Lecture Notes in Mathematics

Edited by A. Dold and B. Eckmann

1014

Complex Analysis – Fifth Romanian-Finnish Seminar

Part 2

Proceedings of the Seminar
held in Bucharest, June 28 – July 3, 1981

Edited by
C. Andreian Cazacu, N. Boboc, M. Jurchescu and I. Suciu

Springer-Verlag
Berlin Heidelberg New York Tokyo 1983

Editors

Cabiria Andreian Cazacu
Nicu Boboc
Martin Jurchescu
Institute of Mathematics
Str. Academiei 14, 70109-Bucharest, Romania

Ion Suciu
Dept. of Mathematics, INCREST
Bdul Păcii 220, 79622 Bucharest, Romania

AMS Subject Classifications (1980): 30-06 (30C60, 30C70, 30C55, 30D45, 30E10, 30F40, 30Fxx); 31-06; 32-06; (58-06); 35-06, 47-06

ISBN 3-540-12683-X Springer-Verlag Berlin Heidelberg New York Tokyo
ISBN 0-387-12683-X Springer-Verlag New York Heidelberg Berlin Tokyo

Library of Congress Cataloging in Publication Data.
Romanian-Finnish Seminar on Complex Analysis (5th: 1981: Bucharest, Romania)
Vth Romanian-Finnish Seminar on Complex Analysis. (Lecture notes in mathematics; 1013–1014)
1. Functions of complex variables–Congresses. 2. Functions of several complex variables–Congresses. 3. Mappings (Mathematics)–Congresses. 4. Functional analysis–Congresses. 5. Potential, Theory of–Congresses. I. Andreian Cazacu, Cabiria. II. Title. III. Series: Lecture notes in mathematics (Springer-Verlag); 1013–1014.
QA3.L28 no. 1013–1014 [QA331] 510s [515.9] 83-20179
ISBN 0-387-12682-1 (v. 1: U.S.)
ISBN 0-387-12683-X (v. 2: U.S.)

Printed in Germany

Printing and binding: Beltz Offsetdruck, Hemsbach/Bergstr.
2146/3140-543210

CONTENTS

Editors' note: for the sake of completeness we reproduce here the Contents of Part I of these proceedings (LNM 1013), as well as the list of lectures not published, the list of participants, and the organizing committee.

VOLUME II

III SECTION - SEVERAL COMPLEX VARIABLES

IV SECTION - POTENTIAL THEORY

V SECTION - FUNCTION THEORETICAL METHODS IN FUNCTIONAL ANALYSIS (OPERATORS AND DIFFERENTIAL OPERATORS)

LECTURES NOT PUBLISHED IN THIS PROCEEDINGS

I SECTION

H. Helling,
Real algebraic models of Teichmüller spaces

Maire Kiikka,
Piecewise linear approximation of quasiconformal maps.

Marjatta Näätänen,
Generic fundamental polygons for surfaces of genus two.

R.Näkki,
Conformal mappings and Lip

S.Rickman,
Omitted values and a defect relation for quasimeromorphic mappings.

M.Seppälä,
Moduli of Klein surfaces

M.Vuorinen,
On the uniqueness of sequential limits of quasiconformal mappings.
Math.Scand. 53 (1983)

II SECTION

I.Bârză,
Fonctions et formes différentielles de type méromorphe sur les surfaces de Riemann non orientables. I,II
Revue Roumaine Math.Pures Appl., 28 , 8 et 7,(1983)

A.Boivin, P.M.Gauthier et W.Hengartner,
Approximation uniforme par des fonctions harmoniques avec singularités
Bull.Math. Soc.Sc.Math.RSRoumanie 28 (76),(1984)

A.Lebowitz and H.Shulman,
Numerical results in degenerating a compact hyper-elliptic Riemann surface of genus 4
An.Univ.Bucureşti Mat., 32 (1982), 35-37.

J.T.Lyons,
A Phragmén-Lindelöf theorem of Fuchs

P.Mocanu,
On starlike functions with respect to symmetric functions

G.Stegbuchner,
Sets of uniqueness for regular functions satisfying an integrated Lipschitz condition
Math.Nachr.,108 (1982), 73-88, An.Univ. Bucureşti Mat., 34 (1984).

H.Wiesler,
On some classes of analytic functions

III SECTION

V.Ancona,
Images directes de faisceaux amples
Trans.Amer.Math.Soc. Vol. 274 (1982), 89-100, An.Univ.Bucuresti Mat., 34 (1984).

A.Baran,
The existence of a subspace connecting given subspaces of a Stein space

P.de Bartolomeis
On complex analyticity of harmonic maps
Bull.Math.Soc.Sc.Math.RSRoumanie,28 (76),(1984)

C.Borcea,
Some remarks on deformations of Hopf manifolds

A.Dimca,
On contact classification of map germs

S.Dimiev,
Faisceaux de fonctions presque-plurisousharmoniques
Bull.Math.Soc.Sc.Math., RSRoumanie, 28 (76), (1984)

O.Dogaru and K.Teleman,
Sur une classe de faisceaux
Revue Roumaine Math.Pures Appl., 28,7,(1983),567-577.

Simone Dolbeault,
Produits de courants et résidus d'après
Mme Laurent-Thiebaut

A.Iordan,
Tubular neighbourhoods for Cauchy-Riemann manifolds

J.Lawrynowicz ,
Condenser capacities connected with a variation of
Kählerian metric in the Dirichlet integral

A.Mihai,
Fibrés vectoriels holomorphes sur des surfaces réglées.

P.Papadopol,
Un théorème de Dolbeault algébrique

P.Skibinski,
The maximum principle for the quotient of plurisub-harmonic functions and some theorems of Noether type

D.Stefănescu,
Algebraic elements over meromorphic power series in
positive characteristic
Bull.Math.Soc.Sc.Math. RSRoumanie, 26 (74),1,(1982),
77-91 and 27 (75),2,(1983), 169-178.

K.Teleman,
Faisceaux céllulaires
Revue Roumaine Math.Pures Appl., 28,7,(1983), 637-641.

IV SECTION

V SECTION

K.Fishman,
Interpolation of linear operators in spaces of analytic functions
Bull.Math.Soc.Sc.Math.RSRoumanie, 28 (76), (1984)

L.Florescu,
Metric bases
Analele Univ.Iaşi, (1984) or (1985)

St.Frunză,
The duality of spectral spaces for operators on Banach spaces

D.Gaspar,
On complex interpolation methods for linear operators

P.Lounesto,
Hypercomplex mapping theorem

M.Puta,
Some remarks on geometric quantization

M.Putinar,
A spectral mapping theorem

I.Suciu,
Exact factorization and complexifications

Wen Guo-chun,
Function theoretic methods for nonlinear elliptic complex equations
Bull.Math.Soc.Sc.Math.RSRoumanie, 28 (76), (1984)

LIST OF PARTICIPANTS

AUSTRIA

Haslinger F.	Univ.Wien - Math.Inst.
Stegbuchner H.	Univ.Salzburg - Math.Inst.

BULGARIA

Dimiev S.	Bulgarian Acad.Sci.-Inst.Math.Sofia
Milev N.	Bulgarian Acad.Sci.-Inst.Math.Sofia
Tonev T.V.	Bulgarian Acad.Sci.-Inst.Math.Sofia

CANADA

Boivin A.	Univ.Montréal - Dépt.Math.Statistique
Gauthier P.	Univ.Montréal - Dépt. Math.Statistique
Gilligan B.	Univ.Regina - Dept.Math.

CHINA

Wen Guo-chun	Univ.Beijing - Inst.Math.

England

Lyons T.J.	Jesus College - Oxford

Finland

Erkama T.	Univ.Joensuu - Dept.Math.
Kahanpää L.	Univ.Jyväskylä- Dept.Math.
Eiikka Maire	Univ.Helsinki - Dept.Math.
Lahtinen A.	Univ.Helsinki - Dept.Math.
Lehto O.	Univ.Helsinki - Dept.Math.
Louhivaara I.S.	Freie Univ.Berlin - Math.Inst.I
Lounesto P.	Helsinki Univ.Techn.- Dept.Math.
Martio O.	Univ.Jyväskylä - Dept.Math.
Näätänen Marjatta	Univ.Helsinki - Dept.Math.
Näkki R.	Univ.Jyväskylä - Dept.Math.
Pirinen A.	Univ.Joensuu - Dept.Math.
Rickman S.	Univ.Helsinki - Dept.Math.
Sarvas J.	Univ.Helsinki - Dept.Math.
Seppälä M.	Helsinki Sch.of Economics
Toppila S.	Univ.Helsinki - Dept.Math.
Vuorinen M.	Univ.Helsinki - Dept.Math.

FRANCE

Abdelkader O.	Univ.Paris VI
Dolbeault P.	Univ.Paris VI
Dolbeault Simone	Univ.Poitiers - Serv.Math.
Ferrand Jacqueline	Univ.Paris VI

D.R.GERMANY

Kühnau R.	Martin Luther Univ. Halle-Wittenberg, Sekt.Math.

F.R.GERMANY

Janssen K.	Univ.Düsseldorff	- Inst.Statistik
Hansen W.	Univ.Bielefeld	- Math.Inst.
Helling H.	Univ.Bielefeld	- Math.Inst.
Spallek K.	Univ.Bochum	- Math.Inst.

ISRAEL

Fishman K.	Univ.Bar-Ilan

ITALIA

Ancona V.	Univ.Ferrara	- Ist.Mat.
de Bartolomeis P.	Univ.Firenze	- Ist.Mat.
Lupacciolu G.	Univ.Roma	- Ist.Mat.
Piccini P.	Univ.Roma	- Ist.Mat.
Rea C.	Univ.Aquilla	- Ist.Mat.
Sbordone C.	Univ.Napoli	- Ist.Mat.
Succi F.	Univ.Roma	- Ist.Mat.
Tomassini G.	Univ.Firenze	- Ist.Mat.

JAPAN

Kuramochi Z.	Univ.Hokkaido	- Dept.Math. Sapporo
Kuribayashi A.	Chuo Univ.	- Dept.Math.

NORWAY

Kari Hag	Univ.Trondheim NTH - Inst.Math.
Martens H.H.	Univ.Trondheim NTH - Inst.Math.

POLAND

Ławrynowicz J.	Inst.Mat. PAN, Łódź	
Skibinski P.	Univ.Łódź	- Inst.Math.
Szafraniec F.H.	Univ.Krakow	- Inst.Math.

ROMANIA

Andreian Cazacu Cabiria	Univ.Bucharest - Fac.Math.
Adam D.	INCREST - Dept.Math.Bucharest
Anghel N.	General School no.14 - Bucharest
Apostol C.	INCREST - Dept.Math.Bucharest
Arsene Gr.	INCREST - Dept.Math.Bucharest
Bally V.	General School - Bucharest
Baran A.	INCREST - Dept.Math.Bucharest
Bănică C.	INCREST - Dept.Math.Bucharest
Bănulescu Martha	Univ.Bucharest - Inst.Math.
Berechet O.	INCREST - Dept.Math.Bucharest
Bîrză I.	Univ.Bucharest - Fac.Math.
Boboc N.	Univ.Bucharest - Fac.Math.
Borcea C.	INCREST - Dept.Math.Bucharest
Borcea Veronica	Univ.Iaşi - Fac.Math.
Bucur Gh.	INCREST - Dept.Math.Bucharest
Caraman P.	Univ.Iaşi - Inst.Math.
Buiculescu Mioara	Statistical Center Bucharest
Ceauşescu Zoia	INCREST - Dept.Math.Bucharest
Colţoiu M.	INCREST - Dept.Math.Bucharest
Cornea F.	General School - Bucharest
Crăciunaş S.	Inst.Research Metallurgy-Bucharest
Deliu Anca	IPGGH - Bucharest
Dimca A.	INCREST - Dept.Math.Bucharest
Dogaru O.	Polytech.Inst. - Bucharest
Fekete O.	Univ.Cluj-Napoca - Fac.Math.
Flondor P.	Polytech.Inst. - Bucharest
Florescu L.	Univ.Iaşi - Fac.Math.
Frunză Monica	Univ.Iaşi - Fac.Math.
Gheondea A.	INCREST - Dept.Math.Bucharest
Godini Gliceria	INCREST - Dept.Math.Bucharest
Gologan R.	INCREST - Dept.Math.Bucharest
Gussi Gh.	INCREST - Dept.Math.Bucharest
Iordan A.	Inst.Constr. - Bucharest
Ivaşcu D.	Polytech.Inst. - Bucharest
Jurchescu M.	Univ.Bucharest - Fac.Math.
Marinaş Marina	Inst.Constr. - Bucharest
Mateescu M.	Polytech.Inst. - Bucharest
Mihai Alexandru	Univ.Bucharest - Fac.Math.
Mihalache N.	INCREST - Dept.Math.Bucharest
Minea Gh.	INCREST - Dept.Math.Bucharest

Mocanu P.	Univ.Cluj-Napoca - Fac.Math.
Pascu E.	INCREST - Dept.Math.Bucharest
Pascu N.N.	Univ.Braşov - Fac.Math.
Podaru V.	Univ.Braşov - Fac.Math.
Popa E.M.	IM Aiud
Popa E.	Univ.Iaşi - Fac.Math.
Potra F.	INCREST - Dept.Math.Bucharest
Puta M.	Univ.Timişoara - Fac.Math.
Putinar M.	INCREST - Dept.Math.Bucharest
Răuţu Gh.	Statistical Center Bucharest
Sălăgean G.	Univ.Cluj-Napoca - Fac.Math.
Sburlan S.	IIS Constanţa
Singer B.	General School Bucharest
Stănăşilă O.	Polytechnic Inst.Bucharest
Stoica L.	INCREST - Dept.Math.Bucharest
Suciu I.	INCREST - Dept.Math.Bucharest
Şabac Mihaela	General School Bucharest
Şabac M.	Univ.Bucharest - Fac.Math.
Şerb I.	Univ.Cluj-Napoca - Fac.Math.
Ştefănescu D.	Univ.Bucharest - Fac.Phys.
Teleman K.	Univ.Bucharest - Fac.Math.
Teleman S.	INCREST - Dept.Math.Bucharest
Timotin D.	INCREST - Dept.Math.Bucharest
Tudor C.	Univ.Bucharest - Fac.Math.
Valuşescu I.	INCREST - Dept.Math.Bucharest
Vernescu A.	Mihai Viteazu High School Bucharest
Voiculescu D.	INCREST - Dept.Math.Bucharest
Wiesler H.	Univ.Cluj-Napoca - Fac.Math.

SWEDEN

Wallin H.	Univ.of Umeá - Dept.Math.

SWITZERLAND

Wohlhauser A.	Ec.Polytechnique Fédérale Lausanne

USSR

Hà Huy Khoái	Moscow Steklov Math.Inst.

USA

Frank Evelyn	Univ.Illinois - Dept.Math.
Knopp M.	Temple Univ.Philadelphia Dept.Math.
Miller S.S.	State Univ.New York -College Brockport - Dept.Math.
Renggli H.	Kent State Univ. - Dept.Math.
Weill G.	Polytechnic Inst. New York

TURKEY

Göktürk Zerrin	Bosphorus Univ. - Dept.Math.

ORGANIZING COMMITTEE

O.LEHTO
Department of Mathematics
University of Helsinki

CABIRIA ANDREIAN CAZACU
Faculty of Mathematics
University of Bucharest

WORKING COMMITTEE

I Section - Quasiconformal and quasiregular mappings, Teichmüller spaces and Kleinian groups.
Cabiria Andreian Cazacu, P.Caraman, D. Ivaşcu

II Section - Function theory of one complex variable.
P.Mocanu

III Section - Several complex variables
C.Bănică, Gh.Gussi, M.Jurchescu, O.Stănăşilă

IV Section - Potential theory
N.Boboc, Gh.Bucur, L.Stoica

V Section - Function theoretical methods in Functional Analysis. (Operators and Differential Operators)
I.Suciu, I.Valuşescu, D.Voiculescu

Secretariat: Simona Pascu, Virginia Zamă

The Editorial Committee expresses its gratitude to Springer-Verlag for kind consent of publishing the Proceedings in the Series Lecture Notes in Mathematics.
It thanks sincerely Simona Pascu for the contribution given in preparing the Proceedings.

UN THÉORÈME D'ANNULATION POUR LES FIBRÉS EN DROITES SEMI-POSITIFS SUR UNE VARIÉTÉ KÄHLÉRIENNE FAIBLEMENT 1-COMPLÈTE

par

Osama ABDELKADER

L'objet de cet article est d'améliorer les résultats de [1]. Une variété analytique complexe munie d'une fonction plurisousharmonique d'exhaustion ψ, de classe C^2, est dite faiblement 1-complète par rapport à ψ. Soient X une variété kählérienne faiblement 1-complète par rapport à ψ et $E \to X$ un fibré hermitien en droites muni d'une métrique hermitienne h, Ω la forme de courbure définie par h. On dit que le fibré E est semi-positif si la forme Ω est semi-positive. Dans [1] on a les résultats suivants :

THÉORÈME 1. - Soient X une variété kählérienne faiblement 1-complète de *dimension complexe* n *et* $E \to X$ *un fibré hermitien en droites, sur* X, *muni d'une structure hermitienne. Supposons la forme* Ω *de courbure de* E *semi-définie positive et de rang* $\geq K$ *en tout point de* X *et soit* m *la plus petite des* K *plus grandes valeurs propres de* Ω. *Supposons* m *minorée sur* X *par une constante* > 0. *Alors*

$H^p(X, \Omega^p(E)) = 0$ *dès que* $p + q \geq 2n - K + 1$.

THÉORÈME 2. - *Supposons satisfaites les hypothèses du théorème 1 à l'exception de la condition sur* m, *alors*

$H^q(X_c, \Omega^p(E)) = 0$ *pour* $p + q \geq 2n - K + 1$ *et* $c \in \mathbb{R}$ *où* $X_c = \{x \in X \; ; \; \psi(x) < c\}$.

Nous allons montrer le théorème 1 sans l'hypothèse sur m i.e. :

THÉORÈME 3. - *Soient* X *une variété kählérienne faiblement* 1-*complète de dimension complexe* n *et* $E \to X$ *un fibré hermitien en droites muni d'une structure hermitienne. Supposons la forme de courbure de* E *semi-définie positive et de rang* $\geq K$ *en tout point de* X. *Alors* $H^q(X, \Omega^p(E)) = 0$ *dès que* $p + q \geq 2n - K + 1$.

Après la parution des théorèmes 1 et 2, H.SKODA [12] a montré le théorème 3, pour $p = n$, en adaptant la méthode de trois poids de HÖRMANDER [5], T.OHSAWA et K.TAKEGOSHI [11] ont montré le théorème 3 par une méthode d'approximation différente.

Dans ce travail on adapte la méthode de HÖRMANDER [4] , théorème 3.4.7. en modifiant la métrique sur la base et la métrique sur les fibres à l'aide de la fonction plurisousharmonique d'exhaustion sur la base pour étendre la procédure d'approximation de KAZAMA [8] pour les (p,q)-formes d''-fermées. On applique ensuite le théorème d'approximation à la démonstration du théorème 3 .

Je tiens à remercier P.DOLBEAULT pour son encouragement continu, ainsi que H.SKODA: ils m'ont fait des remarques et des suggestions qui ont été d'une grande utilité.

§ 1. Notations. On pose $\gamma = \sqrt{-1}\,\Omega$ et on note $e(\gamma)$ le produit extérieur par γ. Soit $(u_j)_{j\in I}$ un recouvrement suffisamment fin de X tel que u_j soit un ouvert de carte de X et de trivialisation de E . La structure hermitienne de E est donnée par un système de fonctions $(b_j^{(o)})$ avec $b_i^{(o)} = |b_{ji}|^2\, b_j^{(o)}$ sur $u_i \cap u_j$ où les b_{ji} sont les fonctions de transition qui définissent E . Pour (u_j) assez fin, il existe $(a_j^{(o)})$ un système de fonctions de classe C^∞ , > 0 , sur u_j tel que la métrique kählérienne de X soit :

$$ds_o^2 = \sum_{\alpha,\beta=1}^{n} g_{j\alpha\beta}^{(o)}\, dz_j^\alpha\, d\bar{z}_j^\beta \quad \text{où} \quad g_{j\alpha\beta}^{(o)} = \frac{\partial^2 \log a_j^{(o)}}{\partial z_j^\alpha\, \partial \bar{z}_j^\beta} .$$

Pour une métrique kählérienne ds^2 sur X et une structure hermitienne h sur E, donnée par un système de fonctions (b_j) sur u_j , sur l'espace $A^{p,q}(X,E)$ des (p,q)-formes C^∞ , à valeurs dans E , nous avons le produit scalaire ponctuel $(\ ,\ )$ définie par : $(\varphi_j, \eta_j)dv = \frac{1}{b_j}\, \varphi_j \wedge \star\bar{\eta}_j$ pour $\eta = (\eta_j)$, $\varphi = (\varphi_j) \in A^{p,q}(X,E)$ où $\star$ est l'opérateur de Hodge définie par ds^2 et où dv désigne l'élément de volume par rapport à ds^2 .

Soit $\mathcal{D}^{p,q}(X,E)$ l'espace des (p,q)-formes C^∞ à support compact de X à valeurs dans E . Considérons le produit scalaire global

$$\langle\varphi, \eta\rangle = \int_X (\varphi, \eta)dv \quad \text{pour} \quad \varphi, \eta \in \mathcal{D}^{p,q}(X,E).$$

L'opérateur $d'' : A^{p,q}(X,E) \to A^{p,q+1}(X,E)$ est défini par $(d''\varphi)_j = (d''\varphi_j)$ et son adjoint formel, par rapport au $\langle\ ,\ \rangle$, est défini par $\delta'' : A^{p,q+1}(X,E) \to A^{p,q}(X,E)$ où $(\delta''\varphi)_j = -\star b_j\, d'((b_j)^{-1} \star \varphi_j)$. Soit ω la forme fondamentale de la métrique kählérienne ds^2 . Posons $L = e(\omega)$ où $e(\omega)$ est le produit extérieur par ω .

On définit l'opérateur $\Lambda : A^{p,q}(X,E) \to A^{p-1,q-1}(X,E)$ par $\Lambda = (-1)^{p+q} \star L \star$.
L'opérateur de Laplace-Beltrami $\Box'' : A^{p,q}(X,E) \to A^{p,q}(X,E)$ est défini par $\Box''=d''\delta''+\delta''d''$.

Posons $\gamma' = \sqrt{-1}\, d'd'' \log b_j$; $j \in I$, donc on a l'égalité :

$$\Box'' - \star^{-1} \Box'' \star : e(\gamma')\Lambda - \Lambda e(\gamma') \quad \text{(voir [9])} .$$

Pour toute $\varphi \in \mathcal{D}^{p,q}(X,E)$ on a :

(1) $\quad <d''\varphi , d''\varphi> + <\delta''\varphi, \delta''\varphi> = <\Box''\varphi, \varphi> \geqslant <(e(\gamma')\Lambda - \Lambda e(\gamma'))\varphi , \varphi>$

car $<\star^{-1} \Box'' \star\varphi , \varphi> = <\Box'' \star \varphi , \star \varphi> \geqslant 0$.

Pour $c \in R$, l'ouvert relativement compact $X_c = \{x \in X ; \psi(x) < c\}$ est une variété faiblement 1-complète par rapport à la fonction d'exhaustion $\Phi_c = \frac{c}{c - \psi}$.
Fixons $c_1, c_2 \in \mathbb{R}$ tels que le bord de X_{c_1} soit lisse et $c_1 < c_2$. Soit $\lambda : \mathbb{R} \to \mathbb{R}$ une fonction C^∞ telle que :

$\lambda(t)$, $\lambda'(t)$ et $\lambda''(t) \geqslant 0$ pour $t \geqslant 0$, $\lambda(t) = \lambda'(t) = \lambda''(t) = 0$ pour $t \leqslant c_3$ et $\lambda''(t) > 0$ pour $t > c_3$.

Etant donné une suite de métriques kählériennes $(d\sigma_\nu^2)$ sur X et une suite de structures hermitiennes données par le système (b_j^ν) ; $\nu \in \mathbb{N}$, sur les fibres de $E|_{X_{c_2}}$. On considère le produit scalaire :

$$<\varphi, \eta>_{\nu'}^{\nu} = \int_{X_{c_2}} (\varphi,\eta)_{\nu'}^{\nu} dv_\nu$$

et la norme $\|\varphi\|_{\nu'}^{\nu} = \sqrt{<\varphi, \varphi>_{\nu'}^{\nu}}$ pour $\varphi, \eta \in \mathcal{D}^{p,q}(X_{c_2},E)$ où l'indice ν' en bas (resp. ν en haut) désigne le produit scalaire par rapport au système $(b_j^{\nu'})$ (resp. à $d\sigma_\nu^2$) et dv_ν désigne l'élément de volume par rapport à $d\sigma_\nu^2$.

Soit $\delta_{\nu'}^{''\nu}$ l'opérateur adjoint de d'' par rapport au $< , >_{\nu'}^{\nu}$ i.e. :

$(\delta_{\nu'}^{''\nu} \varphi)_j = - \star_\nu b_j^{\nu'} d'((b_j^{\nu'})^{-1} \star_\nu \varphi_j)$ où l'opérateur $\star_\nu$ est défini par $d\sigma_\nu^2$.

Nous introduisons la forme sesquilinéaire définie positive

$$a_\nu<\varphi , \eta > = <\varphi,\eta>_\nu^\nu + <d''\varphi , d''\eta>_\nu^\nu + <\delta_\nu^{''\nu} \varphi , \delta_\nu^{''\nu}\eta>_\nu^\nu .$$

Nous désignons par $\mathcal{L}_\nu^{p,q}(X_{c_2},E)^\nu$ le complété de $\mathcal{D}^{p,q}(X_{c_2},E)$ par rapport à la norme $\|\varphi\|_\nu^\nu$ et par $W_\nu^{p,q}(X_{c_2},E)$ le complété de $\mathcal{D}^{p,q}(X_{c_2},E)$ par rapport à la norme $\sqrt{a_\nu <\varphi, \varphi>}$. Une forme φ est dans $W_\nu^{p,q}(X_{c_2},E)$ s'il existe une suite de Cauchy (φ_k) ; $\varphi_k \in \mathcal{D}^{p,q}(X_{c_2},E)$, convergente vers φ dans $\mathcal{L}_\nu^{p,q}(X_{c_2},E)^\nu$, telle que $(d''\varphi_k)$ et $(\delta_\nu^{''\nu} \varphi_k)$ sont encore deux suites de Cauchy dans $\mathcal{L}_\nu^{p,q+1}(X_{c_2},E)^\nu$

et $\mathscr{L}_\nu^{p,q-1}(X_{c_2},E)^\nu$ respectivement. On note encore par d" l'opérateur $\mathscr{L}_\nu^{p,q}(X_{c_2},E)^\nu \to \mathscr{L}_\nu^{p,q+1}(X_{c_2},E)^\nu$ qui étend l'opérateur d" sur $\mathscr{D}^{p,q}(X_{c_2},E)$, alors une forme $\varphi \in \mathscr{L}_\nu^{p,q}(X_{c_2},E)^\nu$ est dans le domaine de d" si et seulement si d" φ, définie au sens des distributions, est dans $\mathscr{L}_\nu^{p,q+1}(X_{c_2},E)^\nu$. Alors l'opérateur d" est un opérateur linéaire fermé de domaine dense, donc on peut définir son adjoint $(d'')_\nu^{*\nu}$. On désigne par $D_\nu^{p,q}(d'')$, $N_\nu^{p,q}(d'')$ et $Im_\nu^{p,q+1}(d'')$ le domaine, le noyau et l'image de d" dans $\mathscr{L}_\nu^{p,q}(X_{c_2},E)^\nu$ et $\mathscr{L}_\nu^{p,q+1}(X_{c_2},E)^\nu$ respectivement. De la même manière on définit $D_\nu^{p,q}((d'')_\nu^{*\nu})$, $N_\nu^{p,q}((d'')_\nu^{*\nu})$ et $Im_\nu^{p,q-1}((d'')_\nu^{*\nu})$ dans $\mathscr{L}_\nu^{p,q}(X_{c_2},E)^\nu$ et $\mathscr{L}_\nu^{p,q-1}(X_{c_2},E)^\nu$ respectivement.

§ 2. Conditions de métrique complète, carré integrable et $W^{p,q}$ ellipticité.

LEMME 1. - *Soient X une variété kählérienne faiblement 1-complète par rapport à ψ et $E \to X$ un fibré hermitien en droites. Soient $c_1, c_2 \in \mathbb{R}$, $c_1 < c_2$ et $\lambda \in C^\infty(\mathbb{R})$ comme dans le § 1. Supposons la forme de courbure de E semi-définie positive et de rang $\geqslant K$ en tout point de X. Etant donnée une forme $f \in A^{p,q}(X_{c_2},E)$, il existe une suite de métriques kählériennes complètes $d\sigma_\nu^2$ sur X_{c_2}, une suite de métriques hermitiennes h^ν sur les fibres de $E|_{X_{c_2}}$ et une constante $c > 0$ indépendante de ν, telles que $\langle f,f\rangle_\nu^\nu < +\infty$ et*

$$\langle d''\varphi, d''\varphi\rangle_\nu^\nu + \langle \delta_\nu''^\nu\varphi, \delta_\nu''^\nu\varphi\rangle_\nu^\nu \geqslant c\langle\varphi,\varphi\rangle_\nu^\nu;\ \nu\in\mathbb{N},$$

pour toute $\varphi \in \mathscr{D}^{p,q}(X_{c_2},E)$ dès que $p + q \geqslant 2n - K + 1$.

Démonstration.

On peut supposer que $\psi \geqslant 0$ sur X, sinon on remplace ψ par $e^\psi > 0$ sur X.

Considérons une fonction $\mu(t) \in C^\infty(\mathbb{R})$ telle que : $\mu(t), \mu'(t), \mu''(t) \geqslant 0$ pour $t \geqslant 0$.

Définissons la métrique h par le système des fonctions $(\frac{1}{b_j})$ où $b_j = e^{\mu(\Phi)}\, b_j^{(0)}$; $\Phi = \Phi_{c_2}$, sur les fibres de $E|_{X_{c_2}}$ et la métrique ds_1^2 par : $ds_1^2 = \sum_{\alpha,\beta=1}^n g_{j\alpha\bar\beta}^{(1)}\, dz_j^\alpha\, d\bar z_j^\beta$ où $g_{j\alpha\bar\beta}^{(1)} = \frac{\partial^2 \log a_j}{\partial z_j^\alpha\, \partial \bar z_j^\beta}$ et $a_j = e^{\mu(\Phi)}\, a_j^{(0)}$. D'après NAKANO [9] et [10], étant donnée une forme $f \in A^{p,q}(X_{c_2},E)$, on peut choisir la fonction $\mu(t)$ telle que $\langle f,f\rangle < +\infty$ et la métrique ds_1^2 soit complète où le produit scalaire est définie par rapport à ds_1^2 et h .

Nous considérons la fonction $\lambda(t)\in C^\infty(\mathbb{R})$ définie dans le § 1 . Nous définissons une suite de structures hermitiennes sur les fibres de $E|_{X_{c_2}}$ par les systèmes $\{b_j^\nu\} = \{e^{\nu\lambda(\Phi)}b_j\}$; $\nu\in\mathbb{N}$. On définit h^ν par : $h^\nu = (h_j^\nu)$ où $h_j^\nu = \frac{1}{b_j^\nu}$. La forme de courbure de $E|_{X_{c_2}}$, par rapport au système $\{b_j^\nu\}$ est donnée par :

$$\Omega^\nu = d'd''\log b_j^\nu = \sum_{\alpha,\beta=1}^{n} \Omega^\nu_{j\alpha\bar\beta}\, dz_j^\alpha\wedge d\bar z_j^\beta \quad , j\in I \quad , \text{ sur } u_j \ .$$

Soient $\delta \geqslant 1$ un nombre réel et $d\sigma_\nu^2$ la suite de métriques kählériennes définies par :

$$d\sigma_\nu^2 = \sum_{\alpha,\beta=1}^{n} (g^{(o)}_{j\alpha\bar\beta} + \delta\Omega^\nu_{j\alpha\bar\beta})dz_j^\alpha\, d\bar z_j^\beta = \sum_{\alpha,\beta=1} g^\nu_{j\alpha\bar\beta}\, dz_j^\alpha\, d\bar z_j^\beta \quad ; \quad \nu\in\mathbb{N} \ .$$

Donc $d\sigma_\nu^2 \geqslant ds_1^2$. Alors $d\sigma_\nu^2$ est complète dès que ds_1^2 l'est.

Soient $\eta_1^1 \geqslant \eta_2^1 \geqslant\ldots\geqslant \eta_n^1 \geqslant 0$ (resp. $\mu_1^\nu \geqslant \mu_2^\nu\geqslant\ldots\geqslant \mu_n^\nu \geqslant 0$) les valeurs propres de $(\Omega_{j\alpha\bar\beta})$ (resp. de $(\Omega^\nu_{j\alpha\bar\beta})$) par rapport à $(g^{(o)}_{j\alpha\bar\beta})$ en tout point de X_{c_2} où $\Omega_{j\alpha\bar\beta} = \dfrac{\partial^2 \log b_j^{(o)}}{\partial z_j^\alpha\,\partial\bar z_j^\beta}$. La compacité de $\bar X_{c_2}$ et le fait que rang $(\Omega_{j\alpha\bar\beta}) \geqslant K$, en tout point de X, impliquent qu'il existe une constante $c_o > 0$ minorant η_K^1 sur X_{c_2} .

En considérant l'inégalité $(\Omega^\nu_{j\alpha\bar\beta}) \geqslant (\Omega_{j\alpha\bar\beta})$ sur u_j, $j\in I$, $\nu\in\mathbb{N}$, d'après le théorème 6.44 de [6] on a : $\mu_K^\nu \geqslant \eta_K^1$; $\nu\in\mathbb{N}$.

On va achever la démonstration par un calcul analogue à celui de [3] . Une forme $\varphi = (\varphi_j)\in A^{p,q}(X,E)$, s'exprime sur u_j par :

$$\varphi_j = \varphi_{j\alpha_1\ldots\alpha_p\bar\beta_1\ldots\bar\beta_q}\, dz_j^{\alpha_1}\wedge\ldots\wedge dz_j^{\alpha_p}\wedge d\bar z_j^{\beta_1}\wedge\ldots\wedge d\bar z_j^{\beta_q}\otimes s_j$$

où s_j est une section de E au-dessus de u_j . Soient $(g_{\nu j}^{\bar\beta\alpha})$ la matrice inverse de $(g^\nu_{j\alpha\bar\beta})$ et $\gamma_\nu = \sqrt{-1}\,\Omega^\nu$. Pour toute forme $\varphi\in A^{p,q}(X_{c_2},E)$ on a la formule (cf. [7], p. 132-133) :

$$(2)\quad ((\Lambda_\nu e(\gamma_\nu) - e(\gamma_\nu)\Lambda_\nu)\varphi)_{\alpha_1\ldots\alpha_p\bar\beta_1\ldots\bar\beta_q} = \frac{1}{b_j^\nu}\Sigma\, g_{\nu j}^{\bar\mu\eta}\, \Omega^\nu_{j\eta\bar\mu}\, \varphi_{j\alpha_1\ldots\alpha_p\bar\beta_1\ldots\bar\beta_q}$$

$$+ \frac{1}{b_j^\nu}\Sigma\sum_{i=1}^{p}(-1)^i g_{\nu j}^{\bar\mu\eta}\,\Omega^\nu_{j\alpha_i\bar\mu}\,\varphi_{j\eta\alpha_1\ldots\hat\alpha_i\ldots\alpha_p\bar\beta_1\ldots\bar\beta_q}$$

$$+ \frac{1}{b_j^\nu}\Sigma\sum_{i=1}^{q}(-1)^i g_{\nu j}^{\bar\mu\eta}\,\Omega^\nu_{j\eta\bar\beta_i}\,\varphi_{j\alpha_1\ldots\alpha_p\bar\mu\bar\beta_1\ldots\hat{\bar\beta}_i\ldots\bar\beta_q}$$

où l'opérateur Λ_ν est défini par $d\sigma_\nu^2$.

On note que rang $(\Omega^\nu_{j\alpha\bar\beta}) \geqslant K$. Soit $x_o \in X$, un point arbitraire fixé, tel que rang $(\Omega^\nu_{j\alpha\bar\beta})(x_o) = \ell \geqslant K$. Alors en x_o il existe un système de coordonnées $(z^1, z^2, \ldots, z^n)$ et une transformation unitaire telles que :

$$ds_o^2(x_o) = \sum_{\alpha=1}^{n} dz^\alpha\, d\bar z^\alpha \quad \text{et} \quad \Omega^\nu(x_o) = \sum_{\alpha=1}^{n}\mu_\alpha^\nu\, dz^\alpha \wedge d\bar z^\alpha .$$

Alors en x_o la formule (2) devient :

$$((\Lambda_\nu e(\gamma_\nu) - e(\gamma_\nu)\Lambda_\nu)\varphi)_{\alpha_1\ldots\alpha_p\bar\beta_1\ldots\bar\beta_q} = \frac{1}{b_j^\nu}\, B\, \varphi_{j\alpha_1\ldots\alpha_p\bar\beta_1\ldots\bar\beta_q} \quad \text{où}$$

$$B = \sum_{i=1}^{\ell}\frac{\mu_i^\nu}{1+\delta\mu_i^\nu} - \sum_{i=1}^{p}\frac{\mu^\nu_{\alpha_i}}{1+\delta\mu^\nu_{\alpha_i}} - \sum_{i=1}^{q}\frac{\mu^\nu_{\beta_i}}{1+\delta\mu^\nu_{\beta_i}} . \text{ Donc}$$

$$B = \frac{1}{\delta}\left[\sum_{i=1}^{\ell}\left(1-\frac{1}{1+\delta\mu_i^\nu}\right) - \sum_{\alpha_i\leqslant\ell}\left(1-\frac{1}{1+\delta\mu^\nu_{\alpha_i}}\right) - \sum_{\beta_i\leqslant\ell}\left(1-\frac{1}{1+\delta\mu^\nu_{\beta_i}}\right)\right].$$

Etant fixé $\alpha_1 < \alpha_2 < \ldots < \alpha_p$ (resp. $\beta_1 < \beta_2 < \ldots < \beta_q$), on suppose que dans l'ensemble des indices $\alpha_1\, \alpha_2 \ldots \alpha_p$ (resp. $\beta_1\, \beta_2 \ldots \beta_q$) il y en a s_1 (resp. s_2) parmi les ℓ premiers. Alors :

$$B = \frac{1}{\delta}\left[\ell - s_1 - s_2 + \sum_{\alpha_i\leqslant\ell}\frac{1}{1+\delta\mu^\nu_{\alpha_i}} + \sum_{\beta_i\leqslant\ell}\frac{1}{1+\delta\mu^\nu_{\beta_i}} - \sum_{i=1}^{\ell}\frac{1}{1+\delta\mu_i^\nu}\right]$$

$$\leqslant \frac{1}{\delta}\left[\ell - s_1 - s_2 + \sum_{i=1}^{\ell}\frac{1}{1+\delta\mu_i^\nu}\right]$$

$$\leqslant \frac{1}{\delta}\left[2\ell - s_1 - s_2 - K + \sum_{i=1}^{K}\frac{1}{1+\delta\mu_K^\nu}\right].$$

On note que pour $p + q \geqslant 2n - K + 1$, $p - (n-\ell) + q - (n-\ell) \geqslant 2\ell - K + 1$.

Alors il existe au moins $2\ell - K + 1$ des α_i et β_i dans les ℓ premiers, donc $s_1 + s_2 \geqslant 2\ell - K + 1$. Alors :

$$B \leqslant \frac{1}{\delta}\left[-1 + \frac{n}{1+\delta\mu_K^\nu}\right] \leqslant -\frac{1}{2\delta} \quad \text{pour} \quad \delta \geqslant \frac{2n-1}{\mu_K^\nu} .$$

Donc pour toute forme $\varphi \in A^{p,q}(X_{c_2},E)$; $p + q \geqslant 2n - K + 1$, en tout point de X_{c_2} , on a :

$$((e(\gamma_\nu)\Lambda_\nu - \Lambda_\nu e(\gamma_\nu))\, \varphi,\varphi)_\nu^\nu \geqslant \frac{1}{2\delta}\,(\varphi,\varphi)_\nu^\nu \ ; \ \delta \geqslant \frac{2n-1}{c_o} .$$ Donc de

l'inégalité de NAKANO (1) résulte :

$$\langle d''\varphi, d''\varphi\rangle_\nu^\nu + \langle \delta_\nu''^\nu\varphi, \delta_\nu''^\nu\varphi\rangle_\nu^\nu \quad \geqslant \quad c\langle\varphi,\varphi\rangle_\nu^\nu \ ; \ \nu \in \mathbb{N}$$

pour toute $\varphi \in \mathscr{D}^{p,q}(X_{c_2},E)$ dès que $p + q \geqslant 2n - K + 1$ avec $c = \frac{1}{2\delta}$. Ce qui démontre le lemme.

D'après le lemme 1 et le théorème 2 de ([2] , page 94) résulte :

COROLLAIRE. - <u>Dans les hypothèses du lemme 1, on a</u> :

$$H^q(X_{c_2}, \Omega^p(E)) = 0 \quad \underline{\text{pour}} \quad p + q \geqslant 2n - K + 1$$

ce qui démontre le théorème 2 de l'introduction.

§ 3. <u>Un théorème d'approximation.</u>

Considérons l'ouvert relativement compact :

$$X_{c_1} = \{x \in X \ ; \ \Psi(x) < c_1\} \ , \quad c_1 < c_2 \ , \text{ alors}$$

$$X_{c_1} = \{x \in X \ ; \ \Phi(x) < c_3\} \text{ où } c_3 = \frac{c_2}{c_2 - c_1}$$ donc X_{c_1} est un ouvert relativement compact de X_{c_2} . On désigne par $\|\varphi\|_{c_1}$ la norme :

$$(4) \quad \|\varphi\|_{c_1} = \left(\int_{X_{c_1}} (\varphi,\varphi)_o^o\, dv_o\right)^{1/2}$$

pour $\varphi \in \mathscr{D}^{p,q}(X_{c_1},E)^1$.

Les opérateurs :

$$d'' : \mathscr{L}_\nu^{p,q}(X_{c_2},E)^\nu \to \mathscr{L}_\nu^{p,q+1}(X_{c_2},E)^\nu$$

$$(d'')_\nu^{\star\nu} : \mathscr{L}_\nu^{p,q}(X_{c_2},E)^\nu \to \mathscr{L}_\nu^{p,q-1}(X_{c_2},E)^\nu$$

sont définis au sens des distributions i.e. : une forme $\varphi \in \mathscr{L}_\nu^{p,q}(X_{c_2},E)^\nu$ telle que : $d''\varphi \in \mathscr{L}_\nu^{p,q+1}(X_{c_2},E)^\nu$ et $(d'')_\nu^{\star\nu}\varphi \in \mathscr{L}_\nu^{p,q-1}(X_{c_2},E)^\nu$ si et seulement si

$$\langle \varphi, \delta_\nu''^\nu\tau\rangle_\nu^\nu = \langle d''\varphi,\tau\rangle_\nu^\nu \quad \text{pour toute} \quad \tau \in \mathscr{D}^{p,q+1}(X_{c_2},E)$$

$$\langle \varphi, d''\rho\rangle_\nu^\nu = \langle (d'')_\nu^{\star\nu}\varphi, \rho\rangle_\nu^\nu \quad \text{pour toute} \quad \rho \in \mathscr{D}^{p,q-1}(X_{c_2},E) .$$

Soit $N_{c_1}^{p,q-1}(d'')$ (resp. $D^{p,q-1}(d'')$) le noyau (resp. le domaine) de l'opérateur $d'' : \mathscr{L}_o^{p,q-1}(X_{c_1},E)^o \to \mathscr{L}_o^{p,q}(X_{c_1},E)^o$ où $\mathscr{L}_o^{p,q}(X_{c_1},E)$ est le complété de $\mathscr{D}^{p,q}(X_{c_1},E)$ par rapport à la norme $\|\varphi\|_{c_1}$.

Théorème d'approximation. Soient $X, E, d\sigma_\nu^2$ et h^ν comme dans le lemme 1. Alors la restriction $j : N_o^{p,q-1}(d'') \to N_{c_1}^{p,q-1}(d'')$ avec $p + q \geqslant 2n - K + 1$ est d'image dense par rapport à la norme $\|\varphi\|_{c_1}$.

Démonstration.

Supposons que le théorème est faux, alors $\overline{j\, N_o^{p,q-1}(d'')} \subsetneqq N_{c_1}^{p,q-1}(d'')$ où $\overline{j\, N_o^{p,q-1}(d'')}$ est l'adhérence de $j\, N_o^{p,q-1}(d'')$ dans $\mathscr{L}_o^{p,q-1}(X_{c_2},E)^o$.
Alors il existe une forme $\hat{\eta} \in N_{c_1}^{p,q-1}(d'')$ telle que $\hat{\eta} \neq 0$ et $\hat{\eta}$ soit orthogonale à $\overline{N_o^{p,q-1}(d'')}$. Donc $< \Psi, \hat{\eta} >_{c_1} = \int_{X_{c_1}} (\Psi, \hat{\eta})_o^o \, dv_o = 0$ pour tout $\Psi \in N_o^{p,q-1}(d'')$.

On prolonge $\hat{\eta}$ sur X_{c_2} par zéro en dehors de X_{c_1} en un élément $\eta \in \mathscr{L}_o^{p,q-1}(X_{c_2},E)^o$.

Donc $<\Psi, \eta>_o^o = \int_{X_{c_2}} (\Psi, \eta)_o^o \, dv_o = 0$ pour toute $\Psi \in N_o^{p,q-1}(d'')$. D'autre part $d\sigma_\nu^2 = d\sigma_o^2$ sur le support de η ; $\nu \in \mathbb{N}$, donc

$$0 = <\Psi, \eta>_o^o = \int_{X_{c_2}} <\Psi, \eta>_o^o \, dv_o = \int_{X_{c_2}} (\Psi, e^{\nu\lambda(\Phi)} \eta)_\nu^\nu \, dv_\nu \quad \text{pour } \psi \in N_o^{p,q-1}(d'').$$

i.e. : $e^{\nu\lambda(\Phi)}\eta$ est orthogonale à $N_o^{p,q-1}(d'') \supset N_\nu^{p,q-1}(d'')$. D'après le théorème 1.1. de [13] et le fait que $d\sigma_\nu^2$, $\nu \in \mathbb{N}$, est complète on a :

$$W_\nu^{p,q}(X_{c_2},E) = \{\varphi \in \mathscr{L}_\nu^{p,q}(X_{c_2},E)^\nu ;\ d''\varphi \in \mathscr{L}_\nu^{p,q+1}(X_{c_2},E)^\nu \text{ et } (d'')_\nu^{*\nu}\varphi \in \mathscr{L}^{p,q-1}(X_{c_2},E)^\nu\}$$
$$= D_\nu^{p,q}(d'') \cap D^{p,q}((d'')_\nu^{*\nu})$$

et $\mathscr{D}^{p,q}(X_{c_2},E)$ est dense dans $W_\nu^{p,q}(X_{c_2},E)$ par rapport à la norme $(< \varphi, \varphi>_\nu^\nu + <d''\varphi, d''\varphi>_\nu^\nu + <(d'')_\nu^{*\nu}\varphi, (d'')_\nu^{*\nu}\varphi>_\nu^\nu)^{1/2}$. Donc du lemme 1 résulte :

(5) $$< \varphi, \varphi>_\nu^\nu \leqslant \frac{1}{c}(<d''\varphi, d''\varphi>_\nu^\nu + <(d'')_\nu^{*\nu} \varphi, (d'')_\nu^{*\nu} \varphi>_\nu^\nu)$$

pour toute $\varphi \in W_\nu^{p,q}(X_{c_2},E)$ dès que $p+q \geqslant 2n-K+1$.

Donc de (5) pour toute $\varphi \in D^{p,q}((d'')^{\star\nu}_{\nu}) \cap N^{p,q}_{\nu}(d'')$ résulte :

(6) $$<\varphi, \varphi>^{\nu}_{\nu} \leq \frac{1}{c} <(d'')^{\star\nu}_{\nu}\varphi \ , \ (d'')^{\star\nu}_{\nu} \varphi >^{\nu}_{\nu}$$

pour $p + q \geq 2n - k + 1$ et $\nu \in \mathbb{N}$.

Du (6) et du lemme 4.1.2. de [5], on déduit qu'il existe $\varphi^{\nu} \in D^{p,q}(d'')^{\star\nu}_{\nu})$ telle que:

$e^{\nu\lambda(\Phi)}\eta = (d'')^{\star\nu}_{\nu}\varphi^{\nu}$ et

(6') $$(\|\varphi^{\nu}\|^{\nu}_{\nu})^2 \leq \frac{1}{c}(\|e^{\nu\lambda(\Phi)}\eta\|^{\nu}_{\nu})^2 .$$

Pour toute $g \in \mathcal{D}^{p,q-1}(X_{c_2}, E)$ on a :

$$<(d'')^{\star\nu}_{\nu}\varphi^{\nu}, g>^{\nu}_{\nu} = <\varphi^{\nu}, d''g>^{\nu}_{\nu} = <e^{-\nu\lambda(\Phi)}\varphi^{\nu}, d''g>^{\nu}_{o} = <\delta''^{\nu}_{o}(e^{-\nu\lambda(\Phi)}\varphi^{\nu}), g>^{\nu}_{o}$$
$$= <e^{\nu\lambda(\Phi)}\delta''^{\nu}_{o}(e^{-\nu\lambda(\Phi)}\varphi^{\nu}), g>^{\nu}_{\nu}$$

i.e : $(d'')^{\star\nu}_{\nu}\varphi^{\nu} = e^{\nu\lambda(\Phi)}\delta''^{\nu}_{o}(e^{-\nu\lambda(\Phi)}\varphi^{\nu})$; $\nu \in \mathbb{N}$.

Posons $y^{\nu} = e^{-\nu\lambda(\Phi)}\varphi^{\nu}$ donc $\eta = \delta''^{\nu}_{o} y^{\nu}$; $\nu \in \mathbb{N}$. Mais supp $\eta \in \bar{X}_{c_1}$, donc $\delta''^{\nu}_{o} y^{\nu} = 0$ sur $X_{c_2} - \bar{X}_{c_1}$; $\nu \in \mathbb{N}$. D'autre part $\lambda(\Phi) = 0$ sur $\bar{X}_{c_1}$ i.e. : $d\sigma^2_{\nu} = d\sigma^2_{o}$ sur $\bar{X}_{c_1}$, donc on peut écrire : $\delta''^{\nu}_{o} y^{\nu} = \delta''^{o}_{o} y^{\nu}$; $\nu \in \mathbb{N}$, donc $\eta = \delta''^{o}_{o} y^{\nu}$; $\nu \in \mathbb{N}$.

La métrique $d\sigma^2_o$ est définie sur u_j par la matrice $\left(g^{(o)}_{j\alpha\bar{\beta}} + \delta \frac{\partial^2\mu(\Phi)}{\partial z^{\alpha}_j \partial \bar{z}^{\beta}_j} + \delta \frac{\partial^2 \log b^{(o)}_j}{\partial z^{\alpha}_j \partial \bar{z}^{\beta}_j}\right)$. Soient $\eta_1 \geq \eta_2 \geq \dots \geq \eta_n \geq 0$ les valeurs propres de $\left(\frac{\partial^2\lambda(\Phi)}{\partial z^{\alpha}_j \partial \bar{z}^{\beta}_j}\right)$ par rapport à cette matrice en tout point de X_{c_2} . Soit $x_o \in u_j$ un point arbitraire fixé ; alors en x_o il existe des coordonnées et une transformation unitaire telle que :

$$d\sigma^2_o(x_o) = \sum_{\alpha=1}^{n} dz^{\alpha} d\bar{z}^{\beta} \quad \text{et} \quad d\sigma^2_{\nu}(x_o) = \sum_{\alpha=1}^{n} (1 + \nu\delta\eta_{\alpha}) dz^{\alpha} d\bar{z}^{\beta} \quad \text{car}$$

$$d\sigma^2_{\nu} = d\sigma^2_o + \nu\delta \sum_{\alpha,\beta=1}^{n} \frac{\partial^2\lambda(\Phi)}{\partial z^{\alpha}_j \partial \bar{z}^{\beta}_j} dz^{\alpha}_j d\bar{z}^{\beta}_j$$

. Une (p,q)- forme $\varphi = (\varphi_j)$; $j \in I$ à valeurs dans E est représentée sur u_j par :

$$\varphi_j = \sum_{\substack{\alpha_1,\dots,\alpha_p \\ \beta_1,\dots,\beta_q}} \varphi_{j\alpha_1\dots\alpha_p\bar{\beta}_1\dots\bar{\beta}_q} dz^{\alpha_1}_j \wedge \dots \wedge dz^{\alpha_p}_j \wedge d\bar{z}^{\beta_1}_j \wedge \dots \wedge d\bar{z}^{\beta_q}_j \otimes s_j$$

où s_j est une section de E au-dessus de u_j . D'après la définition de $(\ ,\ )^{\nu}_{\nu}$

en tout point de X_{c_2}, on a :

$$(\varphi,\varphi)^{\nu}_{\nu} = \frac{1}{b^{\nu}_{j}} \sum_{\substack{\alpha_1,\dots,\alpha_p\\ \beta_1,\dots,\beta_q}} \prod_{i=1}^{p} \frac{1}{1+\nu\delta\eta_{\alpha_i}} \prod_{i=1}^{q} \frac{1}{1+\nu\delta\eta_{\beta_i}} \varphi_{j\alpha_1\dots\alpha_p\bar\beta_1\dots\bar\beta_q} \times \varphi_{j\alpha_1\dots\alpha_p\bar\beta_1\dots\bar\beta_q} \quad \text{i.e.:}$$

(7) $$(\varphi,\varphi)^{\nu}_{\nu} \geqslant \frac{1}{(1+\nu\delta\eta_1)^{2n}} (\varphi,\varphi)^{o}_{\nu} .$$

Posons $z^{\nu} = (1+\nu\delta\eta_1)^{-n}\, y^{\nu}$, donc d'après (7) et la définition de y^{ν} on obtient :

(8) $$\langle z^{\nu}, z^{\nu}\rangle^{o}_{o} \leqslant \int_{X_{c_2}} e^{\nu\lambda(\varphi)} (z^{\nu}, z^{\nu})^{o}_{o}\, dv_o = \int_{X_{c_2}} \frac{e^{-\nu\lambda(\Phi)}}{(1+\nu\delta\eta_1)^{2n}} (\varphi^{\nu},\varphi^{\nu})^{o}_{o}\, dv_o \leqslant \int_{X_{c_2}} (\varphi^{\nu},\varphi^{\nu})^{\nu}_{\nu}\, dv .$$

De (8) et (6') et du fait que $\lambda(\Phi) = 0$ sur le support de η on obtient :

(9) $$\langle z^{\nu}, z^{\nu}\rangle^{o}_{o} \leqslant \frac{1}{c} \int_{X_{c_1}} (\eta,\eta)^{o}_{o}\, dv_o = a < +\infty .$$

Donc la suite (z^{ν}) est fortement bornée dans $\mathcal{L}^{p,q}_{o}(X_{c_2},E)^{o}$; $p+q \geqslant 2n-K+1$, alors il existe une sous-suite, qu'on note encore (z^{ν}) qui converge faiblement vers $z \in \mathcal{L}^{p,q}_{o}(X_{c_2},E)^{o}$.

Pour tout $\varepsilon > 0$, il existe une constante $c_{\varepsilon} > 0$ telle que :

$\lambda(x) \geqslant c_{\varepsilon} > 0$ pour $x \in X_{c_2} \setminus \bar{X}_{c_1+\varepsilon}$;

$$e^{\nu c_{\varepsilon}} \int_{X_{c_2}\setminus \bar X_{c_1+\varepsilon}} (z^{\nu},z^{\nu})^{o}_{o}\, dv_o \leqslant \int_{X_{c_2}\setminus \bar X_{c_1+\varepsilon}} e^{\nu\lambda(\Phi)} (z^{\nu}, z^{\nu})^{o}_{o}\, dv_o \leqslant \int_{\bar X_{c_2}} (\varphi^{\nu},\varphi^{\nu})^{\nu}_{\nu}\, dv_{\nu} \leqslant a \quad \text{i.e. :}$$

(10) $$\int_{X_{c_2}\setminus \bar X_{c_1+\varepsilon}} (z^{\nu}, z^{\nu})^{o}_{o}\, dv_o \leqslant e^{-\nu c_{\varepsilon}}\, a .$$

Le premier membre de (10) converge vers 0 quand $\nu \to +\infty$, donc (z^{ν}) converge vers 0 presque partout sur $X_{c_2} \setminus \bar X_{c_1+\varepsilon}$. Pour toute $\alpha \in \mathcal{L}^{p,q}_{o}(X_{c_2}\setminus \bar X_{c_1+\varepsilon},E)^{o}$, on a :

(11) $$\left|\int_{X_{c_2}\setminus \bar X_{c_1+\varepsilon}} (z^{\nu},\alpha)^{o}_{o}\, dv_o\right| \leqslant \int_{X_{c_2}\setminus \bar X_{c_1+\varepsilon}} (z^{\nu},z^{\nu})^{o}_{o}\, dv_o \int_{X_{c_2}\setminus \bar X_{c_1+\varepsilon}} (\alpha,\alpha)^{o}_{o}\, dv_o .$$

Le premier terme du deuxième membre de (11) converge vers 0 quand $\nu \to +\infty$, donc :

$\int_{X_{c_2}\setminus \bar X_{c_1+\varepsilon}} (z,\alpha)^{o}_{o}\, dv_o = 0$, en particulier $\int_{X_{c_2}\setminus \bar X_{c_1+\varepsilon}} (z,z)^{o}_{o}\, dv_o = 0$ i.e. : $z = 0$ presque partout sur $X_{c_2} \setminus \bar{X}_{c_2+\varepsilon}$ pour tout $\varepsilon > 0$; cela entraîne : $z = 0$ presque partout sur $X_{c_2} \setminus \bar{X}_{c_1}$ i.e. : $\operatorname{supp} z \subset \bar{X}_{c_1}$.

On a : $\eta = \delta_o''^o y^\nu$ et $z^\nu = y^\nu$ sur le support de η ; alors

$$\langle \eta, g\rangle_o^o = \langle \delta_o''^o z^\nu, g\rangle_o^o = \langle z^\nu, d''g\rangle_o^o \xrightarrow{\nu\to\infty} \langle z, d''g\rangle_o^o = \langle \delta_o''^o z, g\rangle_o^o$$

pour toute $g \in \mathcal{D}^{p,q-1}(X_{c_2}, E)$ i.e. : $\eta = \delta_o''^o z$.

D'autre part $\bar{X}_{c_1}$ est compact et son bord est lisse, alors d'après la proposition 1.2.3. de [4] il existe une suite (g_μ) ; $g_\mu \in \mathcal{D}^{p,q}(X_{c_2}, E)$, $\operatorname{supp} g_\mu \subset X_{c_1}$ telle que $\delta_o''^o g_\mu$ (resp. g_μ) converge vers $\delta_o''^o z$ dans $\mathcal{L}_o^{p,q-1}(X_{c_1}, E)^o$ (resp. z dans $\mathcal{L}_o^{p,q}(X_{c_1}, E)^o$) . Donc pour toute $f \in D_{c_1}^{p,q-1}(d'')$:

$$\int_{X_{c_1}} (f,\eta)_o^o \, dv_o = \int_{X_{c_1}} (f, \delta_o''^o z)_o^o \, dv_o = \lim_{\mu\to\infty} \int_{X_{c_1}} (f, \delta_o''^o g_\mu)_o^o \, dv_o$$

$$= \lim_{\mu\to\infty} \int_{X_{c_1}} (d''f, g_\mu)_o^o \, dv_o = \int_{X_{c_1}} (d''f, z)_o^o \, dv_o \quad .$$

En particulier si $f \in N_{c_1}^{p,q-1}(d'')$: $\int_{X_{c_1}} (f,\eta)_o^o \, dv_o = 0$ i.e. : $\hat{\eta} = 0$ en contradiction avec l'hypothèse que $\hat{\eta} \neq 0$ et $\hat{\eta} \in N_{c_1}^{p,q-1}(d'')$.

§ 4. Démonstration du théorème 3.

Cas où $p + q \geqslant 2n - K + 2$:

La démonstration est formelle et n'utilise pas le théorème d'approximation. Soient X et E comme dans l'énoncé du théorème 3. Soit Ψ la fonction plurisousharmonique d'exhaustion définie sur X . Posons $X_c = \{x \in X \ ; \ \Psi(x) < c\}$, $c \in \mathbb{R}$. Soient $\{c_\mu\}$ et $\{\varepsilon_\mu\}$, $\mu = 1, 2, \ldots$ deux suites de nombres réels positifs telles que :

$$(12) \quad \begin{cases} \text{(i)} & X = \bigcup_{\mu=1}^{\infty} X_{c_\mu} \\ \text{(ii)} & \bar{X}_{c_\mu} \subset X_{c_\mu + \varepsilon_\mu} \quad \text{et} \quad \bar{X}_{c_\mu + \varepsilon_\mu} \subset X_{c_{\mu+1}} \end{cases} .$$

Soit $\varphi \in A^{p,q}(X,E)$ telle que $d''\varphi = 0$. On montre par récurrence, sur μ , qu'il existe une forme η_μ définie sur un voisinage ouvert $X_{c_\mu+\varepsilon_\mu}$ de $\bar{X}_{c_\mu}$ telle que :

(13) $$\varphi|_{X_{c_\mu+\varepsilon_\mu}} = d''\tilde{\eta}_\mu \quad \text{et} \quad \eta_{\mu+1}|_{X_{c_\mu+\varepsilon_\mu}} = \eta_\mu .$$

Pour $\mu = 1$, il suffit de montrer que :

$$\varphi|_{X_{c_1+\varepsilon_1}} = d''\,\eta_1 .$$

Mais $\varphi|_{X_{c_1+\varepsilon_1}} \in A^{p,q}(X_{c_1+\varepsilon_1},E)$ alors l'existence de η_1 est une conséquence du théorème 2. Supposons les formes $\eta_1,\ldots,\tilde{\eta}_\mu$ satisfaisant la condition (13) ; d'après le théorème 2 il existe une forme $\eta'_{\mu+1} \in A^{p,q-1}(X_{c_{\mu+1}+\varepsilon_{\mu+1}}, E)$ telle que :

$$\varphi|_{X_{c_{\mu+1}+\varepsilon_{\mu+1}}} = d''\tilde{\eta}'_{\mu+1} .$$

Sur $X_{c_\mu+\varepsilon_\mu}$ la forme $\eta'_{\mu+1} - \eta_\mu$, de type $(p,q-1)$, est d''-fermée car :

$$d''(\eta'_{\mu+1} - \eta_\mu)|_{X_{c_\mu+\varepsilon_\mu}} = \varphi|_{X_{c_\mu+\varepsilon_\mu}} - \varphi|_{X_{c_\mu+\varepsilon_\mu}} = 0$$

par hypothèse $p + q \geqslant 2n - K + 2$ et $K \leqslant n - 1$, donc $q - 1 > 1$. Alors d'après le théorème 2, il existe une forme $\beta \in A^{p,q-2}(X_{c_\mu+\varepsilon_\mu},E)$ telle que :

$\eta'_{\mu+1} - \eta = d''\beta$ dès que $p + q - 1 \geqslant 2n - K + 1$.

Soit α une fonction de classe C^∞ réelle telle que $\alpha = 1$ sur $\bar{X}_\mu$ et $\alpha = 0$ sur $X - \bar{X}_{c_\mu+\varepsilon_\mu}$. Considérons la forme $\eta_{\mu+1} = \eta'_{\mu+1} - d''(\alpha\beta) \in A^{p,q-1}(X_{c_{\mu+1}+\varepsilon_{\mu+1}},E)$. Donc $d''\,\eta_{\mu+1} = d''\,\eta'_{\mu+1} = \varphi|_{X_{c_{\mu+1}+\varepsilon_{\mu+1}}}$ et sur $X_{c_\mu+\varepsilon_\mu}$ on a : $\eta_{\mu+1} = \eta'_{\mu+1} - d''\beta = \eta_\mu$. Ce qui établit la récurrence sur μ . Alors il existe une forme η sur X telle que:

$\eta|_{X_{c_\mu+\varepsilon_\mu}} = \eta_\mu$ satisfaisant l'équation $d''\eta = \varphi$.

Ce qui démontre le théorème 3 dans le cas $p + q \geqslant 2n - K + 2$.

Cas où $p + q \geqslant 2n - K + 1$:

Soient $\langle\varphi,\varphi\rangle = \int_X (\varphi,\varphi)dv$ le produit scalaire défini par la métrique kählérienne initiale sur X et la métrique initiale sur les fibres de E et

$\|\varphi\| = \sqrt{\langle\varphi,\varphi\rangle}$ la norme associée. Dans les notations du théorème d'approximation, il existe une constante $A > 0$ telle que pour toute $\varphi \in \mathscr{L}^{p,q}_{loc}(X_{c_2},E)$, où $\mathcal{L}^{p,q}_{loc}(X_{c_2},E)$ est l'espace des (p,q)-formes mesurables de carré intégrable sur tout compact de X_{c_2}, on a :

(14) $$\|\varphi\|^2_{X_{c_1}} = \int_{X_{c_1}} (\varphi,\varphi)\, dv \leq A \int_{X_{c_1}} (\varphi,\varphi)^o_o\, dv_o$$

car sur X_{c_1} la fonction $\lambda(\Phi) = 0$ et les fonctions $\mu(\Phi)$, $\mu'(\Phi)$, $\mu''(\Phi)$ sont bornées. D'après le théorème de Sard, on peut choisir une suite de nombres réels (a_m), $m \in N$, telle que :

(15) $$\begin{cases} \text{(i)} \quad a_{m+1} > a_m > 0 \text{ et } a_m \text{ tend vers l'infini quand } m \text{ tend vers l'infini ;} \\ \qquad m \in \mathbb{N}\,. \\ \text{(ii)} \quad \text{le bord de l'ouvert relativement compact } X_m = \{x \in X,\ \Psi(x) < a_m\} \text{ est lisse.} \end{cases}$$

Pour tout couple (a_{m+1}, a_{m-1}), $m \geq 1$ on définit :

(a) Une fonction plurisousharmonique Φ_{m+1} d'exhaustion sur X_{m+1} par

$$\Phi_{m+1} = \frac{a_{m+1}}{a_{m+1} - \Psi}\,.$$

(b) Une fonction $\mu_{m+1}(\Phi_{m+1})$ telle que $\mu_{m+1}(t) \geq 0$, $\mu'_{m+1}(t) \geq 0$, $\mu''_{m+1}(t) \geq 0$ pour $t \geq 0$.

(c) Une fonction $\lambda_{m+1}(t)$ telle que : $\lambda_{m+1}(t)$, $\lambda'_{m+1}(t)$ et $\lambda''_{m+1}(t) \geq 0$ pour $t \geq 0$, $\lambda_{m+1}(t) = \lambda'_{m+1}(t) = \lambda''_{m+1}(t) = 0$ pour $t \leq (1 - \frac{a_{m-1}}{a_{m+1}})^{-1}$ et $\lambda''_{m+1}(t) > 0$ pour $t > (1 - \frac{a_{m-1}}{a_{m+1}})^{-1}$.

On définit une métrique kählérienne $ds^2_{1,m+1}$ sur X_{m+1} par

$$ds^2_{1,m+1} = \sum_{\alpha,\beta=1}^{n} g^{1,m+1}_{j\alpha\bar\beta}\, dz^\alpha_j\, d\bar z^\beta_j \quad \text{où} \quad g^{1,m+1}_{j\alpha\bar\beta} = \frac{\partial^2 \log a^{m+1}_j}{\partial z^\alpha_j\, \partial \bar z^\beta_j} \quad \text{avec}$$

$a^{m+1}_j = e^{\mu_{m+1}(\Phi_{m+1})}\, a^{(o)}_j$. On définit une nouvelle structure hermitienne sur les fibres de E par le système $(b^{\nu,m+1}_j) = (b^{m+1}_j\, e^{\nu\lambda_{m+1}(\Phi_{m+1})})$ où $b^{m+1}_j = e^{\mu_{m+1}(\Phi_{m+1})}\, b^{(o)}_j$; $\nu \in \mathbb{N}$. De la même façon que dans le lemme 1 on peut choisir la fonction $\mu_{m+1}(\Phi_{m+1})$ de sorte que $ds^2_{1,m+1}$ soit complète et pour une forme donnée $\varphi \in A^{p,q}(X_{m+1},E)$ on a $< \varphi , \varphi > < + \infty$ où le produit scalaire est défini par $ds^2_{1,m+1}$ et $(b^{o,m+1}_j)$. Nous définissons une suite de métriques kählériennes complètes sur X_{m+1} par :

$$d\sigma^2_{\nu,m+1} = \sum_{\alpha,\beta=1}^{n} (g^{(o)}_{j\alpha\bar\beta} + \delta_{m+1}\,\Omega^{\nu,m+1}_{j\alpha\bar\beta})dz_j\,d\bar z_j \quad \text{où} \quad \Omega^{\nu,m+1}_{j\alpha\bar\beta} = \frac{\partial^2 \log b^{\nu,m+1}_j}{\partial z^\alpha_j\,\partial \bar z^\beta_j} \quad \text{et}$$

$\delta_{m+1} \geqslant \max(1, \frac{2n-1}{C_{m+1}})$ où C_{m+1} est une constante > 0 minorant la plus petite des K plus grandes valeurs propres de $(\Omega^{o,m+1}_{j\alpha\bar\beta})$ par rapport à $(g^{(o)}_{j\alpha\bar\beta})$. Nous considérons le produit scalaire :

$$<\varphi\,,\eta>^{\nu,m+1}_{\nu,m+1} = \int_{X_{a_{m+1}}} (\varphi,\eta)^{\nu,m+1}_{\nu,m+1} dv_{\nu,m+1}$$

pour $\varphi,\ \eta \in \mathcal{D}^{p,q}(X_{a_{m+1}},E)$ où l'indice ν, m + 1 en bas (resp. en haut) désigne le produit scalaire par rapport au système $(b^{\nu,m+1}_j)$ (resp. à $d\sigma^2_{\nu,m+1}$) et où $dv_{\nu,m+1}$ désigne l'élément de volume par rapport à $d\sigma^2_{\nu,m+1}$. Soit $\|\varphi\|^{\nu,m+1}_{\nu,m+1}$ la norme définie par : $\|\varphi\|^{\nu,m+1}_{\nu,m+1} = \sqrt{<\varphi,\,\varphi>^{\nu,m+1}_{\nu,m+1}}$. Soient $\mathcal{L}^{p,q}_{\nu,m+1}(X_{a_{m+1}},E)^{\nu,m+1}$ le complété de $\mathcal{D}^{p,q}(X_{a_{m+1}},E)$ par rapport à la norme $\|\ \|^{\nu,m+1}_{\nu,m+1}$ et $\mathcal{L}^{p,q}_{o,m+1}(X_{a_{m-1}},E)^{o,m+1}$ le complété de $\mathcal{D}^{p,q}(X_{a_{m-1}},E)$ par rapport à la norme $\|\varphi\|_{a_{m-1}} = (\int_{X_{a_{m-1}}} (\varphi\,,\varphi)^{o,m+1}_{o,m+1}\,dv_{o,m+1})^{1/2}$. On désigne par $N^{p,q-1}_{o,m+1}(d'')$ (resp. $N^{p,q-1}_{o,m-1}(d'')$) le noyau de l'opérateur

$d'' : \mathcal{L}^{p,q-1}_{o,m+1}(X_{a_{m+1}},E)^{o,m+1} \to \mathcal{L}^{p,q}_{o,m+1}(X_{a_{m+1}},E)^{o,m+1}$

(resp. $d'' : \mathcal{L}^{p,q-1}_{o,m+1}(X_{a_{m-1}},E)^{o,m+1} \to \mathcal{L}^{p,q}_{o,m+1}(X_{a_{m-1}},E)^{o,m+1}$).

Dans les notations ci-dessus, d'après (14) pour $c_1 = a_{m-1}$ et $c_2 = a_{m+1}$, il existe une constante A_{m-1} telle que, pour toute $\varphi \in \mathcal{L}^{p,q}_{loc}(X_{m+1},E)$ on ait :

$$(16)\quad \|\varphi\|^2_{X_{a_{m-1}}} = \int_{X_{a_{m-1}}} (\varphi,\varphi)\,dv \leqslant A_{m-1} \int_{X_{a_{m-1}}} (\varphi,\varphi)^{o,m+1}_{o,m+1}\,dv_{o,m+1}.$$

Pour montrer le théorème 3, il suffit de démontrer que, pour toute $f \in \mathcal{L}^{p,q}_{loc}(X,E)$, $d''f = 0$, il existe $g \in \mathcal{L}^{p,q}_{loc}(X,E)$ telle que $f = d''g$ si $p + q \geqslant 2n - K + 1$, car d'après le théorème de Dolbeault (cf. [4], théorème 2.2.4. et 2.2.5.) on a l'isomorphisme suivant :

$$H^{p,q}_{loc}(X,E) = \frac{\{f \in \mathcal{L}^{p,q}_{loc}(X,E)\ ;\ d''f = 0\}}{\{\mathcal{L}^{p,q}_{loc}(X,E)\} \cap \{d''g ; g \in \mathcal{L}^{p,q-1}_{loc}(X,E)\}} \approx H^q(X,\,\Omega^p(E)),\ \text{pour } q \geqslant 1.$$

Posons $f_m = f|_{X_{a_m}}$. Nous allons montrer par récurrence sur m qu'il existe une suite (g_m) , $m \geq 1$ telle que :

$$(17)\quad \begin{cases} \text{(i)} & g_m \in \mathscr{L}^{p,q-1}_{loc}(X_{a_m},E) \\ \text{(ii)} & f_m = d'' g_m \\ \text{(iii)} & \|g_{m+1} - g_m\|^2_{X_{a_{m-1}}} < \frac{1}{2^m} \end{cases}$$

pour $m = 1$, l'existence de g_1 est une conséquence du théorème 2 et de l'isomorphisme ci-dessus car $f_1 \in \mathscr{L}^{p,q}_{loc}(X_{a_1},E)$ et $d''f_1 = 0$. Supposons les formes $g_1,\dots,g_m$ satisfaisant (17) déjà choisies. D'après le théorème 2 et le fait que $f_{m+1} \in \mathscr{L}^{p,q}_{loc}(X_{a_{m+1}},E)$, $d''f_{m+1} = 0$, il existe une forme $g'_{m+1} \in \mathscr{L}^{p,q-1}_{loc}(X_{a_{m+1}},E)$ telle que $d''g'_{m+1} = f_{m+1}$. La forme $(g'_{m+1} - g_m)|_{X_{a_{m-1}}}$ est dans $\mathscr{L}^{p,q-1}_{o}(X_{a_{m-1}},E)^o$, d''-fermée car $d''(g'_{m+1} - g_m)|_{X_{a_{m-1}}} = (f_{m+1} - f_m)|_{X_{a_{m-1}}} = 0$.

D'après le théorème d'approximation, il existe une forme $h \in N^{p,q-1}_{o,m+1}(X_{a_{m+1}},E)$ telle que :

$$(18)\quad \int_{X_{a_{m-1}}} (h - (g'_{m+1} - g_m) , h - (g'_{m+1} - g_m))^{o,m+1}_{o,m+1}\, dv_{o,m+1} < 1/2^m A_{m-1} .$$

Posons $g_{m+1} = g'_{m+1} - h \in \mathscr{L}^{p,q-1}_{loc}(X_{a_{m+1}},E)$. D'après la définition de $N^{p,q-1}_{o,m+1}(X_{a_{m+1}},E)$ on a $d''h = 0$. Donc d'après la définition de g'_{m+1} et de (18) on a :

$$(19)\quad \begin{cases} d'' g_{m+1} = d'' g'_{m+1} = f_{m+1} \\ \| g_{m+1} - g_m\|^2_{X_{a_{m-1}}} < 1/2^m . \end{cases}$$

D'après (19) , la suite (g_ν), $\nu \geq m+1$ converge uniformément sur X_m , par rapport à la norme $\| \ \|$, vers une forme g . Donc sur X_m on a :
$g - g_m = \lim_{\nu\to\infty}(g_\nu - g_m)$ et la forme $(g_\nu - g_m)$ $(\nu \geq m+1)$ est d''-fermée. D'autre part l'opérateur d'' est fermé. Alors
$d''g = d''g_m + d''(\lim_{\nu\to\infty}(g - g_m)) = d''g_m = f_m = f|_{X_m}$, $m \in \mathbb{N}$, ce qui démontre le théorème 3 dans le cas $p + q \geq 2n - K + 1$.

<u>Remarque</u>.

De l'isomorphisme de Le Potier (voir [1]) et du théorème 3 on déduit le théorème 2 de [1] sans l'hypothèse sur la minoration de m_o .

B I B L I O G R A P H I E

[1] ABDELKADER (O.). - Annulation de la cohomologie d'une variété kählérienne faiblement 1-complète à valeurs dans un fibré vectoriel holomorphe semi-positif. C.R.A.S. Paris, t. 290, série A, p. 75, 1980.

[2] ANDREOTTI (A.) and VESENTINI (E.). - Carleman estimates for the Laplace-Beltrami equation on complex manifolds. Publ. I.H.E.S., n° 25, p. 81-130, 1965.

[3] GIRBAU (J.). - Sur le théorème de Le Potier d'annulation de la cohomologie. C.R.A.S., Paris, t. 283, série A, p. 355, 1976.

[4] HÖRMANDER (L.). - L^2 estimates and existence theorems for the $\bar{\partial}$ operators. Acta Math., 113, pp. 89-152, 1965.

[5] HÖRMANDER (L.). - An introduction to complex analysis in several variables. North-Holland, 1973.

[6] KATO (T.). - Purturbation theory for linear operators, Springer-Verlag, Berlin, vol. 132, 1966.

[7] KODAÏRA (K.) and MORROW (J.). - Complex manifolds, Halt. Rinehart, and Winston, inc., 1971.

[8] KAZAMA (H.). - Approximation theorem and application to Nakano's vanishing theorem for weakly 1-complete manifolds, Memoir of the Faculty of Science, Kyushu University, ser. A, vol. 27, n° 2, 1973.

[9] NAKANO (S.). - On the inverse of monoidal transformation. Publ. R.I.M.S., vol. 6, pp. 483-502, 1970-1971.

[10] NAKANO (S.). - Vanishing theorems for weakly 1-complete manifolds II .Publ.R.I.M.S., Kyoto University 10 , pp. 101-110, 1974.

[11] OHSAWA (T.) and TAKEGOSHI (K.). - A vanishing theorem for $H^p(X, \Omega^q(B))$ on weakly 1-complete manifolds. A paraître.

[12] SKODA (H.). - Remarques à propos des théorèmes d'annulation pour les fibrés semi-positifs. Séminaire P.Lelong,H.Skoda (Analyse), année 1978-1980, Lecture Notes in Math., n° 822, Springer Verlag, Berlin-Heidelberg-New York, 1980, p. 252.

[13] VESENTINI (E.). - Lecture on Levi convexity of complex manifolds and cohomology vanishing theorems. Tata Institute of Fundamental Research, Bombay, 1967.

SOME RESULTS ON MIXED MANIFOLDS

by

Paul FLONDOR and Eugen PASCU

In [J,1], professor M.Jurchescu considered the notion of mixed manifold, generalizing the notion of C^∞-family of complex manifolds. Then, in [J.2], the notion of mixed-space is also considered. Some important theorems about mixed spaces (manifolds) have been proved.

In this paper, we shall deal only with mixed manifolds. The corresponding results for the more delicate case of the mixed spaces will be treated in the future.

Let us recall some of the notions and results concerning the mixed manifolds (for more details, see [J,1] and [J,2]). First, consider the category $\mathcal{L}$ (local models) whose objects are open subsets of spaces of type $\mathbb{R}^m \times \mathbb{C}^n$ and where the morphisms are C^∞- maps which are holomorphic with respect to the complex variables. Localizing by standard procedures the category $\widetilde{\mathcal{L}}$, one obtains the category $\mathcal{M}$ of mixed manifolds. The structure sheaf of such a manifold X is the sheaf of germs of complex valued morphisms on X and will be denoted by $\mathcal{O}_X(\mathbb{C})$. We shall denote by $\mathcal{O}_X(\mathbb{R})$ the sheaf of germs of real valued morphisms on X. Manifolds are paracompact, Hausdorff and with countable base.

DEFINITION. A mixed manifold is a Cartan manifold iff the following conditions are fulfilled:

1) X is $\mathcal{O}_X(\mathbb{C})$-convex, namely for any compact subset K of X, the set $\hat{K}=\{x\in X \mid |f(x)| \le \sup_{y\in K} |f(y)|\ ,\ f\in\Gamma(X, \mathcal{O}_X(\mathbb{C}))\}$ is a compact subset of X (or equivalently, for any discrete subset A of X, there exists $f\in\Gamma(X, \mathcal{O}_X(\mathbb{C}))$ such that $\sup_{x\in A}|f(x)| =+\infty$).

2) For any point $x\in X$, there exist global coordinates in x.

3) For any $x\in X$, $y\in X$, $x\neq y$, there exists $f\in\Gamma(X, \mathcal{O}_X(\mathbb{C}))$ such that $f(x)\neq f(y)$.

Note that C^∞-manifolds and complex Stein manifolds are Cartan manifolds.

We remind that for any Cartan manifold X and for any separate coherent $\mathcal{O}_X(\mathbb{C})$-module $\mathcal{F}$, one has:

A) For each $x\in X$, $\mathcal{F}_x$ is $\mathcal{O}_{X,x}(\mathbb{C})$-generated by global sections.

B) $H^q(X,\mathcal{F})=0$, $q\geq 1$.

Also, the following embedding theorem holds:

Any Cartan manifold X of type (m,n) may be embedded in a certain $\mathbb{R}^M \times \mathbb{C}^N$ (for sufficiently large M and N).

There are some examples which show that even in the case of C^∞-families of complex manifolds, "good" properties of the fibres cannot be "well" extended.

Example 1. Let X be the (mixed) open subset of $\mathbb{R} \times \mathbb{C}$ given by $X = \mathbb{R} \times \mathbb{C} \setminus \{(0,0)\}$. It is a C^∞-family of complex open sets. Its fibres are Stein (open sets of $\mathbb{C}$). But, it is not a Cartan manifold (open subset of $\mathbb{R} \times \mathbb{C}$) as one can easily see, by noticing that for $K = \{(t,z) \in X \mid |t| \leq 1 \text{ , } |z| = 1\}$ one gets $\hat{K} = \{(t,z) \in X \mid |t| \leq 1, \ |z| \leq 1\}$, hence not a compact subset of X.

Example 2. Let $X = \mathbb{R} \times \mathbb{C} \setminus \{(t,0) \mid |t| \leq 1\}$. It is a C^∞-family of Stein open sets, and it is not a Cartan open subset of $\mathbb{R} \times \mathbb{C}$. Let $\eta : \mathbb{R} \to \mathbb{R}$ a C^∞-map whose support is equal to $[-1,1]$.

Let us consider $f \in \Gamma(X, \mathcal{O}_X(\mathbb{C}))$ defined by:

$$f(t,z) = \begin{cases} \dfrac{\eta(t)}{z} & \text{if } |t| \leq 1 \\ 0 & \text{otherwise} \end{cases}$$

This function cannot be extended over any point of the boundary of X (in the sense that for any $(t,z) \in \partial X$, there exists no open mixed polydisc $U \ni (t,z)$ and no function $F \in \Gamma(U, \mathcal{O}_U(\mathbb{C}))$ with the property $F|_{X \cap U} = f|_{X \cap U}$

However, it is simple enough to characterize cohomologically the open $\mathcal{O}_X(\mathbb{C})$-convex subsets of Cartan manifolds. One has the following (more or less known):

THEOREM 1. Let Y be a Cartan manifold of type (m,n) X an open subset of Y. Then, X is $\mathcal{O}_X(\mathbb{C})$-convex iff $H^q(X, \mathcal{O}_X(\mathbb{C})) = 0$ for each $q = 1, 2, \ldots, n$.

Another result consists of characterizing the Cartan manifolds X by means of the algebra of the global sections of $\mathcal{O}_X(\mathbb{C})$.

DEFINITION. A $\mathbb{C}$-algebra, algebraically isomorphic to $\Gamma(X, \mathcal{O}_X(\mathbb{C}))$ for a certain Cartan manifold X of type (m,n) is called a Cartan algebra.

Cartan algebras and their morphisms (defined naturally as unitary, nonzero, $\mathbb{C}$-algebra morphisms) form a category, denoted by $\mathcal{AC}$.

DEFINITION. Let A be a Cartan algebra. An element of the set $\mathrm{Hom}_{\mathcal{AC}}(A,\mathbb{C})$ is called a character of A.

<u>Example</u>. For each $x \in X$, $\chi_x : \Gamma(X, \mathcal{O}_X(\mathbb{C})) \longrightarrow \mathbb{C}$ defined by $\chi_x(f)=f(x)$ is a character of $\Gamma(X, \mathcal{O}_X(\mathbb{C}))$. It is called point-character.

Due to the embedding theorem, to A and B theorems and to [N], each character of a Cartan algebra $\Gamma(X, \mathcal{O}_X(\mathbb{C}))$ is a point-character. If one considers the canonical topology on $\Gamma(X, \mathcal{O}_X(\mathbb{C}))$ (see [J,1]) characters are continuous maps. It is not difficult to prove that if $u \in \mathrm{Hom}_{\mathcal{AC}}(\Gamma(X, \mathcal{O}_X(\mathbb{C})), \Gamma(Y, \mathcal{O}_Y(\mathbb{C})))$, then it carries $\Gamma(X, \mathcal{O}_X(\mathbb{R}))$ into $\Gamma(Y, \mathcal{O}_Y(\mathbb{R}))$.

Let $\mathcal{C}$ be the category of Cartan manifolds. One can define the functor:

$$\mathcal{G} : \mathcal{C} \longrightarrow \mathcal{AC}$$

by:

$$\mathcal{G}(X)=\Gamma(X,\mathcal{O}_X(\mathbb{C}));$$

for each $f\in \mathrm{Hom}_{\mathcal{C}}(X,Y)$

$$\mathcal{G}_{XY}(f):\Gamma(Y,\mathcal{O}_Y(\mathbb{C}))\longrightarrow\Gamma(X,\mathcal{O}_X(\mathbb{C}))$$ is given by

$$\mathcal{G}_{XY}(f)(h)=h\circ f\ .$$

Then:

THEOREM 2. $\mathcal{G}$ is an antiequivalence of categories.

Proof. We show that

$$\mathcal{G}_{XY}:\mathrm{Hom}_{\mathcal{C}}(X,Y)\to \mathrm{Hom}_{\mathcal{AC}}(\Gamma(Y,\mathcal{O}_Y(\mathbb{C})),\ \Gamma(X,\mathcal{O}_X(\mathbb{C})))$$

is onto when

a) $Y=\mathbb{R}^m \times \mathbb{C}^n$

b) Y is any Cartan manifold of type (m,n).

(Note that, as $\Gamma(Y,\mathcal{O}_Y(\mathbb{C}))$ separates points, $\mathcal{G}_{XY}$ is one-to-one).

For a) let us notice that if

$$u:\Gamma(\mathbb{R}^m \times \mathbb{C}^n,\mathcal{O}_{\mathbb{R}^m\times\mathbb{C}^n}(\mathbb{C}))\to\Gamma(X,\mathcal{O}_X(\mathbb{C}))$$

then $u(\pi_i)=f_i\in\Gamma(X,\mathcal{O}_X(\mathbb{C}))$ (for each canonical projection π_i, $i=1,\dots,m+n$). Note that $f_i\in\Gamma(X,\mathcal{O}_X(\mathbb{R}))$ for $i=1,\dots,m$.

We obtain

$f=(f_1,\dots,g_{m+n}):X\to\mathbb{R}^m\times\mathbb{C}^n$ with $f\in \mathrm{Hom}_{\mathcal{C}}(X,\mathbb{R}^m\times\mathbb{C}^n)$.

One can show then:

$$\mathcal{G}_{X,\mathbb{R}^m \times \mathbb{C}^n}(f)=u.$$

b) We consider the mixed embedding $\varphi : Y \longrightarrow \mathbb{R}^M \times \mathbb{C}^N$.

By a) there exists $f : X \longrightarrow \mathbb{R}^M \times \mathbb{C}^N$ such that

$$\mathcal{G}_{X,\mathbb{R}^M \times \mathbb{C}^N}(f)=u \circ \mathcal{G}_{Y,\mathbb{R}^M \times \mathbb{C}^N}(\varphi)$$

Now f factorizes through Y i.e. one gets then the commutative diagram

$$\begin{array}{ccc} X & \xrightarrow{\ f\ } & \mathbb{R}^M \times \mathbb{C}^N \\ & {\scriptstyle \tilde{f}}\searrow \quad \nearrow{\scriptstyle \varphi} & \\ & Y & \end{array}$$

for a $\tilde{f} \in \mathrm{Hom}_{\mathcal{C}}(X,Y)$.

We obtain then

$$\mathcal{G}_{XY}(\tilde{f})=u .$$

Moreover, let us note that one may refind the manifold X starting from $\Gamma(X, \mathcal{O}_X(\mathbb{C}))$ as follows: One takes $\mathfrak{X}$ = set of characters of $\Gamma(X, \mathcal{O}_X(\mathbb{C}))$ and endows it with the topology induced by the weak topology on the dual of $\Gamma(X, \mathcal{O}_X(\mathbb{C}))$. The map $\theta : X \longrightarrow \mathfrak{X}$ defined by $\theta(x)=\chi_x$ is a homeomorphism. One has $\Gamma(X, \mathcal{O}_X(\mathbb{R})) = \bigcap_{\chi \in \mathfrak{X}} \chi^{-1}(\mathbb{R})$.

The mixed structure on $\mathfrak{X}$ may be obtained by lifting (in the continuous functions) the mixed structures on $\mathbb{R}^p \times \mathbb{C}^q$ by all $f=(f_1,\dots,f_p,f_{p+1},\dots,f_{p+q})$ with $f_i \in \Gamma(X, \mathcal{O}_X(\mathbb{R}))$, $i=1,\dots,p$, $f_{p+j} \in \Gamma(X, \mathcal{O}_X(\mathbb{C}))$ $j=1,\dots,q$, and p,q positive integers. The map θ considered above becomes then an isomorphism.

THEOREM 3. (suggested by C.Bănică). Let X be a mixed manifold of type (m,n), let $\mathcal{F}$ be a coherent $\mathcal{O}_X(\mathbb{C})$-module. Then:

$$H^q(X,\mathcal{F})=0, \text{ for } q \geq n+1$$

Proof. By a fact similar to that in [R] one can show that the problem is "local". As a consequence of [J,1], one gets for each $x \in X$, for a sufficiently small open Cartan neighbourhood U of x a resolution:

$$0 \to \mathcal{O}_U^{r_{m+n}}(\mathbb{C}) \to \ldots \to \mathcal{O}_U^{r_o}(\mathbb{C}) \to \mathcal{F}|_U \to 0 \quad .$$

Then one obtains

$H^q(V,\mathcal{F}|_V)=H^{q+m+n}(V, \mathcal{O}_V^{r_{m+n}}(\mathbb{C}))$, for each V open subset of U and for each $q \geq n+1$, by mixed Dolbeault resolution ([J,1]). As $H^k(V, \mathcal{O}_V(\mathbb{C}))=0$, for $k \geq n+1$, one gets

$$H^q(V,\mathcal{F}|_V)=0 \quad \text{for any } q \geq n+1$$

and the conclusion follows.

We shall now consider vector bundles.

DEFINITION. A trivial complex vector bundle of rank k over the mixed manifold X is the mixed manifold $X \times \mathbb{C}^k$. A morphism between the trivial complex vector bundles on X, $X \times \mathbb{C}^k$ and $X \times \mathbb{C}^j$ is given by a mixed morphism

$$h=(h_1,h_2):X \times \mathbb{C}^k \to X \times \mathbb{C}^j$$

which satisfies:

a) h_1 is the canonical projection on X.

b) There exists a (uniquely determined) mixed morphism $g: X \to \mathcal{L}(\mathbb{C}^k, \mathbb{C}^j)$ such that the diagram

$$\begin{array}{ccc} X \times \mathbb{C}^k & \xrightarrow{g \times 1_{\mathbb{C}^k}} & \mathcal{L}(\mathbb{C}^k, \mathbb{C}^j) \times \mathbb{C}^k \\ & {\scriptstyle h_2} \searrow \quad \swarrow {\scriptstyle \delta} & \\ & \mathbb{C}^j & \end{array}$$

where $\delta(A,z) = A \cdot z$, is commutative.

Note that such a morphism is an isomorphism iff $j=k$ and the values of g are in $GL(k,\mathbb{C})$.

DEFINITION. A (locally trivial)' complex vector bundle of rank k over the mixed manifold X is a mixed manifold E together with a mixed morphism $\pi: E \to X$ which is onto, such that there exists an open covering $\mathcal{U} = (U_i)_{i \in I}$ of X and the following properties are satisfied:

1) There exist mixed isomorphisms $f_i : \pi^{-1}(U_i) \to U_i \times \mathbb{C}^k$ and the diagram

$$\begin{array}{ccc} \pi^{-1}(U_i) & \xrightarrow{f} & U_i \times \mathbb{C}^k \\ & {\scriptstyle \pi} \searrow \quad \swarrow {\scriptstyle \pi_1} & \\ & U_i & \end{array}$$

is commutative.

2) $f_i \circ f_j^{-1}|_{U_i \cap U_j \times \mathbb{C}^k} : U_i \cap U_j \times \mathbb{C}^k \to U_i \cap U_j \times \mathbb{C}^k$ are isomorphisms of trivial complex vector bundles of rank k over $U_i \cap U_j$ for each $(i,j) \in I \times I$.

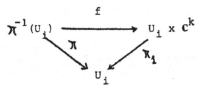

One can show that there exists a one-to-one and onto correspondence between (classes of isomorphisms of) locally trivial complex vector bundles of rank k over a mixed manifold X and (classes of isomorphisms of) locally free sheaves of rank k on X.

The usual canonical operations between such bundles may be performed in the category of (locally trivial) complex vector bundles of finite rank over a mixed manifold X. The existence of the dual E^* and the fact that each element $h^* \in H^0(X,E^*)$, determines a mixed morphism $h:E \to \mathbb{C}$ whose restrictions to each fibre $\pi^{-1}(x)$ are $\mathbb{C}$-linear, enables us with a technique, by means of which we may prove:

THEOREM 4. A (locally trivial) complex vector bundle of finite rank over a Cartan manifold, is itself a Cartan manifold.

We remind that all these results will be extended to spaces.

We want to express all our gratitude to professor M.Jurchescu for his constant help.

Added in Proof

Professor B.Gilligan from Regina University informed us that as a consequence of Theorem 4, and by means of a technique similar to that from [M,M] he succeeded into proving the analogous of Theorem 4 for principal bundles which have a complex Lie-structure group and Stein fibres over Cartan manifolds.

REFERENCES

[B,S] Bănică C., Stănăşilă O., Metode algebrice în teoria globală a spaţiilor complexe, Ed.Academiei RSR, Bucureşti, 1974.

[G,R] Grauert H., Remmert R., Theorie des Steinschen Räume, Springer, Grundlehren der Mathematischen Wissenschafften, no.227, 1977.

[J,1] Jurchescu M., Variétés mixtes, Proceedings of the IIIrd Romanian-Finnish Seminar, Springer Lectures Notes in Mathematics, no. 743 (1979), pp.431-448.

[J,2] Jurchescu M., Espaces mixtes, Proceedings of the Vth Romanian-Finnish Seminar, Springer Lectures Notes in Mathematics (this volume).

[M,M] Matsushima Y., Morimoto A., Sur certains espaces fibrés holomorphes sur une variété de Stein, Bull. Soc.Math.France, vol.88, no.2, 1960, p.137-155.

[N] Nagel, A., Cohomology, maximal ideals and point evaluations, Proc.Amer.Math.Soc.vol.42, no.1, 1974, pp. 47-50.

[R] Reiffen, H.J., Riemannsche Hebbarkeitssätze für Cohomologieklassen mit kompaktem Träger, Math.Ann., vol. 164, 1966, pp.272-279.

HOLOMORPHIC REDUCTIONS OF HOMOGENEOUS SPACES

Bruce Gilligan*

1. INTRODUCTION.

Suppose X is a compact complex analytic space of dimension n. If one defines the algebraic dimension k of X to be the transcendence degree of its function field $M(X)$ over $\mathbb{C}$, then it is known that $k \leq n$ [31],[26]. But given any $n \geq 2$ and k with $0 \leq k \leq n$, one can find a space X of dimension n having algebraic dimension k. Analogous statements can also be made about the rank of the function algebras of non-compact spaces.

Now it may turn out that there exists a complex space X* and a holomorphic mapping $\pi: X \to X^*$ of X onto X* which induces an isomorphism $\pi^*: \mathcal{O}(X^*) \to \mathcal{O}(X)$ of $\mathcal{O}(X^*)$ onto $\mathcal{O}(X)$ (resp. and also induces an isomorphism $\pi^*: M(X^*) \to M(X)$ of $M(X^*)$ onto $M(X)$). Such couples (X^*, π) were called *holomorphic* (resp. *meromorphic*) *reductions* by Remmert [27]. Ideally one would like to find a reduction where X* is even holomorphically (resp. meromorphically) separable. An important example is the Remmert reduction: If X is an irreducible holomorphically convex analytic space, then the natural map $\pi: X \to X/\sim$, where $x_1 \sim x_2$ for $x_1, x_2 \in X$ if and only if $f(x_1) = f(x_2)$ for every $f \in \mathcal{O}(X)$, is a proper holomorphic mapping and $X/\sim$ is a Stein analytic space [27]. The essential point in this case is that the holomorphic separation (the partition given by the equivalence classes of $\sim$) is an analytic decomposition of X in the sense of K. Stein [34]. But this is certainly not always so! For, Grauert [16] has given examples of complex manifolds containing non-compact divisors on which every holomorphic function is constant, but off of which the holomorphic functions separate points and give local coordinates. However, if X is a complex manifold which is homogeneous under the action of a Lie group, then it is quite easy to see that the *holomorphic* (resp. *meromorphic*) *separation reduction* of X actually does exist (cf. sections 2 and 4).

This note is a survey of some results relating to reductions of homogeneous spaces with the main emphasis on the holomorphic reduction where a good deal of recent progress has been made. In particular, we point out the role that ends play in this theory.

*Partially supported by NSERC Grants A-3494 & T1365

2. REDUCTIONS OF COMPLEX HOMOGENEOUS SPACES.

If a complex Lie group G is acting holomorphically and transitively on the connected complex manifold X, then X is called a *complex homogeneous space*. As is well-known, X is biholomorphic to the complex coset space G/H, where $H := \{g \in G : gx_0 = x_0\}$ for some fixed $x_0 \in X$. Holomorphic and meromorphic reductions of complex homogeneous spaces always exist. For, if one defines to be equivalent those points which the function algebra (resp. the function field) cannot separate, then it is straightforward to show that the resulting partition of X is G-invariant. Hence it follows from a remark of Remmert and van de Ven [28] that the subgroup J of G which stabilizes the equivalence class of x_0 is closed, complex and contains H. From this it is easy to conclude that the fibers of the homogeneous fibration $\pi: G/H \to G/J$ are exactly the analytic sets of this partition. Clearly π is surjective and induces an isomorphism of $\mathcal{O}(G/J)$ onto $\mathcal{O}(G/H)$ (resp. $M(G/J)$ onto $M(G/H)$) and is called the *holomorphic* (resp. *meromorphic*) *separation fibration*.

First let us look at meromorphic reductions. For complex tori it is classical that the meromorphic separation fibration exists and its base is an abelian variety. And Chow [11] showed that a compact homogeneous *algebraic* manifold whose algebraic dimension equals its complex dimension is projective algebraic. The case of an arbitrary compact complex homogeneous space was handled by Grauert and Remmert [17]. By looking at the intersection of divisors they showed the existence of the meromorphic separation fibration and then using analytic techniques proved that its base is projective algebraic. An example is any homogeneous Hopf surface which fibers equivariantly over $\mathbb{P}^1$ with an elliptic curve as fiber. But even though the fiber of the separation fibration is always complex parallelizable, it need not be a torus or tower of tori. For, as noted in [17], it follows from a theorem of Tits [36] that $M(S/\Gamma) = \mathbb{C}$ for any uniform discrete subgroup Γ of a semi-simple complex Lie group S.

In the case of non-compact homogeneous spaces not so much is known. For example, it is not clear what meromorphic separability implies about the global geometry. However, Ahiezer [1] has recently shown that for certain non-uniform Zariski dense discrete subgroups Γ of a semi-simple complex Lie group S one has $M(S/\Gamma) = \mathbb{C}$. As well Vogt [37] has investigated divisors on reduced abelian groups, i.e. non-compact quotients $T = \mathbb{C}^n/\Gamma$ with $\mathcal{O}(T) = \mathbb{C}$, and has given a characterization in terms of the lattice Γ for when every divisor is the divisor of a theta function.

The holomorphic function algebra of a complex Lie group is closely related to the Cauchy-Riemann structure of a maximal compact subgroup. Recall that any connected Lie group G has an Iwasawa decomposition, i.e. G is homeomorphic to $K \times \mathbb{R}^s$, where K is a maximal compact subgroup of G [19]. In particular, any such K is connected. Now if G is a connected complex Lie group with Lie algebra $\mathfrak{g}$ and $\mathfrak{k}$ is the Lie algebra of K, then the complex subalgebra $\mathfrak{k} \cap i\mathfrak{k}$ is not unique. However, the condition $\mathfrak{k} \cap i\mathfrak{k} = (0)$, which following Matsushima and Morimoto [23] we call condition (P), is independent of the choice of K, for any two maximal compact subgroups of G are conjugate. Clearly condition (P) is equivalent to the fact that K contains no complex subgroup of positive dimension. The remarkable fact is the following result of Matsushima-Morimoto [23]:

THEOREM. *A connected complex Lie group* G *satisfies condition* (P) *if and only if* G *is Stein.*

Let us sketch the proof. It is easy to see that a holomorphically separable complex Lie group satisfies condition (P). For, if we set $\mathfrak{k}' := \mathfrak{k} \cap i\mathfrak{k}$ and let K' denote the corresponding complex subgroup of K, then given $f \in \mathcal{O}(G)$, the composition of $f|K'$ with $\exp: \mathfrak{k}' \to K'$ is bounded on $\mathfrak{k}'$ and hence constant. Since G is holomorphically separable and exp is locally biholomorphic, it follows that $K' = (e)$, i.e. $\mathfrak{k}' = (0)$. To prove the other direction one first observes that any semi-simple complex Lie group can be faithfully represented onto a closed subgroup and thus is Stein and satisfies condition (P). Also if an abelian group satisfies condition (P), then it is not hard to see that it is biholomorphic to $\mathbb{C}^m \times (\mathbb{C}^*)^n$. For the general case one needs the following (for the proof see [23]):

PROPOSITION. *Suppose* G *is a connected complex Lie group which satisfies condition* (P) *and is not simple. Then there exists a proper connected normal closed complex subgroup* A *of* G, *which is isomorphic to a closed complex subgroup of* $GL(n,\mathbb{C})$, *such that the quotient group* G/A *satisfies condition* (P).

The proof of the Theorem is now completed by using induction on the dimension of G and the fact that the total space of a principal bundle with Stein base whose structure group is a closed complex subgroup of some $GL(n,\mathbb{C})$ is Stein. As noted by Serre [30] this follows easily since the total space of a holomorphic vector bundle with Stein base is Stein and $GL(n,\mathbb{C})$ and thus any closed complex subgroup of it are also Stein.

Now suppose one wants to form the holomorphic separation of a connected complex Lie group G. To do this, let

$$G_0 := \{g \in G : f(g) = f(e) \text{ for every } f \in \mathcal{O}(G)\}.$$

Clearly G/G_0 is holomorphically separable. What is not so clear is that G_0 is a normal subgroup of G and therefore the *complex Lie group* G/G_0 is Stein. For this reason Morimoto called G_0 the *Steinizer* of G. He in fact proved the following [24]:

THEOREM. *Suppose G is a connected complex Lie group and let G_0 be its Steinizer. Then G_0 is a connected central closed complex subgroup of G and $\mathcal{O}(G_0) = \mathbb{C}$.*

Only under additional assumptions is there anything known about the function algebras of complex homogeneous spaces $X = G/H$ and these assumptions are, for the most part, about the pair (G,H) not about the underlying complex manifold X. One such assumption is that the group G is reductive and then assuming H is connected and using Lie algebra techniques Matsushima [21] proved that G/H is Stein precisely if H is also reductive. Also Barth and Otte [3] showed that if the quotient G/H of a reductive group G, linearly algebraic over $\mathbb{C}$, is holomorphically separable, then H is an algebraic subgroup of G. It then follows from earlier work of Bialynicki-Birula, Hochschild and Mostow [5] that such a G/H is quasi-affine, i.e. biregularly equivalent to a Zariski open subset of an affine algebraic variety.

If G is nilpotent, then G/H is called a nil-manifold. For these spaces with H connected Matsushima [22] gave a Lie algebra condition which ensures that G/H is Stein. And in a joint paper with Huckleberry [14], we showed that if $\pi: G/H \to G/J$ is the holomorphic separation fibration, then G/J is Stein and J/H is connected with $\mathcal{O}(J/H) = \mathbb{C}$. The idea underlying this result is a condition (P) à la Matsushima-Morimoto! The proof involves factoring out by the center of G (it is connected and has positive dimension) and using induction along with Matsushima-Morimoto's solution of the Serre Conjecture for holomorphic fiber bundles with Stein base and fiber having a *connected* complex Lie group as structure group [23].

Many questions are still open. For instance, there is much evidence to suggest that a holomorphically separable complex homogeneous space has a Stein envelope of holomorphy. But we simply do not know if this is always so! One of the main difficulties is whether the

Serre Conjecture, for which there exist counterexamples (e.g. [32]), holds for homogeneous fibrations of solv-manifolds (see [33]). Clearly the resolution of the conjecture in this setting would be a big step towards the complete understanding of the function algebras of homogeneous spaces.

3. THE ROLE OF ENDS.

In this section we turn our attention to the role that the ends of a complex homogeneous space play in determining its function theory. First let us recall the definition which may be formulated in the following way. An *end* (in the sense of Freudenthal [12]) of a non-compact Hausdorff space X is a function E which assigns to each compact subset K of X an arc component E(K) of $X \setminus K$ subject to the condition $E(K) \supseteq E(L)$ if $K \subseteq L$.

Now in complex analysis one does not usually consider either the ends or the one-point compactifications of complex spaces, for there is generally no natural way to extend the complex structure. As an example, if $\mathbb{C}^n \cup \{\infty\}$ for $n > 1$ had a complex structure, then every holomorphic function on $\mathbb{C}^n$ would extend and thus would be constant. However, in 1953 Serre [30] observed, generalizing a result of Behnke [4], that a Stein space of dimension greater than one has one end. Thus the possibility exists that "the ends of a complex space form some sort of obstruction to that space having a Stein envelope of holomorphy."

What is known about the ends of homogeneous spaces? Freudenthal showed that a connected topological group which is locally connected, locally compact and second countable has at most two ends [12]. For a connected Lie group G one can see this directly from the Iwasawa decomposition [19], G having two ends precisely if $G \sim K \times \mathbb{R}$, where K is a maximal compact subgroup of G. Using the Iwasawa decomposition together with a spectral sequence argument, Borel [9] proved that a homogeneous space G/H with H connected has at most two ends. Moreover, G/H has two ends precisely if $G/H \sim K/L \times \mathbb{R}$, where K (resp. L) is a maximal compact subgroup of G (resp. of H, contained in K). Also Borel pointed out that for every integer $k > 2$ there exist discrete subgroups Γ_k of $SL(2,\mathbb{R})$ such that $SL(2,\mathbb{R})/\Gamma_k$ has k ends. Such subgroups Γ_k exist in $SL(2,\mathbb{C})$ as well (e.g. [15]). In fact Bianchi [6] found explicit examples for certain k by constructing fundamental domains; e.g. for $k = 3$ one can take $SL(2,R) < SL(2,\mathbb{C})$, where R is the ring of integers of the quadratic imaginary number field $\mathbb{Q}(\sqrt{-7})$.

Now in order to illustrate the connection with function theory, we will construct a class of complex homogeneous spaces which are holomorphically separable and have two ends. Suppose $Q = S/P$ is a homogeneous rational manifold, i.e. S is a semi-simple complex Lie group and P is a parabolic subgroup. Since $\pi_1(Q) = 0$, there is an equivariant embedding of Q into some projective space [7], which we take to be the hyperplane at infinity $\{z_0 = 0\} = \mathbb{P}^{N-1} \subset \mathbb{P}^N$. Now connect each point of Q with the point $[1:0:\ldots:0]$ by a complex line in $\mathbb{P}^N$. The resulting space $\hat{X}$ is the total space of a $\mathbb{P}^1$-bundle over Q. Let $\mathbb{C}^N = \mathbb{P}^N \setminus \mathbb{P}^{N-1}$ and let X be the *affine cone* $\hat{X} \cap \mathbb{C}^N$ with its vertex $[1:0:\ldots:0]$ removed. Then X is a $\mathbb{C}^*$-bundle over Q, for the zero and infinity sections of the first bundle have been removed. There is an induced action of S on X, since $H^1(Q,\mathcal{O}) = 0 = H^2(Q,\mathcal{O})$ and $H^2(Q,\mathbb{Z})$ is discrete. Moreover, S acts transitively on X since S-orbits in X lie over S-orbits in Q and the bundle has no sections. That affine homogeneous cones minus their vertices are the only possibilities is the content of the next result [13].

THEOREM. *Suppose G is a connected complex Lie group and H is a closed complex subgroup such that G/H has more than one end and $\mathcal{O}(G/H) \neq \mathbb{C}$. Let $\pi: G/H \to G/J$ be the holomorphic separation fibration. Then G/J is an affine homogeneous cone with its vertex removed and J/H is connected and compact.*

The proof involves essentially three steps. First one has to note that if the fiber and base of a locally trivial fiber bundle are connected and non-compact, then its total space has one end. This can be seen directly by using the definition of ends, or else by using a spectral sequence argument pointed out to us by Abels. As an immediate consequence, if the total space has more than one end then either the fiber is compact and the base has the same number of ends as the total space or the base is compact and the fiber has at least as many ends as the total space. This allows one to proceed by induction provided that there really do exist enough fibrations. Now one always has the normalizer fibration $G/H \to G/N$ (see [36],[10]), where $N := N_G(H^0)$ is the normalizer in G of the connected component of the identity H^0 of H. Its base G/N is an orbit in some $\mathbb{P}^k$ which can be further fibered because the G'-orbits in G/N are closed [13]. But the main technical difficulty is to show the existence of a fibration of the fiber $N/H = N/H^0/H/H^0 = \hat{N}/\Gamma$, where $\Gamma := H/H^0$ is a discrete subgroup of the complex Lie group $\hat{N} := N/H^0$, whenever $\hat{N}$ is neither semi-simple nor solvable. A natural idea is to fiber by the radical orbits, but it is not true that the radical orbits are always closed (e.g. [25]). However, using the Zassenhaus Lemma of Auslander [2], one can show that in the case of discrete isotropy there exists a "hull" which contains the radical orbits. The third step is to apply analytic arguments in order to show that the only homogeneous $\mathbb{C}^*$-bundles with compact base whose total spaces have function algebras of maximal rank are the affine homogeneous cones with their vertices removed.

Thus a holomorphically separable complex homogeneous space X which has more than one end does have a Stein envelope of holomorphy, namely the cone together with its vertex. Ironically X is even quasi-affine, for except in the case $X = \mathbb{C}^*$, the semi-simple group S acting transitively on the base is acting transitively on X. Thus the Theorem follows *a posteriori* from Borel's Theorem [9]. But in fact the proof does not depend on Borel's Theorem and uses only the analytic assumption $\mathcal{O}(X) \neq \mathbb{C}$ and not some assumption about the pair (G,H).

In passing we note that Steinsiek [35] has recently classified the semi-simple complex Lie groups which can act transitively and effectively on homogeneous rational manifolds and that for some of these spaces the group is not unique.

4. OTHER FORMS OF HOMOGENEITY.

If X is a compact complex manifold, then the group Aut(X) of all biholomorphic maps of X onto itself is a complex Lie group [8]. But for X non-compact this is no longer so. A standard example is $\mathbb{C}^2$, since $\mathrm{Aut}(\mathbb{C}^2)$ contains automorphisms of the form

$$\alpha_f : \mathbb{C}^2 \to \mathbb{C}^2, \quad (z,w) \mapsto (z, w + f(z)),$$

where $f \in \mathcal{O}(\mathbb{C})$, and thus is not finite dimensional. However, $\mathbb{C}^2$ itself is a complex Lie group. A class of examples which are homogeneous under the full automorphism group but on which no Lie group can act transitively has been given by W. Kaup [20]. Using his idea, one can construct such spaces which also have an arbitrary number of ends.

Suppose k is a positive integer and $z_1, \ldots, z_{k-1} \in \mathbb{C}$ are distinct points. Let $X := \mathbb{C}^2 \setminus \{(z_1,0), \ldots, (z_{k-1},0)\}$. Then X is holomorphically separable and has k ends. Moreover, X is homogeneous, i.e. given $x_1, x_2 \in X$ there exists $g \in \mathrm{Aut}(X)$ with $g(x_1) = x_2$. To see this, first note that $\mathbb{C} \times \mathbb{C}^*$ is homogeneous under complex linear maps. Hence we are done once we show that given any point $(z,0) \in X$, where $z \neq z_i$, $1 \leq i \leq k-1$, there exists $g \in \mathrm{Aut}(X)$ such that $g(z,0) = (z',w)$ with $w \neq 0$. But given any $f \in \mathcal{O}(\mathbb{C})$ such that $f(z) = 0$ precisely if $z = z_i$, $1 \leq i \leq k-1$, then the associated automorphism α_f of $\mathbb{C}^2$ defined above leaves X invariant, i.e. $\alpha_f \in \mathrm{Aut}(X)$, and does the job.

Now if $k > 2$ then we claim that no Lie group G can act transitively on X. Of course the theorem in the previous section rules out complex Lie groups since X has more than two ends, but we can also eliminate real Lie groups. For, suppose $X = G/H$ for some G and H. Since $\pi_1(X) = 0$, it follows from the exact homotopy sequence of the fibration $G \to G/H$ that H is connected. But then X can have at most two ends by Borel's Theorem [9], which is a contradiction. Note that this means that the subgroup of Aut(X) generated by non-singular complex linear maps of $\mathbb{C} \times \mathbb{C}^*$ and by any automorphism α_f of the form described above is not a Lie group.

The question then arises whether one can say anything about the structure of complex manifolds homogeneous under the action of a real Lie group. Using the remark of Remmert-van de Ven [28], together with a result of Holmann [18], one can again show the existence of the *holomorphic* (resp. *meromorphic*) *separation fibration*. But very little is known even in the holomorphically separable case. Examples are provided by bounded homogeneous domains in $\mathbb{C}^n$. They have one end and it follows from Liouville's Theorem that no connected complex Lie group acts transitively on them. However, we know of no examples having more than one end. Indeed, we propose the following!

CONJECTURE. *Suppose X is a connected complex manifold homogeneous under the action of a real Lie group such that X has more than one end and $\mathcal{O}(X)$ has maximal rank. Then X is also homogeneous under the action of a connected complex Lie group. In particular, X is an affine homogeneous cone with its vertex removed.*

If $X = G/H$ with H connected, then the proof follows easily from Borel's Theorem [9]. For, G/H then has two ends and the orbit of any maximal compact subgroup of G is a compact real hypersurface whose Levi form thus has the same signature at every point. Because $\mathcal{O}(G/H)$ has maximal rank, it is easy to see (cf. [15, Theorem 2]) that these hypersurfaces are in fact strongly pseudoconvex. Thus G/H is a $\mathbb{C}^*$-bundle over a homogeneous projective algebraic manifold Y [29]. Hence $\mathrm{Aut}(\hat{X})$ is a complex Lie group, where $\hat{X}$ is the associated $\mathbb{P}^1$-bundle over Y. The stabilizer in $\mathrm{Aut}(\hat{X})$ of X is then a complex Lie group which is acting transitively on X.

However, H may have an infinite number of connected components and then the above proof does not work.

5. REFERENCES.

1. Ahiezer,D.N., Invariant meromorphic functions on semi-simple complex Lie groups, Invent. Math. (to appear).

2. Auslander,L., On radicals of discrete subgroups of Lie groups, Amer. J. Math. 85 (1963), 145-150.

3. Barth,W. and M.Otte, Invariante holomorphe Funktionen auf reduktiven Liegruppen, Math. Ann. 201 (1973), 97-112.

4. Behnke,H., Über die Fortsetzbarkeiten analytischer Funktionen mehrerer Veränderlichen und den Zusammenhang der Singularitäten, Math. Ann. 117 (1940/41), 89-97

5. Bialynicki-Birula,A., G. Hochschild and G. Mostow, Extensions of representations of algebraic linear groups, Amer. J. Math. 85 (1963), 131-144.

6. Bianchi,L., Sui gruppi di sostituzioni lineari con coefficienti appartenenti a corpi quadratici imaginari, Math. Ann. 40 (1892), 332-412.

7. Blanchard,A., Espaces fibrés kählériens compacts, C.R. Acad. Sc. Paris 238 (1954), 2281-2283.

8. Bochner,S. and D.Montgomery, Groups on analytic manifolds, Ann. of Math. 48 (1947), 659-669.

9. Borel,A., Les bouts des espaces homogènes de groupes de Lie, Ann. of Math. 58 (1953), 443-457.

10. Borel,A. and R.Remmert, Über kompakte homogene Kählersche Mannigfaltigkeiten, Math. Ann. 145 (1962), 429-439.

11. Chow,W.L., On the projective embeddings of homogeneous varieties. *Algebraic Geometry and Topology*, (A symposium in honor of S. Lefschetz), 122-128, Princeton Univ. Press, 1957.

12. Freudenthal,H., Über die Enden topologischer Räume und Gruppen, Math. Z. 33 (1931), 692-713.

13. Gilligan,B., Ends of complex homogeneous manifolds having non-constant holomorphic functions, Arch. Math. (to appear).

14. Gilligan,B. and A.Huckleberry, On non-compact complex nil-manifolds, Math. Ann. 238 (1978), 39-49.

15. Gilligan,B. and A.Huckleberry, Complex homogeneous manifolds with two ends, Michigan Math. J. (to appear).

16. Grauert,H., Bemerkungswerte pseudokonvexe Mannigfaltigkeiten, Math. Z. 81 (1963), 377-391.

17. Grauert,H. and R.Remmert, Über kompakte homogene komplexe Mannigfaltigkeiten, Arch. Math. 13 (1962), 498-507.

18. Holmann,H., Holomorphe Blätterungen komplexer Räume, Comment. Math. Helv. 47 (1972), 185-204.

19. Iwasawa,K., On some types of topological groups, Ann. of Math. 50 (1949), 507-558.

20. Kaup,W., Reelle Transformationsgruppen und invariante Metriken auf komplexen Räumen, Invent. Math. 3 (1967), 43-70.

21. Matsushima,Y., Espaces homogènes de Stein des groupes de Lie complexes, I, Nagoya Math. J. 16 (1960), 205-218.

22. Matsushima,Y., Espaces homogènes de Stein des groupes de Lie complexes, II, Nagoya Math. J. 18 (1961), 153-164.

23. Matsushima,Y. and A.Morimoto, Sur certaines espaces fibrés holomorphes sur une variété de Stein, Bull. Soc. Math. France 88 (1960), 137-155.

24. Morimoto,A., Non-compact complex Lie groups without non-constant holomorphic functions. *Proceedings of the Conference on Complex Analysis*, Minneapolis, 1964, 256-272.

25. Otte,M. and J.Potters,Beispiele homogener Mannigfaltigkeiten, Manuscripta Math. 10 (1973), 117-127.

26. Remmert,R., Meromorphe Funktionen in kompakten komplexen Räumen, Math. Ann. 132 (1956), 277-288.

27. Remmert,R., Reduction of Complex Spaces. *Seminar on Analytic Functions*, Princeton, 1958, 190-205.

28. Remmert,R. and A.van de Ven, Zur Funktionentheorie homogener komplexer Mannigfaltigkeiten, Topology 2 (1963), 137-157.

29. Rossi,H., Homogeneous strongly pseudoconvex hypersurfaces, Rice Univ. Studies 59 (1972), 131-145.

30. Serre,J-P., Quelques problèmes globaux relatifs aux variétés de Stein, *Colloque sur les fonctions de plusieurs variables*, Bruxelles, 1953, 57-68.

31. Siegel,C.L., Meromorphe Funktionen auf kompakten analytischen Mannigfaltigkeiten, Nachr. Akad. Wiss. Göttingen, Math-Phys. Kl. IIa (1955), 71-77.

32. Skoda,H., Fibrés holomorphes à base et à fibre de Stein, Invent. Math. 43 (1977), 97-107.

33. Snow,J.E., Complex solv-manifolds of dimension two and three, Univ. of Notre Dame, Ph.D. Thesis, 1979.

34. Stein,K., Analytische Zerlegungen komplexer Räume, Math. Ann. 132 (1956), 63-93.

35. Steinsiek,M., Universität Münster, Dissertation (to appear).

36. Tits,J., Espaces homogènes complexes compacts, Comment. Math. Helv. 37 (1962), 111-120.

37. Vogt,C., Geradenbündel auf toroiden Gruppen, Universität Düsseldorf, Dissertation, 1981.

University of Regina,
Regina, Canada S4S 0A2

Espaces mixtes

par

M. Jurchescu

Cet exposé contient les faits fondamentaux d'une théorie des faisceaux cohérents sur des espaces mixtes.

La notion d'espace mixte considerée ici renferme celle d'espace analytique complexe et aussi une notion possible d'espace différentiable et, en particulier, fournit le cadre naturel pour une théorie des familles différentiables d'espaces complexes.

Rappelons qu'une théorie des faisceaux cohérents sur les variétés mixtes a été déjà présentée dans [9.b].

§1. Catégorie des modèles lisses. Espaces $\mathcal{L}$-fonctés locaux.

Un modèle lisse (d'espace mixte) est un ouvert D d'un espace vectoriel mixte $R^m \times C^n$; plus précisement, dans cette situation, on dira que le modèle lisse D est de type (m,n).

Il y a un seul modèle lisse maximal de type (m,n), à savoir $D = R^m \times C^n$. L'unique modèle lisse de type (0,0) est maximal et sera désigné par e.

Si D est un modèle lisse, disons de type (m,n), on appelle fonction morphe complexe sur D toute fonction $f \in C^\infty(D,C)$ telle que, pour tout point $s \in pr_1(D)$, la fonction

$$D(s) \ni z \longmapsto f(s,z) \in C$$

soit holomorphe sur l'ouvert

$$D(s) := \{ z \in C^n \mid (s,z) \in D \}$$

de C^n.

De même, on appelle fonction morphe réelle toute fonction $f \in C^\infty(D,C)$ telle que, pour tout point $s \in pr_1(D)$, la fonction

$$D(s) \ni z \longmapsto f(s,z) \in \mathbb{R}$$

soit localement constate. Pour qu'une application $f: D \longrightarrow \mathbb{R}$ soit une fonction morphe réelle il faut et il suffit que $j \circ f$ soit une fonction morphe complexe, où $j : \mathbb{R} \longrightarrow \mathbb{C}$ est l'inclusion canonique.

Si D et D' sont deux modèles lisses, on appelle <u>application morphe</u> du premier dans le second toute application $\varphi : D \longrightarrow D'$ dont les composantes réelles et complexes sont des fonctions morphes.

On désigne par $\mathcal{L}$ la catégorie dont les objets sont les modèles lisses et dont les morphismes sont les applications morphes: c'est une catégorie petite avec produits directs finis.

Pour tout espace topologique X, soit Ens(X) la catégorie des faisceaux d'ensembles sur X.

<u>Définition.</u> Un espace <u>$\mathcal{L}$-foncté</u> est un espace topologique X muni d'un foncteur

$$\mathcal{O}_X : \mathcal{L} \longrightarrow \text{Ens}(X).$$

Rappelons que la notion d'espace K-foncté, pour K une catégorie quelconque est due à Douady [4] qui l'a utilisé pour sa définition des espaces analytiques banachiques.

Si X est un espace $\mathcal{L}$-foncté, on désignera par $O_X(D)$ et $O_X(\varphi)$ les images d'un objets D et d'un morphisme φ de $\mathcal{L}$ respectivement. De même, pour tout point $x \in X$ la fibre au point x du faisceau $O_X(D)$ sera désignée $O_{X,x}(D)$ et la fibre au point x du morphisme de faisceaux $O_X(\varphi)$ sera désignée par $O_{X,x}(\varphi)$.

Si X et Y sont deux espaces $\mathcal{L}$-fonctés, un morphisme d'espaces $\mathcal{L}$-fonctés du premier dans le second est un couple

$\varphi = (\varphi_0, \varphi_1)$, où φ_0 est une application continue de l'espace topologique X dans l'espace topologique Y et où φ_1 est un morphisme fonctoriel de O_Y dans le foncteur composé $\varphi_{0*} \circ O_X$.

Par adjonction, la donnée de φ_1 équivaut à la donnée d'un morphisme fonctoriel $\varphi^1 : \varphi_0^* \circ O_Y \longrightarrow O_X$; pour tout objet D de $\mathcal{L}$ et tout point $x \in X$, on désignera par

$$\varphi^*_{x,D} : O_{Y,\varphi_0(x)}(D) \longrightarrow \mathcal{O}_{X,x}(D)$$

la fibre au point x du morphisme de faisceaux d'ensembles

$$\varphi^1(D) : \varphi_0^* O_Y(D) \longrightarrow O_X(D).$$

Les espaces $\mathcal{L}$-fonctés et les morphismes de tels espaces forment une catégorie, la composition des morphismes dans cette catégorie ayant une définition évidente; on la désignera par $\mathcal{L}'$.

<u>Exemples.</u> 1) Tout modèle lisse D est un espace $\mathcal{L}$-foncté avec le foncteur structural

$$\mathcal{O}_D := \mathcal{H}om_{\mathcal{L}}(D, .),$$

c'est-à-dire que

$$\Gamma(U, \mathcal{O}_D(D')) := \mathrm{Hom}_{\mathcal{L}}(U, D')$$

pour tout ouvert U de D et tout objet D' de $\mathcal{L}$. De même, toute application morphe de modèles lisses est la composante topologique d'un morphisme d'espaces $\mathcal{L}$-fonctés évident.

On voit, d'ailleurs, que de cette manière $\mathcal{L}$ se réalise comme sous-catégorie pleine de $\mathcal{L}'$.

2) Même assertion vaut pour la catégorie $\mathcal{M}_0$ des variétés mixtes de dimension finie [9.b].

3) Soit X un espace $\mathcal{L}$-foncté. Tout ouvert U de X est un espace $\mathcal{L}$-foncté avec le foncteur structural

$$\mathcal{O}_U := \mathcal{O}_{X|U} ,$$

et on dira alors que U est un **sous-espace** $\mathcal{L}$-**foncté ouvert** de X. On a un morphisme canonique évident $i: U \longrightarrow X$ et, pour tout espace $\mathcal{L}$-foncté T, tout morphisme $\varphi: T \longrightarrow X$ tel que $\varphi_0(T) \subset U$ se factorise d'une manière unique à travers i.

Pour tout espace $\mathcal{L}$-foncté X, le foncteur

$$\mathcal{H}om_{\mathcal{L}'}(X,.): \mathcal{L} \longrightarrow \mathrm{Ens}(X)$$

définit une nouvelle structure d'espace $\mathcal{L}$-foncté sur X, et on a un morphisme fonctoriel évident

$$\theta_X: \mathcal{H}om_{\mathcal{L}'}(X,.) \longrightarrow \mathcal{O}_X ,$$

donc un morphisme d'espaces $\mathcal{L}$-fonctés

$$(\mathrm{id}, \theta_X): (X, \mathcal{O}_X) \longrightarrow (X, \mathcal{H}om_{\mathcal{L}'}(X,.)).$$

Définition. On dit qu'un espace $\mathcal{L}$-foncté X est **semi-local** si le morphisme précedent est un isomorphisme.

Lorsque X est un espace $\mathcal{L}$-foncté semi-local, on identifiera $\mathcal{O}_X$ avec $\mathcal{H}om_{\mathcal{L}'}(X,.)$ via l'isomorphisme θ_X.

Exemple. Pour tout espace $\mathcal{L}$-foncté $(X, \mathcal{O}_X)$, l'espace $\mathcal{L}$-foncté associé

$$(X, \mathcal{H}om_{\mathcal{L}'}(X,.))$$

este semi-local.

Définition. On dit qu'un espace $\mathcal{L}$-foncté X est **local** s'il satisfait aux conditions suivantes :

a) X est semi-local.

b) Pour tout couple d'objets D',D" de $\mathcal{L}$ et tout sous-espace $\mathcal{L}$-foncté ouvert U de X, l'application canonique

$$\mathrm{Hom}_{\mathcal{L}'}(U, D' \times D'') \longrightarrow \mathrm{Hom}_{\mathcal{L}'}(U, D') \times \mathrm{Hom}_{\mathcal{L}'}(U, D'')$$

est bijective.

c) Il existe un morphisme $X \longrightarrow e$.

La catégorie des espaces $\mathcal{L}$- fonctés locaux $\mathcal{L}''$ est, par définition, la sous-catégorie pleine de $\mathcal{L}'$ dont les objets sont les espaces $\mathcal{L}$-fonctés locaux. Il est clair que e est un objet final dans la catégorie $\mathcal{L}''$.

<u>**Exemples**</u>. $\mathcal{L}$ et M_o sont des sous-catégories pleines de $\mathcal{L}''$.

Désignons par α l'addition et par μ la multiplication dans le modèle lisse $\mathbb{C}$; α et μ sont des applications (holo) morphes de $\mathbb{C} \times \mathbb{C}$ dans $\mathbb{C}$.

Pour tout espace $\mathcal{L}$-foncté local X, $\mathcal{O}_X(\alpha)$ et $\mathcal{O}_X(\mu)$ sont des morphismes de faisceaux de $\mathcal{O}_X(\mathbb{C}) \times \mathcal{O}_X(\mathbb{C})$ dans $\mathcal{O}_X(\mathbb{C})$; d'autre part on a une application $\mathbb{C} \longrightarrow \Gamma(X, \mathcal{O}_X(\mathbb{C}))$ fournie par l'unique morphisme $X \longrightarrow e$. Ces trois opérations définissent une structure de faisceau de $\mathbb{C}$-algèbres (associatives unifères et commutatives) sur $\mathcal{O}_X(\mathbb{C})$.

Pour tout espace $\mathcal{L}$-foncté local X et tout point x de X, l'anneau $\mathcal{O}_{X,x}(\mathbb{C})$ est local et son corps résiduel est isomorphe à $\mathbb{C}$ en tant que $\mathbb{C}$-algèbre; on désignera par $\mathfrak{m}_x$ son idéal maximal. De plus, pour tout morphisme d'espaces $\mathcal{L}$-fonctés locaux $\varphi : X \longrightarrow Y$, et tout point x de X, l'application $\varphi^*_{x,\mathbb{C}}$: $\mathcal{O}_{Y,\varphi_o(x)}(\mathbb{C}) \longrightarrow \mathcal{O}_{X,x}(\mathbb{C})$ est un morphisme de $\mathbb{C}$-algèbres locales.

Considérations analogues valent pour le modèle lisse $\mathbb{R}$. Notons que, pour tout espace $\mathcal{L}$-foncté local X, le morphisme canonique $\mathcal{O}_X(\mathbb{R}) \longrightarrow \mathcal{O}_X(\mathbb{C})$ provenant de l'inclusion canonique $\mathbb{R} \longrightarrow \mathbb{C}$ est un morphisme de faisceaux de $\mathbb{R}$-algèbres.

On utilisera les structures algébriques précédentes pour la définition de l'espace tangent à un espace $\mathcal{L}$-foncté local X en un point x.

Rappelons d'abord qu'un <u>espace vectoriel mixte</u> est un espace vectoriel réel E muni d'un sous-espace vectoriel E_1 de E et d'une structure vectorielle complexe sur l'espace vectoriel réel E_1; on dit que E_1 est la <u>composante complexe</u> de l'espace vectoriel mixte E et que l'espace vectoriel réel $E_2 := E/E_1$ est la <u>composante réelle</u> de E.

Si E et F sont deux espaces vectoriels réels, on appelle <u>morphisme</u> d'espaces vectoriels mixtes du premier dans le second toute application $\mathbb{R}$-linéaire $\phi : E \longrightarrow F$ telle que $\phi(E_1) \subset F_1$ et que l'application $\phi_1 : E_1 \longrightarrow F_1$ induite par ϕ, soit $\mathbb{C}$-linéaire; on dit que ϕ_1 est la <u>composante complexe</u> de ϕ et que l'application $\mathbb{R}$-linéaire $\phi_2 : E_2 \longrightarrow F_2$, coinduite par ϕ, est la <u>composante réelle</u> de ϕ.

Par exemple, $\mathbb{R}^m \times \mathbb{C}^n$ est un espace vectoriel mixte avec la composante complexe $\mathbb{C}^n$ et la composante réelle $\mathbb{R}^m$. Tout espace vectoriel mixte de dimension finie (en tant qu'espace vectoriel réel) est isomorphe à un (et à un seul) $\mathbb{R}^m \times \mathbb{C}^n$; on dira alors que c'est un espace vectoriel mixte de <u>type</u> (m,n).

Soient maintenant X un espace $\mathcal{L}$-foncté local, x un point de X et $j_{X,x} : \mathcal{O}_{X,x}(\mathbb{R}) \longrightarrow \mathcal{O}_{X,x}(\mathbb{C})$ l'application canonique. Considérons l'espace vectoriel complexe $\mathrm{Der}_{\mathbb{C}}(\mathcal{O}_{X,x}(\mathbb{C}))$ des dérivations complexes sur l'algèbre $\mathcal{O}_{X,x}(\mathbb{C})$. Les dérivations $u \in \mathrm{Der}_{\mathbb{C}}(\mathcal{O}_{X,x}(\mathbb{C}))$ qui prennent des valeurs réelles sur $\mathcal{O}_{X,x}(\mathbb{R})$, c'est-à-dire pour lesquelles l'application $u \circ j_{X,x}$ se factorise à travers $\mathbb{R}$, forment un sous-espace vectoriel réel de $\mathrm{Der}_{\mathbb{C}}(\mathcal{O}_{X,x}(\mathbb{C}))$; de même les dérivations $u \in \mathrm{Der}_{\mathbb{C}}(\mathcal{O}_{X,x}(\mathbb{C}))$ qui s'annulent sur $\mathcal{O}_{X,x}(\mathbb{R})$ forment un sous-espace vectoriel complexe de $\mathrm{Der}_{\mathbb{C}}(\mathcal{O}_{X,x}(\mathbb{C}))$. On désignera le premier par $T(X)_x$ et le second par $T_1(X)_x$.

Ainsi $T(X)_x$ est un espace vectoriel mixte avec la composante complexe $T_1(X)_x$, et sera appelé <u>l'espace tangent</u> (mixte) à X au point x; sa composante réelle sera désignée par $T_2(X)_x$.

Soit, enfin, $\varphi : X \longrightarrow Y$ un morphisme d'espaces $\mathcal{L}$-fonctés locaux. Pour tout point $x \in X$, l'application $\mathbb{C}$-linéaire

$$\mathrm{Der}_{\mathbb{C}}(\varphi^*_{x,X}) : \mathrm{Der}_{\mathbb{C}}(\mathcal{O}_{X,x}(\mathbb{C})) \longrightarrow \mathrm{Der}_{\mathbb{C}}(\mathcal{O}_{Y,\varphi_o(x)}(\mathbb{C}))$$

induit un morphisme d'espaces vectoriels mixtes

$$d\varphi_x : T(X)_x \longrightarrow T(Y)_{\varphi_o(x)}$$

qu'on appelle l'<u>application linéaire tangente</u> à X au point x; on désigne par $d_1\varphi_x$ sa composante complexe et par $d_2\varphi_x$ sa composante réelle.

Par exemple, pour D un modèle lisse de type (m,n) et pour tout point x de D, on a $TD_x = \mathbb{R}^m \times \mathbb{C}^n$; en outre, si φ: $: D \longrightarrow F$ est une application morphe de modèles lisses, F maximal, alors $d\varphi_x$ est la dérivée usuelle.

§2. <u>Espaces mixtes</u>.

<u>Lemme 2.1</u>. Soit $f : D \longrightarrow F$ une application morphe de modèles lisses, F maximal. Alors le produit fibré $X := D \times_F e$ existe dans la catégorie $\mathcal{L}''$, l'espace topologique sous-jacent à X est $f^{-1}(o)$ avec la topologie induite et, pour tout point $x \in X$, $T(X)_x = \mathrm{Ker}\, df_x$.

<u>Définition.</u> On appelle <u>modèle</u> (d'espace mixte) tout espace $\mathcal{L}$-foncté local $X = D \times_F e$ comme ci-dessus. Un <u>espace mixte</u> est un espace $\mathcal{L}$-foncté X localement isomorphe à des modèles.

La catégorie des espaces mixtes est, par définition, la sous-catégorie pleine $\mathcal{M}$ de $\mathcal{L}'$ dont les objets sont les espaces mixtes. Il s'ensuit du lemme 1 que $\mathcal{M} \subset \mathcal{L}''$, et d'autre part on a $\mathcal{L} \subset \mathcal{M}_o \subset \mathcal{M}$, toujours comme sous-catégories pleines.

D'après le lemme 1 on voit aussi que, pour tout espace mixte X et tout point x de X, l'espace tangent $T(X)_x$ est un espace vectoriel mixte de dimension finie.

Un espace mixte X est dit de type (m,n) au point $x \in X$ si l'espace tangent est un espace vectoriel mixte de type (m,n). X est dit purement complexe au point x si $T_2(X)_x=0$ et purement réel au point x si $T_1(X)_x = 0$; X est dit purement complexe (resp. purement réel) s'il est purement complexe (resp.purement réel)en tout point $x \in X$.

On a un foncteur canonique

$$\{\text{Esp.complexes}\} \longrightarrow \mathcal{M}$$

et on voit que ce foncteur induit une équivalence entre la catégorie des espaces analytiques complexes et la sous-catégorie pleine de $\mathcal{M}$ constituée par les espaces mixtes purement complexes.

Soit $\mathcal{L}_2$ la sous-catégorie pleine de $\mathcal{L}$ dont les objets sont les modèles lisses purement réels. On a alors la notion d'espace $\mathcal{L}_2$-foncté local, analogue a celle d'espace $\mathcal{L}$ -foncté local. Le lemme 1 vaut encore dans ce cas, et un espace $\mathcal{L}_2$-foncté local de la forme $X := D \times_F e$, avec D,F objets de $\mathcal{L}_2$, sera appelé modèle d'espace différentiable .Un espace différentiable est un espace $\mathcal{L}_2$-foncté localement isomorphe à des modèles d'espaces différentiables. (Pour un autre point de vue sur la notion d'espace différentiable voir Spallek [12]).

On a un foncteur canonique

$$\{\text{Esp. différentiables}\} \longrightarrow \mathcal{M}$$

et on voit encore que ce foncteur induit une équivalence entre la catégorie des espaces différentiables et la sous-catégorie pleine de $\mathcal{M}$ constituée par les espaces mixtes purement réels.

On identifiera les espaces analytiques complexes avec les espaces mixtes purement complexes et les espaces différentiables avec les espaces mixtes purement réels via les foncteurs précédents.

__Théorème__ 2.2. $\mathcal{M}$ est une catégorie avec produits fibrés, et le produit fibré dans $\mathcal{M}$ commute avec le produit fibré des espaces topologiques sous-jacents. En outre, les espaces analytiques complexes et les espaces différentiables sont stables par rapport aux produits fibrés.

__Définition.__ Soit X un espace mixte. Un __sous-espace mixte__ de X est un espace mixte X' muni d'un morphisme i: :X' $\longrightarrow$ X vérifiant la propriété suivante :

pour tout point x de X' il existe un sous-epsace mixte ouvert U de X contenant le point x, un modèle lisse maximal F et un morphisme d'espaces mixtes φ: $U \longrightarrow$ F tels que

$$U \cap X' = U \times_F e$$

dans la catégorie $\mathcal{M}$.

L'espace topologique X' est alors un sous-espace topologique de X; on dit que X' est un sous-espace mixte __fermé__ de X s'il est fermé en tant qu'espace topologique.

Si φ: X $\longrightarrow$ Y est un morphisme d'espaces mixtes et Y' un sous-epsace mixte de X, alors le produit fibré X': = = X $\times_Y$Y' est un sous-espace mixte de X, même fermé si Y' l'est.

__Définition__. On dit qu'un morphisme d'espaces mixtes φ : X $\longrightarrow$ Y est un __plongement__ si φ induit un morphisme de X sur un sous-espace mixte Y' de Y; on dira alors que Y' est l'__image__ du plongement φ . Le plongement φ est dit __fermé__ si l'image Y' de φ est un sous-espace mixte fermé.

<u>Exemples</u> . 1) Pour tout espace mixte X et tout point x de X on a un plongement fermé unique φ : $e \longrightarrow X$ tel que $\varphi_o(o)=x$; l'image de ce plongement sera noté par e_x.

2) Pour tout morphisme d'espaces mixtes $\varphi : X \longrightarrow Y$ et tout point $y \in Y$, la fibre

$$X(y) := X \times_Y e_y =: \varphi^{-1}(e_y)$$

est un sous-espace mixte fermé de X; on désignera par j_y le morphisme canonique de $X(y)$ dans X et, pour tout $\mathcal{O}_X(\mathbb{C})$-module $\mathcal{F}$ on posera

$$\mathcal{F}(y) := j_y^*(\mathcal{F}),$$

il s'agit naturellement de l'image inverse annelée.

<u>Définition</u>. On dit qu'un morphisme d'espaces mixtes $\varphi : X \longrightarrow Y$ est une <u>immersion locale</u> au point $x \in X$ s'il existe un sous-espace mixte ouvert U de X contenant le point x tel que la restriction $\varphi|_U : U \longrightarrow Y$ soit un plongement.

<u>Théorème</u> 2.3. Pour tout morphisme d'espaces mixtes $\varphi : X \longrightarrow Y$ et tout point x de X, les conditions suivantes sont équivalentes :

i) φ est une immersion locale au point x.

ii) L'application $\varphi^*_{x,D}$ est surjective pour tout objet D de $\mathcal{L}$.

iii) Les applications $\varphi^*_{x,\mathbb{C}}$ et $\varphi^*_{x,\mathbb{R}}$ sont surjectives.

iv) Les applications $d_1\varphi_x$ et $d_2\varphi_x$ sont injectives.

<u>Définition</u>. Un morphisme d'espaces mixtes $\varphi : X \longrightarrow Y$ est dit $\mathbb{C}$ - <u>analytique</u> au point $x \in X$ s'il existe un sous-espace mixte ouvert U de X contenant le point x et un morphisme $\psi : U \longrightarrow \mathbb{C}^N$ tels que le morphisme

$$(\varphi|_U, f): U \longrightarrow Y \times \mathbb{C}^N$$

soit un plongement.

Théorème 2.4. Pour tout morphisme d'espaces mixtes $\varphi: X \longrightarrow Y$ et tout point $x \in X$, les conditions suivantes sont équivalentes :

1) φ est $\mathbb{C}$ - analytique au point x .

2) L'application $\varphi^{*}_{x,D}$ est surjective pour tout modèle lisse purement réel D.

3) L'application $\varphi^{*}_{x,\mathbb{R}}$ est surjective.

4) L'application $d_2\varphi_x$ est injective.

On dit que φ est $\mathbb{C}$ - analytique s'il l'est en tout point de X; il est clair alors que les sous-espaces mixtes $X(y) := \varphi^{-1}(e_y)$, $y \in Y$, sont tous purement complexes.

Si $\varphi: X \longrightarrow Y$ est $\mathbb{C}$ -analytique on dira aussi que X est un espace analytique (complexe) relativement à Y et, lorsque Y est un espace différentiable, que (X, φ) est une famille différentiable d'espaces complexes .

Enfin, on dira qu'un espace X admet suffisamment de morphismes réels si, pour tout point x de X, il existe un morphisme $\varphi: X \longrightarrow \mathbb{R}^N$, avec un entier N dépendant de x, qui soit $\mathbb{C}$ - analytique au point x.

Par exemple, si (X, φ) est une famille différentiable d'espaces complexes, alors X admet suffisamment de morphismes réels.

§3. Faisceaux $\mathcal{L}$-cohérents.

Tout espace $\mathcal{L}$-foncté local X a une structure sous-jacente d'espace $\mathbb{C}$-annelé avec le faisceau structurel $\mathcal{O}_X(\mathbb{C})$;

ceci est vrai, en particulier, si X est un espace mixte.

<u>Definition</u>. Soit X un espace mixte et soit $r \in \mathbb{N} \cup \{\infty\}$. Un $\mathcal{O}_X(\mathbb{C})$-module $\mathcal{F}$ est dit $(\mathcal{F},r)$-<u>cohérent</u> (ou <u>lissement r-cohérent</u>) si, pour tout point x de X, il existe un sous-espace mixte ouvert U de X contenant le point x, un modèle lisse D et un plongement fermé $j: U \longrightarrow D$ tels que l'image directe $j_*(\mathcal{F}_{|U})$ soit un $\mathcal{O}_D(\mathbb{C})$-module r-cohérent [9.b]

On dira "$\mathcal{L}$-<u>cohérent</u>" (ou <u>"lissement cohérent"</u>) pour "$(\mathcal{L},\infty)$-<u>cohérent</u>". Notons que, si le faisceau $\mathcal{O}_X(\mathbb{C})$ est $\mathcal{L}$-cohérent, alors "$\mathcal{L}$-cohérent" coïncide avec "cohérent".

<u>Théorème</u> 3.1. Soit $\varphi: X \longrightarrow Y$ un morphisme d'espaces mixtes, $\mathbb{C}$-analytique en un point $a \in X$, et soit $\mathcal{F}$ un $\mathcal{O}_X(\mathbb{C})$-module $(\mathcal{L},r)$-cohérent, pour un $r \in \mathbb{N} \cup \{\infty\}$, tel que l'espace vectoriel $\mathcal{F}_a/\underline{m}_{\varphi_0(a)}\mathcal{F}_a$ soit de dimension finie. Alors il existe un sous-espace mixte ouvert U de X contenant le point a et un sous-espace mixte ouvert V de Y, avec $\varphi_0(U) \subset V$, vérifiant les propriétés suivantes :

a) Pour tout point $x \in U$, $\mathcal{F}_x$ est un $\mathcal{O}_{Y,\varphi_0(x)}$-module de type fini.

b) L'application $U \cap \operatorname{supp} \mathcal{F} \longrightarrow V$ induite par φ_0 est propre et à fibres finies.

c) $\psi_*(\mathcal{F}_{|U})$ est un $\mathcal{O}_V(\mathbb{C})$-module $(\mathcal{L},r)$-cohérent, où $\psi: U \longrightarrow V$ est le morphisme induit par φ.

<u>Définition</u>. Un morphisme d'espaces analytiques φ: $: X \longrightarrow Y$ est dit <u>fini</u> au point $x \in X$ s'il satisfait aux conditions suivantes :

a) φ est $\mathbb{C}$ - analytique au point x.

b) x est un point isolé dans la fibre $\varphi_0^{-1}(\varphi_0(x))$.

<u>Définition</u>. On dit qu'un espace mixte X, supposé séparé et à base dénombrable, est <u>$\mathcal{O}$-complet</u> s'il possède les propriétés suivantes :

C_1) X est <u>$\mathcal{O}$-convexe</u>, i.e., pour tout compact $K \subset X$, l'ensemble

$$\hat{K} := \left\{ x \in X \;\middle|\; |f(x)| \leq \sup_K |f|, \; \forall f \in \Gamma(X, \mathcal{O}_X(\mathbb{C})) \right\}$$

est compact.

C_2) X est <u>$\mathcal{O}$-séparé</u>, i.e., pour tout point $x \in X$, il existe un morphisme $\varphi : X \longrightarrow \mathbb{R}^M \times \mathbb{C}^N$ (M et N étant des entiers convenables dépendant de x) qui soit fini au point x.

On dit d'un ouvert U de X qu'il est <u>$\mathcal{O}$-complet</u> si le sous-espace mixte ouvert de X associé à U est $\mathcal{O}$-complet.

<u>Exemples</u>.1) Un espace analytique complexe X est un espace de Stein si et seulement si X est un espace $\mathcal{O}$-complet (purement complexe).

2) Tout espace différentiable séparé et à base dénombrable est $\mathcal{O}$- complet.

3) Pour tout espace mixte séparé X, les ouverts $\mathcal{O}$-complets de X forment une base, stable par intersection finie, pour la topologie de X.

4) Tout sous-espace mixte fermé d'un espace mixte $\mathcal{O}$-complet est $\mathcal{O}$-complet.

5) Si $\varphi : X \longrightarrow S$ et $\psi : Y \longrightarrow S$ sont des morphismes d'espaces mixtes et si X,Y et S sont $\mathcal{O}$-complets, alors le produit fibré $X \times_S Y$ est $\mathcal{O}$-complet.

Notons que tout espace mixte $\mathcal{O}$-complet possède suffisamment de morphismes réels.

Définition. Soit X un espace mixte et $\mathcal{F}$ un $\mathcal{O}_X(\mathbb{C})$-module de type fini. Soit $\tilde{\mathcal{F}}$ le faisceau de jets de $\mathcal{F}$, i.e.

$$\tilde{\mathcal{F}}(U) := \prod_{(x,k)\in U\times\mathbb{N}^*} \mathcal{F}_x/\mathfrak{m}_x^k \mathcal{F}_x$$

pour tout ouvert U de X ; $\tilde{\mathcal{F}}$ est un faisceau d'espaces vectoriels topologiques séparés. Soit $j: \mathcal{F} \longrightarrow \tilde{\mathcal{F}}$ le morphisme canonique. On dit que $\mathcal{F}$ est séparé (s'il est de type fini et) si j est un monomorphisme, et quasi-séparé s'il est isomorphe à un quotient global sur X d'un faisceau séparé.

Définition. Soit X un espace mixte, supposé séparé et à base dénombrable, et soit $\mathcal{F}$ un $\mathcal{O}_X(\mathbb{C})$-module séparé. Pour tout ouvert U de X, la topologie canonique de $\mathcal{F}(U)$ est l'unique topologie de Fréchet sur $\mathcal{F}(U)$ pour laquelle l'application $j_U : \mathcal{F}(U) \longrightarrow \tilde{\mathcal{F}}(U)$ est continue.

Notons que la topologie canonique existe dans les deux cas suivants :

1) X est un espace différentiable;

2) $\mathcal{F}$ est $\mathcal{L}$-cohérent.

On peut le voir en utilisant le théorème B pour le polydisque mixte [9.b] .

Théorème 3.2. Soit X un espace mixte $\mathcal{O}$-complet et soit $\mathcal{F}$ un $\mathcal{O}_X(\mathbb{C})$-module $\mathcal{L}$-cohérent quasi-séparé. Alors:

A) pour tout point $x \in X$, $\mathcal{F}_x$ est $\mathcal{O}_{X,x}(\mathbb{C})$-engendré par l'image de $\mathcal{F}(X)$.

B) $H^q(X,\mathcal{F}) = 0$ pour $q \geq 1$.

§4. Espaces de Cartan.

Pour mettre en valeur les théorèmes A et B il est nécessaire de faire des hypothèses supplémentaires sur X.

Définition. On dit qu'un espace mixte X est un espace de Cartan s'il est $\mathcal{O}$-complet et si le faisceau $\mathcal{O}_X(\mathbb{C})$ est $\mathcal{L}$-cohérent séparé.

Notons que pour les variétés mixtes la seconde condition ci-dessus est automatiquement vérifiée de sorte que toute variété mixte $\mathcal{O}$-complète est une variété de Cartan [9.b].

Théorème 4.1. Soit X un espace de Cartan. Alors $\Gamma(X, \mathcal{O}_X(\mathbb{C}))$ sépare les points de X et, pour tout point $x \in X$, il existe un morphisme $\varphi : X \longrightarrow \mathbb{R}^M \times \mathbb{C}^N$ qui soit une immersion locale au point x.

Le théorème B caractérise les espaces de Cartan dans le sens suivant.

Théorème 4.2. Soit X un espace mixte séparé à base dénombrable. On suppose que X admet suffisamment de morphismes réels et que $\mathcal{O}_X(\mathbb{C})$ est $\mathcal{L}$-cohérent séparé. Si le théorème B est vrai pour tout idéal cohérent $\mathcal{J}$ de $\mathcal{O}_X(\mathbb{C})$, alors X est un espace de Cartan.

Théorème 4.3. Soit X un espace de Cartan et soit U un ouvert de Cartan de X. Les conditions suivantes sont équivalentes :

i) Pour tout compact $K \subset U$, on a

$$\hat{K}_U = \hat{K}_X.$$

ii) Pour tout $\mathcal{O}_X(\mathbb{C})$-module cohérent séparé $\mathcal{F}$, l'application de restriction

$$\rho_U^X : \mathcal{F}(X) \longrightarrow \mathcal{F}(U)$$

est d'image dense pour la topologie canonique de $\mathcal{F}(U)$.

iii) L'application

$$\rho_U^X : \Gamma(X, \mathcal{O}_X(\mathbb{C})) \longrightarrow \Gamma(U, \mathcal{O}_X(\mathbb{C}))$$

est d'image dense pour la topologie canonique de $\Gamma(U, \mathcal{O}_X(\mathbb{C}))$.

<u>Définition</u>. Dans les conditions du théorème 4.3, on dira que U est un <u>ouvert de Runge de X</u>.

Nous considérons maintenant le cas d'une famille différentiable d'espaces complexes.

Soit donc S un espace différentiable, X un espace mixte et $\pi : X \longrightarrow S$ un morphisme $\mathbb{C}$-analytique d'espaces mixtes. On supposera que l'espace X est séparé à base dénombrable et $\mathcal{O}$- convexe et que le faisceau $\mathcal{O}_X(\mathbb{C})$ est $\mathcal{L}$-cohérent séparé.

<u>Théorème</u> 4.4. Dans les conditions explicités ci-dessus on a les assertions suivantes :

a) X est un espace de Cartan si et seulement si X(s) est un espace de Stein pour tout $s \in S$.

b) Si X est un espace de Cartan et K un compact de X, alors

$$\hat{K} \cap X(s) = (K \cap X(s))\hat{}_{X(s)} .$$

c) Si X est un espace de Cartan et U un ouvert de X, alors U est un ouvert de Runge de X si et seulement si $U(s)$ est un ouvert de Runge de X(s) pour tout point $s \in S$.

§5. <u>Pseudo-convexité</u>

<u>Définition.</u> Soit D un modèle lisse. Une fonction $\varphi \in C^\infty(D) := C^\infty(D, \mathbb{R})$ est dite <u>strictement plurisousharmonique</u> si, pour tout point $s \in pr_1(D)$, la fonction

$$D(s) \ni z \longmapsto \varphi(s,z) \in \mathbb{R}$$

est strictement plurisousharmonique; on désignera par $SC^{\infty}(D)$ l'ensemble des fonctions strictement plurisousharmoniques sur D.

Définition . Soit X un espace mixte. On dit qu'une fonction réelle u sur X est (de classe C^{∞} et) strictement plurisousharmonique si, pour tout point $x \in X$, il existe un sous-espace ouvert U de X contenant le point x, un plongement fermé j de U dans un modèle lisse D et une fonction $u' \in SC^{\infty}(D)$ tels que $u = u' \circ j_o$.

On désignera par $SC^{\infty}(X)$ l'ensemble des fonctions de classe C^{∞} strictement plurisousharmoniques sur X.

Théorème 5.1. Soit X un espace mixte séparé à base dénombrable. On suppose que X est $\mathcal{O}$-convexe et que $\mathcal{O}_X(\mathbb{C})$ est $\mathcal{L}$-cohérent séparé. Alors les conditions suivantes sont équivalentes:

i) X est un espace de Cartan

ii) Il existe $\varphi \in SC^{\infty}(X)$ telle que

$$X_{\alpha} := \{x \in X \mid \varphi(x) < \alpha\} \Subset X$$

pour tout nombre réel α.

iii) L'ensemble $SC^{\infty}(X)$ n'est pas vide.

Définition. Un ouvert D d'un espace mixte X est dit strictement pseudo-convexe au point $a \in \overline{D} \setminus D$ s'il existe un sous-espace mixte ouvert U de X contenant le point a et une fonction $\varphi \in SC^{\infty}(U)$ tels que

$$D \cap U = \{x \in U \mid \varphi(x) < 0\} .$$

On dit que D est strictement pseudo-convexe dans X s'il l'est en tout point de $\overline{D} \setminus D$.

Dans la suite S sera un espace différentiable, X un espace mixte et $\pi : X \longrightarrow S$ un morphisme $\mathbb{C}$-analytique. On supposera que les espaces X et S sont séparés à base dénombrable et que les faisceaux $\mathcal{O}_X(\mathbb{C})$ et $\mathcal{O}_S(\mathbb{C})$ sont $\mathcal{L}$-cohérents séparés.

On utilisera la notion suivante introduite par Schneider [11] qui d'ailleurs l'a donnée dans un contexte plus général (cf. aussi Douady [4]).

Définition. Un $\mathcal{O}_X(\mathbb{C})$-module $\mathcal{F}$ est dit transplat relativement à S au point $x \in X$ si

$$\mathrm{Tor}_i^{\mathcal{O}_{S,s}(\mathbb{C})}(\mathbb{C}, \mathcal{F}_x) = 0, \quad s = \varphi_o(x),$$

pour $i \geq 1$. On dit que $\mathcal{F}$ est transplat relativement à S s'il l'est en tout point de X; on dit que X est transplat relativement à S si le faisceau $\mathcal{O}_X(\mathbb{C})$ l'est.

En prenant X=S et $\pi = \mathrm{id}_S$ on obtient aussi la notion de $\mathcal{O}_S(\mathbb{C})$-module transplat.

On a alors un théorème d'images directes que voiçi:

Théorème 5.2. Soit $D \subset X$ un ouvert strictement pseudoconvexe, relativement propre sur S, et soit $\mathcal{F}$ un $\mathcal{O}_X(\mathbb{C})$-module cohérent. Alors, pour tout point $s \in S$, il existe un sous-espace différentiable ouvert S' de S contenant le point s et un complexe $C^\bullet$ de $\mathcal{O}_{S'}(\mathbb{C})$-modules avec les propriétés suivantes:

a) $C^\bullet$ est borné, C^q est un $\mathcal{O}_{S'}(\mathbb{C})$-module transplat pour tout q et $C^q = 0$ pour $q < 0$.

b) $H^q(C^\bullet) = R^q\pi_{D*}(\mathcal{F})|_{S'}$ pour $q \geq 0$, où $\pi_D := \pi|_D$.

c) Si $\mathcal{F}$ est transplat relativement à S dans les points $x \in \pi^{-1}(s)$, alors

$$H^q(C^\bullet(s)) = H^q(D(s), \mathcal{F}(s))$$

pour $q \geq 0$.

d) Il existe un complexe

$$\mathcal{L}^{\bullet} : \quad \ldots 0 \to \mathcal{L}^{0} \to \ldots \to \mathcal{L}^{N} \to 0 \to \ldots$$

de $\mathcal{O}_{S'}(\mathbb{C})$-modules libres de type fini et un morphisme de complexes $\theta : \mathcal{L}^{\bullet} \longrightarrow C^{\bullet}$ tels que $H^q(\theta)$ soit un épimorphisme pour q=1 et un isomorphisme pour $q \geqslant 2$ (donc $C^{\bullet}$ est 1-pseudo-cohérent).

Théorème 5.3. Dans les conditions du théorème précédent, supposons en outre que $\mathcal{F}$ est transplat relativement à S et que D(s) est un espace de Stein pour un $s \in S$. Alors il existe un voisinage ouvert S' de s dans S tel que

$$R^q \pi_{D*}(\mathcal{F}) |_{S'} = 0$$

pour $q \geqslant 1$.

Dans les corollaires suivants on supposera vérifiées les hypothèses du théorème 5.2 et en outre que X est transplat relativement à S. Le premier de ces corollaires est un théorème de stabilité et le troisième donne la solution du problème de Levi mixte.

Corollaire 1. L'ensemble des points $s \in S$ tels que D(s) soit un espace de Stein est ouvert.

Corollaire 2. Si D(s) est un espace de Stein pour tout $s \in S$, D est un ouvert de Cartan.

Corollaire 3. X est un espace de Cartan si et seulement si il existe une fonction $\varphi \in SC^{\infty}(X)$ telle que $X_c \Subset X$ pour tout réel c.

Notons qu'une solution du problème de Levi dans le cas des variétés mixtes a été obtenu par M.Colţoiu [3] .

Bibliographie

[1] A.Andreotti et H.Grauert : Théorèmes de finitude pour la cohomologie des espaces complexes. Bull.Soc.Math.de France, 90(1962),193-259.

[2] H.Cartan : Variétés analytiques complexes et cohomologie. Coll.sur les fonct.de plus.var., Bruxelles, 1953, 41-55.

[3] M.Coltoiu: The Levi problem . . .(à paraître).

[4] A.Douady : Le problème des modules pour les sous-espaces analytiques compacts d'un espace analytique donné. Ann.Inst.Fourier, 16,1 (1966) 1-98.

[5] O.Forster und K.Knorr : Relativ-analytische Räume und die Kohärenz von Bildgarben. Inventiones Math., 16(1972), 113-160.

[6] H.Grauert : a) Charakterisierung der holomorph vollständigen Räume. Math.Ann., 129 (1955), 233-255.

b)Über Modifikationene und exzeptionelle analytische Mengen. Math.Ann., 146(1962),331-368.

[7] A.Grothendieck :Techniques de construction en Géométrie analytique.II Séminaire Cartan, 13ᵉ année, 1960-61, exposé 9.

[8] C.Houzel : Espaces analytiques relatifs et théorème de finitude. Math.Ann., 205(1973), 13-54.

[9] M.Jurchescu: a)Espaces annelés transcendants et morphismes analytiques. Séminaires de l'Institut de Mathématique, Bucureşti, Editura Academiei, 1971.

b) Variétés mixtes. Proceedings of the IIIrd Romanian-Finnish Seminar on Complex Analysis. Springer Lecture Notes in Math., 743(1979), 431-448.

[10] R.Kiehl : Relativ analytische Räume. Inventiones Math., 16 (1972), 40-112.

[11] M.Schneider : Halbstätigkeitssätze für relativ analytische Raüme. Inventiones Math., 16 (1972), 161-176.

[12] K.Spallek : Differenzierbare Räume. Math.Ann.180 (1969), 269-296.

EQUIVALENCE INDEFINIMENT DIFFERENTIABLE ET EQUIVALENCE ANALYTIQUE REELLE POUR LES GERMES D'ENSEMBLES ANALYTIQUES

N. Milev

On sait que la notion d'équivalence indéfiniment différentiable (C^∞-équivalence) pour les germes d' énsembles analytiques coincide avec la notion d'équivalence analytique réelle /3/. La situation est beaucoup plus compliquéé quand on exige seulement la C^k-équivalence. Ainsi Tougerons a donné un exemple /8/ de germe d'une hypersurface analytique réelle, telle que pour tout entier positif k on a que cette surface est C^k-équivalence àu germe d'une surface algébrique, mais qui n'est pas C^∞-équivalente à aucun germe d'ensemble algébrique. Becker /3/ a prouvé que si V est une hypersurface analytique complexe et p, $p \in V$, est un point singulier isolé, on peut affirmer qu'il existe un entier positif k tel que l'implication suivante est vraie: si (V',p') et (V,p) sont des germes C^k-équivalents, on a qu'ils sont aussi équivalents analytiquement réels.

Dans cet article on étudie la même question pour les ensembles analytiques de codimension arbitraire. On introduit les notions de dimension du plongement, du rang et de l'ordre d'un germe d'un ensemble analytique, qui sont des invariants locaux par rapport à l'équivalence analytique réelle forte. On prouve que, si X est un germe d'ensemble analytique réel avec un point singulier algébriquement isolé (i.e. il s'agit d'un germe finiment determiné), alors il existe un entier positif k, tel que si le germe Y est C^k-équivalent à X et sa dimension du plongement, son rang et son ordre sont les mêmes que ceux de X, on a que X et Y sont équivalents analytiquement réelles.

1. NOTATIONS ET PRELIMINARES

Par O_n est noté l'anneau local des germes des fonctions R-analytiques; par $\mathfrak{m}$ ou $\mathfrak{m}(n)$ est noté l'idéal maximal de l'anneau local O_n; par E_n^k, $k = 1,2,\dots,\infty$, est noté l'anneau local des germes dans l'origine des fonctions différentiables de classe C^k; $O(n,r)$ est l'ensemble des germes $f:(R^n,0)\longrightarrow R^r$ d'applications R-analytiques; $B(n,n)$ est le groupe multiplicatif des germes dans l'origine d'applications inversibles de $O(n,n)$; $E^k(n,r)$ est l'ensemble des germes dans l'origine des applications $f:(R^n,0)\longrightarrow R^r$ de classe C^k; $M(r)$ est l'algèbre des $r\times r$ - matrices sur O_n; $G(r)$ est le groupe multiplicatif des éléments inversibles de $M(r)$;$\omega(f)$ est l'ordre de $f\in O_n$; $T_r(f)$ le polynôme de Taylor de degré r, correspondant au germe $f\in E_n^k$, $r\leq k$; $T_r(f_1,\dots,f_s) = (T_r(f_1),\dots, T_r(f_s))$; $I(X)$ est l'idéal des germes $f\in O_n$ gui s'annulent sur X; $I_k(X)$ est l'idéal des germes $f\in E_n^k$ qui s'annulent sur X.

Soit S un germe d'applications de type suivant

a) analytiques faibles d'après Remmert

b) analytiques fortes, i.e. analytique dans tout point de l'ensemble analytique considéré

c) C^k-différentiables, $k = 1,2,\dots, \infty$.

Les germes X, Y de l'ensembles analytiques $\bar{X}\subset R^n$, $\bar{Y}\subset R^r$, sont applelés S-équivalent (faiblement analytique, fortement analytique, C^k-équivalents), s'il existent $f,g\in S$, tels qu'on a

$$f: X\longrightarrow Y, \quad g: Y\longrightarrow X, \quad g\circ f = 1_X, \quad f\circ g = 1_Y .$$

Pour les ensembles analytiques $\bar{X}$ et $\bar{Y}$ on dira qu'ils sont localement S-équivalents.

<u>Dimension de plongement</u>. Soit X un germe d'ensemble analytique $\bar{X}\subset R^n$ et soit $I(X)$ l'idéal engendré par les germes $f_1,\dots,f_r$, i.e. $I(X) = (f_1,\dots,f_r)$. Le quotient $I(X) +\mathfrak{M}^2/\mathfrak{M}^2$ est un espace vectoriel sur R de dimension finie. Les parties linéaires de $f_1,\dots,f_r$ engendrent cet espace vectoriel. On a

$$\operatorname{rang} \frac{D(\bar{f}_1,\dots,\bar{f}_r)}{D(x_1,\dots,x_n)}(0) = \dim_R(I(X) + \mathfrak{m}^2/\mathfrak{m}^2),$$

$\bar{f}_1,\dots,\bar{f}_r$ étant des représentants des germes $f_1,\dots,f_r$.

Supposons que le rang de la matrice de Jacobi est égal à p et qu'on a

$$\det \frac{D(\bar{f}_1,\dots,\bar{f}_p)}{D(x_1,\dots,x_p)}(0) \neq 0 .$$

D'après le théorème des fonctions implicites le système

$$\bar{f}_j(x_1,\dots,x_p,x_{p+1},\dots,x_n) = 0, \quad j=1,\dots,p ,$$

a la solution analytique suivante $x_j=\bar{g}_j(x_{p+1},\dots,x_n)$, $j=1,\dots,p$. Alors l'ensemble analytique

$$\bar{Y}=\{(y_{p+1},\dots,y_n): \bar{f}_j(\bar{g}_1(y_{p+1},\dots,y_n),\dots,\bar{g}_p(y_{p+1},\dots,y_n),y_{p+1},\dots,y_n) = 0,\ j=p+1,\dots,r\} \subset R^{n-p}$$

est localement équivalent dans le sens analytique fort à $\bar{X}$. De même on a $I(Y)\subset \mathfrak{m}^2$ et $\dim_R(I(Y)+\mathfrak{m}^2/\mathfrak{m}^2) = 0$. Le germe Y est appelé germe canonique de X et on le notera par $\hat{X}$.

Le nombre $\operatorname{emdim} X := n - \dim_R(I(X)+\mathfrak{m}^2/\mathfrak{m}^2)$ sera appelé dimension du plongement de germe X.

On peut prouver que la dimension du plongement est un invariant par rapport à l'équivalence analytique forte. L'espace $R^{\operatorname{emdim} X}$ est l'espace de dimension minimal dans lequel l'ensemble analytique X est plongé localement.

Rang d'un germe d'ensemble analytique. On dit que les germes $f_1,\dots,f_r \in O_n$ sont linéairement indépendants si toute fois quand on a $a_1f_1 + a_2f_2 + \dots + a_rf_r = 0$ avec $a_1,a_2,\dots,a_r \in O_n$, on a aussi

$a_1, a_2, \ldots, a_r \in \mathfrak{M}$. Si de plus les germes linéairement indépendants $f_1, f_2, \ldots, f_r$ engendrent l'idéal J on dit qu'ils forment une base pour J. On sait bien qu'il existent des bases d'un nombre fini d'éléments. De tout sousensemble engendrant J, on peut construir une base en omettant ces éléments qui sont linéairements des autres éléments de J. Le nombre r est dit rang de l'idéal J ($r = \mathrm{rg} J$). La définition est correcte puisque si

$$f = \begin{pmatrix} f_1 \\ \cdots \\ f_r \end{pmatrix} \quad \text{et} \quad g = \begin{pmatrix} g_1 \\ \cdots \\ g_k \end{pmatrix}$$

sont deux bases pour l'idéal J, alors on a $r = k$ et $g = Af$ où $A \in G(r)$. Donc toute base de J est de la forme suivante: Af, où $A \in G(r)$.

On dit que le nombre

$$\mathrm{rg}\, X := \mathrm{rg}\, I(X) - \dim_R (I(X) + \mathfrak{M}^2 / \mathfrak{M}^2)$$

est le rang du germe X. Si $\hat{X}$ est le germe canonique correspondat à X on a $\mathrm{rg}\, X = \mathrm{rg}\, I(\hat{X}) = \mathrm{rg}\, \hat{X}$.

On peut prouver que le rang est invariant par rapport à l'équivalence analytique forte.

Soit $A(h)$ l'orbite du germe $h \in O(n,r)$, définie sous l'action du groupe $G(r) \times B(n,n)$ sur $O(n,r)$:

$$A(h) = \left\{ A(h \circ F) : A \in G(r),\ F \in B(n,n) \right\} .$$

Proposition 1.1. Soient X et Y deux germes canoniques, tels qu'on a $\mathrm{emdim}\, X = \mathrm{emdim}\, Y = n$. Alors, si $f = (f_1, \ldots, f_r)$ est une base pour $I(X)$ et $g = (g_1, \ldots, g_r)$ est une base $I(Y)$, les assertions suivantes équivalentes:

i) les germes X et Y sont équivalents analytiques forts;

ii) $A(f) = A(g)$.

Ainsi le groupe $G(r) \times B(n,n)$ correspond naturellement à la notion d'équivalence analytique forte.

<u>Ordre d'un germe d'ensemble analytique.</u> Soient J un idéal dans O_n et g un germe qui n'appartient à J, $g \notin J$. Alors, on a

$$\max_{h \in J} \omega(g + h) < \infty .$$

Si $f_1, \ldots, f_r$ est une base pour l'idéal J considérons l'idéal $J_j = (f_1, \ldots, f_{j-1}, f_{j+1}, \ldots, f_r)$, $j = 1, \ldots, r$.

La base $f_1, \ldots, f_r$ est appelée ω-base, si pour tout

$$j = 1, \ldots, r \qquad \omega(f_j) = \max_{h \in J_j} \omega(f_j + h).$$

Ayant une base arbitraire pour l'idéal J on peut construir toujours un ω-base.

Soit $f_1, f_2, \ldots, f_r$ une ω-base pour l'idéal J. On peut supposer qu'on a $\omega(f_1) \leq \omega(f_2) \leq \ldots \leq \omega(f_r)$. Le r-tuplet ordonné non-décroissant de nombres entiers positifs $(\omega(f_1), \omega(f_2), \ldots \omega(f_r))$ sera appelé ordre de l'idéal J, note $\omega(J)$. La définition est correcte, i.e. toute ω-base a le même ordre.

Lemme 1.2. Soit $f_1, \ldots, f_r$ une ω-base pour l'idéal J et $\omega(f_1) \leq \ldots \leq \omega(f_r)$. Alors, si $u_{s+1}, \ldots, u_r$ et v sont des éléments de O_n et encore $\omega(v) = 0$, on a

$$\omega(vf_s + u_{s+1}f_{s+1} + \ldots + u_r f_r) = \omega(f_s).$$

On peut prouver l'assertion suivante: si $X \subset Y$, rg $I(X) =$ $=$ rg $I(Y)$ et $\omega(I(X)) = \omega(I(Y))$, alors $X = Y$.

L'ordre du germe X (noté $\omega(X)$) est par définition l'ordre de l'idéal $I(\hat{X})$, i.e. $\omega(X) := \omega(I(\hat{X}))$.

Remarque. Soit $f_1, \ldots, f_r$ une ω-base pour l'idéal $I(X)$. Si $\omega(f_1) = \ldots = \omega(f_k) = 1$ et encore $2 \leq \omega(f_{k+1}) \leq \ldots \leq \omega(f_r)$, alors on a $\omega(X) = (\omega(f_{k+1}), \ldots, \omega(f_r))$.

On peut prouver que l'ordre est invariant par rapport à l'équivalence analytique forte. Comme la C^∞-équivalence implique l'équivalence analytique réelle forte, on voit que la dimension du plongement, le rang et l'ordre sont des invariants par rapport à la C^{∞}-équivalence. On remarquera qu'ils ne sont pas invariants par rapport à la C^k-équivalence.

Exemple. (Becker /3/). Soit $k>0$, $q>k+1$, $r = k(q+1)+1$. Les germes d'ensembles analytiques suivants

$$X = \{(x,y) : x^{q+1} - y^q = 0\} \subset R^2,$$

$$Y = \{(x,y,z) : x^{q+1} - y^q = 0,\ xz - y^{q+1} = 0,\ x^r - z^q = 0\} \subset R^3,$$

sont C^k-équivalents, mais on a $\operatorname{emdim} X = 2$, $\operatorname{rg} X = 1$, $\omega(X) = q$ et d'autre part $\operatorname{emdim} Y = 3$, $\operatorname{rg} Y = 3$, $\omega(Y) = (2,q,q)$.

2. GERMES FINIMENT DETERMINES

Le germe $f \in O(n,r)$ est applelé k-déterminé, si pour tout autre germe $g \in O(n,r)$, avec $T_k(g) = T_k(f)$, on a la même orbite par rapport au groupe $G(r)\times B(n,n)$, i.e. $A(g) = A(f)$.

Considérons l'ensemble

$$H(f) = \{Mf+u_1\dot{f}_{x_1}+\dots+u_n\dot{f}_{x_n} : M \in M(r),\ u_1,\dots,u_n \in O_n\},$$

où $f = \begin{pmatrix} f_1 \\ \cdots \\ f_r \end{pmatrix}$, $\dot{f}_{x_j} = \begin{pmatrix} \partial f_1/\partial x_j \\ \cdots\cdots \\ \partial f_r/\partial x_j \end{pmatrix}$, $j = 1,\dots,n$.

L'ensemble $H(f)$ est un module finiment engendré sur O_n, qui est de même un espace vectoriel sur R. Le germe f est finiment engendré si est seulemeny si $\dim_R O(n,r)/H(f) < \infty$. Plus précisement, si $\mathcal{M}^k(n)O(n,r) \subset \mathcal{M}(n)H(f)$, alors le germe f est k-déterminé.

L'idéal $J \subset O_n$ est appelé k-déterminé, si sa base est k-déterminée. Si $(p_1,\dots,p_s)$ est une base pour J, on a $\omega(p_j) \leq k$, $j = 1,\dots,s$.

Proposition 2.1. Soit l'idéal J_1 un idéal k-déterminé. Si l'on $J_2 \subset J_1 + \mathfrak{M}^{k+1}$ et rg J_2 = rg J_1, $\omega(J_1) = \omega(J_2)$, alors on a $J_1 + \mathfrak{M}^{k+1} = J_2 + \mathfrak{M}^{k+1}$, i.e. $J_1 \equiv J_2 \pmod{\mathfrak{M}^{k+1}}$ et l'idéal est k-déterminé.

Démonstration. Soient $(p_1,\dots,p_r)$ une ω-base pour J_1, et $(q_1,\dots,q_r)$ une ω-base pour J_2, tels que

×) $\omega(p_1) \leq \dots \leq \omega(p_r) \leq k$ et $\omega(p_j) = \omega(q_j)$, $j = 1,\dots,r$.

En vertu de $J_2 \subset J_1 + \mathfrak{M}^{k+1}$, on obtient

××) $q_j = a_{j1}p_1 + \dots + a_{jr}p_r + h_{j1}$, $j = 1,\dots,r$.

où $h_{11},\dots,h_{r1} \in \mathfrak{M}^{k+1}$. Mais $\omega(p_1) = \omega(q_1) = \omega(a_{11}p_1 + \dots + a_{1r}p_r + h_{11}) \geq$
$\geq \min(\omega(a_{11}) + \omega(p_1),\dots,\omega(a_{1r}) + \omega(p_r), \omega(h_{11})) \geq \omega(p_1)$.

Si $\omega(p_1) = \dots = \omega(p_s) < \omega(p_{s+1}) \leq \dots \leq \omega(p_r)$, alors il exist un j, $j = 1,\dots,s$, tel que $\omega(a_{1j}) = 0$. Eventualement en changeant la numeration $1,2,\dots,s$, on pent supposer que $\omega(a_{11}) = 0$. On remarquera qu'après le chanqement de la numeration les conditions (×) restent valables. Mais $\omega(p_2) = \omega(q_2) = \omega(-a_{21} + a_{11}q_2) =$

$$= \omega\left(\begin{vmatrix} a_{11} & a_{12} \\ a_{21} & a_{22} \end{vmatrix} p_2 + \dots + \begin{vmatrix} a_{11} & a_{1r} \\ a_{21} & a_{2r} \end{vmatrix} p_r + h_2\right) \geq \min\left(\omega\left(\begin{vmatrix} a_{11} & a_{12} \\ a_{21} & a_{22} \end{vmatrix}\right) + \omega(p_2), \dots\right.$$

$$\left.\dots, \omega\left(\begin{vmatrix} a_{11} & a_{1r} \\ a_{21} & a_{2r} \end{vmatrix}\right) + \omega(p_r), \omega(h_2)\right) \geq \omega(p_2), \quad h_2 \in \mathfrak{M}^{k+1} .$$

(Dans la deuxième équalité on emploit que $\omega(a_{11}) = 0$ et le leme 1.2 encore.) Par consequence il existe un j, $j=2,\dots,s$, tel que

$$\omega\left(\begin{vmatrix} a_{11} & a_{1j} \\ a_{21} & a_{2j} \end{vmatrix}\right) = 0 .$$

A l'aide de raisonnement semblables on obtient l'équalité suivante

$$A_{1r}q_1 + \dots + A_{rr}q_r = \det A p_r + h_r,$$

où $h_r \in \mathfrak{M}^{k+1}$, A_{jr} étant les complements algebriques des élements a_{jr}, $j = 1,\dots,r$, A est la matrice $\|a_{ij}\|$, $i,j = 1,\dots,r$. On a encore $\omega(Arr) = 0$. Alors $\omega(p_r)=\omega(q_r)\geq \omega(A_{1r}p_1+\dots+A_{rr}p_r)\geq$ $\geq \min(\omega(\det A)+\omega(p_r), \omega(h_r)) \geq \omega(p_r)$, c'est-à-dire $\omega(\det A)=0$, qui signifie que A est une matrice inversible.

Designons la condition (**) dans la forme vectorielle suivante: $q \equiv Ap(\mathrm{mod}\, \mathfrak{M}^{k+1})$. Mais alors on a $A^{-1}q \equiv p(\mathrm{mod}\, \mathfrak{M}^{k+1})$, i.e. $J_1 \subset J_2 + \mathfrak{M}^{k+1}$ et $J_1 \equiv J_2(\mathrm{mod}\, \mathfrak{M}^{k+1})$. Le germe $A^{-1}q$ est aussi k-déterminé. Comme $A^{-1}q$ est une base pour J_2, l'idéal J_2 est aussi k-déterminé.

Corollaire 2.2. Ayant en vue la proposition ci-dessus on obtient que les orbites des bases des deux idéals coincident, i.e. $A(p) = A(q)$.

3. LE RESULTAT

Le germe est dit k-déterminé, si l'idéal $I(X)$ est k-déterminé. Le propriété d'être k-déterminé est invariante par rapport à l'équivalence analytique forte.

Théorème 2.3. Soit le germe X un germe k-déterminé. Alors il existe un nombre entier positif r, tel que pour tout germe Y, C^r-équivalent à X avec les mêmes dimensions du plongement, rang et ordre, on peut affirmer qu'il est équivalent à X par rapport à l'équivalence analytique réelle forte.

Démonstration. Soit $\hat{X}$ le germe canonique correspondant au germe X. Le germe $\hat{X}$ est aussi k-déterminé. Du théorème de Risler /6/,/7/, il suit qu'il existe un nombre entier positif r tel que

$$*)\quad T_k(I(\hat{X})) \subset I(\hat{X}) + \mathfrak{M}^{k+1}.$$

Soit Y un germe C^r-équivalent au germe X avec les mêmes dimensions du plongement, rang et ordre. Considérons le germe canonique $\hat{Y}$ du germe Y . Supposons que la C^r-équivalence de X et Y est réalisée par des germes f et g . Comme emdim $\hat{X}$ = emdim X = = emdim Y = emdim $\hat{Y}$ = n, on a $f, g \in E^r(n,n)$ et $g \circ f = 1_{\hat{X}} \equiv$ $\equiv 1_{R^n} (\mathrm{mod}\ I_r(\hat{X}))$, i.e. $1_{\hat{X}} = 1_{R^n} + h$, où les composantes de h appartiennent à l'idéal $I_r(\hat{X})$. Mais comme $\hat{X}$ est un germe canonique, $I(\hat{X}) \subset \mathfrak{M}^2$, on a $T_k(I_r(\hat{X})) \subset I(\hat{X}) + \mathfrak{M}^{k+1} \subset \mathfrak{M}^2$, i.e. $T_1(I(\hat{X})) = 0$. Par conséquent $T_1(g) \circ T_1(f) = T_1(g \circ f) = T_1(1_{\hat{X}}) = 1_{R^n}$, qui signifie que f est un germe d'une application inversible de classe C^r, i.e. $T_k(f) \in B(n,n)$.

Soit $p = (p_1, \ldots, p_s)$ une ω-base pour l'idéal $I(\hat{X})$ et $q = (q_1, \ldots, q_s)$ une ω-base pour l'idéal $I(\hat{Y})$. Mais $q_j \circ f \in I_r(\hat{X})$, $j = 1, \ldots, s$, et d'après (*) on a

$$q_j \circ T_k(f) \in I(\hat{X}) + \mathfrak{M}^{k+1}.$$

Considérons l'idéal

$$I^*(\hat{Y}) = \{h \circ T_k(f) \ : \ h \in I(\hat{Y})\} \subset I(\hat{X}) + \mathfrak{M}^{k+1} .$$

Mais $q_j \circ T_k(f)$, $j = 1, \ldots, s$, est une ω-base pour l'idéal consideré, i.e. $\mathrm{rg}\ I^*(\hat{Y}) = \mathrm{rg}(I(\hat{Y})) = \mathrm{rg}(\hat{X})$, $\omega(I^*(\hat{Y})) = \omega(I(\hat{Y})) = \omega(\hat{X})$. Il suit de 2.2 qu'on a $A(p) = A(q)$, qui signifie que les germes $\hat{X}$ et $\hat{Y}$ sont équivalents analytiquement réels forts (proposition 1.1). Par conséquent la même chose est vraie et pour les germes X et Y .

Références

1. Abhyancar S., Local analytic geometry, New Jork, Academic Press, 1964.

2. Artin M., On the solutions of analytic equations, Invent. Math., 5, 1968, 277 - 297.

3. Becker J., C^k and analytic equivalence of complex analytic varieties, Math. Ann.,1977, 225, 1, 57 - 67.

4. Mather J., Stability of C^∞ mappings.III.Finitely determined mapderms, Pub. Math. I.H.E.S. 35, 127 - 156, 1968.

5. Mather J., Stability of C^∞ mappings.VI.The Nice Dimensions. Springer Lecture Notes in Math., 192, 1971, 207 - 253.

6. Risler J., Sul la divisibilite des fonctions de classe C^r par les fonctions analytiques reeles, Bull. Soc. Math. France, 1977, 105, 1, 97 - 112.

7. Risler J., Division des fonctions de classe C^r par les fonctions analytiques reelles. C.r.Acad.Sci.,1977, A285, 4, A 237 - 239.

8. Tougeron J., Ideaux de fonctions differentiables, Ann. Inst. Fourier 18, 1, 1968, 177 - 240.

FORMES DE LEVI D'ORDRE SUPERIEUR ET REDUCTION DES VARIABLES DANS L'EQUATION D'UNE HYPERSURFACE REELLE.

Claudio Rea (Italie, l'Aquila) (°)

§1. Une hypersurface réelle S, de classe C^{∞}, appartenante à $\mathbb{C}^n$, peut avoir comme ambiance naturelle un espace avec moins de dimensions.

Cela arrive lorsque S, au voisinage de son point O, peut être transformée en un cylindre $S^{\circ}\times\mathbb{C}^{n-k}$ par une transformation bi-holomorphe, S° étant une hypersurface de $\mathbb{C}^k$.

En d'autres mots il se peut que, par un choix convenable des coordonnées $\zeta_1, \ldots, \zeta_n$ près de O, S, qui avait une équation du type $F(z_1,\ldots,z_n)=0$, en ait une autre $\phi(\zeta_1,\ldots,\zeta_k)=0$ où n'apparaissent que certaines coordonnées.

Nous donnerons dans cet exposé des moyens qui permettent de reconnaître cette possibilité directement de l'équation $F(z_1,\ldots,z_n)=0$ originelle.

Au point de vue de l'analyse il s'agit de trouver les fonctions $\zeta_1,\ldots,\zeta_k$.

On est donc vis-à-vis d'un système linéaire aux dérivées partielles de n équations complèxes en k inconnues avec deux conditions non linéaires données sur S où le jacobien complexe $\partial(\zeta_1,\ldots,\zeta_k)/\partial(z_1,\ldots,z_n)$ doit avoir rang maximal alors que celui des $\zeta_1,\ldots,\zeta_k$ et F ne doit pas l'avoir.

Le nombre k des inconnues n'est pas connu non plus.

Cet étude est purement local. Nous sousentendrons donc toujours un germe S

(°) Rechèrche souténue par le groupe GNASAGA du C.N.R.

d'hypersurface à l'origine et un voisinage de cette dernière qui pourra être rétréci sans qu'on le dise.

Si S est biholomorphement équivalente à $S^{\circ}\times\mathbb{C}^{n-k}$, alors S est appelée (n-k)-cylindre qui sera dit irréductible si k est le plus petit entier jouissant de cette propriété. L'hypersurface réelle S° de $\mathbb{C}^{k}$ est déterminée à un biholomorphisme près et s'appelle directrice du cylindre.

Nous envisagerons aussi des cylindres unilatéraux. S est un cylindre unilatéral lorsqu'il existe un difféomorphisme d'un voisinage de l'origine dans un autre qui transforme S en $S^{\circ}\times\mathbb{C}^{n-k}$ et est holomorphe d'un côté de S.

Nous allons présenter une méthode constituée par une suite de conditions suffisantes chacune desquelles ne s'applique qu'au cas où la précédente n'ait pas donné de réponse.

Il y a toutefois des cylindres qui échappent à cette chaîne de conditions. Le procédé en fait ressemble formellement de près à celui qui permet de trouver les maximums locaux des fonctions d'une variable réelle et qui reste insensible à ces maximums qui ne se révèlent pas par des symptômes algébriques.

Dans le cas où k soit plus grand que un et la forme de Levi de la directrice soit non dégénérée le premier pas de notre méthode donne déjà une réponse assez satisfaisante.

THEOREME 1. Soit $k\geq 2$ un entier donné. La condition qui suit est nécessaire et surfisante pour que S soit un (n-k)-cylindre irréductible, si S n'est pas pseudoconvexe et un (n-k)-cylindre unilatéral, si S est pseudoconvexe, et pour *que la forme de Levi de sa directrice ne dégénère pas*

(i) *Le rang de la forme de Levi de* S *est égal à* k-1 *près de* 0,

(ii) *La fonction* F *définissant* S *satisfait l'équation différentielle*

$$\sum_{1}^{n}{}_{jht} \, F_{j\bar{h}\bar{t}}F_{\bar{s}} - F_{j\bar{s}}F_{\bar{h}\bar{t}})v_j\bar{u}_t\bar{w}_h = 0 \tag{1.1}$$

pour tout $v \in N, u, w \in T$.

Ici T dénote l'espace complexe tangent à S,N le noyau de la forme de Levi

Il faut ici remarquer que l'on peut remplacer dans (1.1) les vecteurs u,w, et v par ceux d'une base de T et de N de sorte que dans l'équation n'apparaîtrons que les dérivées de F. L'équation (1.1) est donc une condition différentielle de troisième degré quasi linéaire ne portant que sur F.

Il est facile de construire un cylindre convexe unilatéral comme celui du théorème qui n'est pas un cylindre des deux côtés.([7]).

Nous définirons dans la suite des formes de Levi d'ordre supérieur L^h.

A l'aide de cette notion on peut donner des conditions suffisantes pour que S soit un cylindre dont la forme de Levi de la directrice pourra dégénérer.

THEOREME 2. *Soit* $k \geq 2$ *un entier et la forme de Levi de* S *ne soit pas nulle en* 0. *Supposons que*

(+) *Le noyau de l'h-ème forme de Levi* L^h *ait dimension* n-k *au voisinage de* 0 *et coïncide avec celui de* L^{h+1}

alors,

si S *n'est pas pseudoconvexe,* S *est un* (n-k)-*cylindre,*

si S *est pseudoconvexe,* S *est un* (n-k)-*cylindre du côté pseudoconvexe.*

Un point d'une hypersurface réelle est dit de <u>type fini</u> s'il existe un entier m tel que toute hypersurface analytique complexe passant par ce point ait un contact d'ordre plus petit que m avec l'hypersurface réelle ([3],[4]).

THEOREME 3. <u>Soit $k \geq 2$ un entier, la forme de Levi de S en 0 soit nulle et 0 soit un point de type fini.</u>

<u>Sous l'hypothèse (+) du théorème 2 on a que</u>

<u>si S est pseudoconvexe alors S est un (n-k)-cylindre unilatéral du côté pseudoconvexe.</u>

L'hypothèse (+) peut s'exprimer par une équation différentielle quasi linéaire de degré h+1 qui, pour h=2 est l'équation (1,1).

Nous concluons cette partie par quelques mots sur le cas k=1 et le cas analytique réel.

La forme de Levi L^2 d'un (n-1)-cylindre est identiquement nulle.

Si S est analytique réelle et $L^2=0$ alors S est un hyperplan et peut s'écrire $\mathrm{Re}\,\zeta_1=0$. ([6]). Dans le cas non analytique S n'est pas nécessairement un cylindre ([6]), même pas unilatéral ([1]).

Il y a peu d'espoir de trouver des conditions suffisantes pour des (n-1)-cylindres différentiables locaux.

Dans le cas analytique réel toute condition de pseudoconvexité est inutile.

Les théorèmes 1 et 2 deviennent:

<u>Si l'hypothèse (+) du théorème 2, ou bien les hypothèses (i) et (ii) sont remplies, alors S est un (n-k)-cylindre.</u>

La deuxième partie de cet énoncé a été prouvée par Freeman ([5]).

§2 Quelques rappels élémentaires.

On dit que le vecteur $v=(v_1,\dots,v_n)$ de $\mathbb{C}^n$ est tangent à l'hypersurface S d'équation $F=0$, au point $z°$ si la droite complexe $z°+\zeta v$, de paramètre $\zeta\in\mathbb{C}$, est tangente à S en $z°$. Si nous présentons les vecteurs sous forme d'opérateurs $v=\sum_1^n v_j \partial/\partial z_j$, alors la condition de tangence devient

$$(2,1) \qquad vF=0.$$

On notera par $T_{z°}$ l'espace tangent complexe de S en $z°$.

Toute fonction f de classe C^1 sur S peut être dérivée en $z°$ le long d'une telle droite sans se soucier de la prolonger au voisinage. Cela a donc un sens de se demander si f est holomorphe sur la droite au point $z°$ ce qui équivaut à l'équation $[\partial f(z°+\zeta v)/\partial\bar{\zeta}]_{\zeta=0}=0$, ou bien $\bar{v}f=0$. Si f se prolonge au voisinage de S, même d'un seul côté, on a certainement

$$(2,2) \qquad \bar{v}f=0, \qquad \text{pour tout } v\in T.$$

Toute fonction de classe C^1 sur S jouissant de cette propriété, nécessaire (mais non pas suffisante) pour son prolongement holomorphe dans le voisinage, s'appelle <u>fonction</u> CR <u>sur</u> S. Une application de S dans une variété est dite CR si ses composantes sont des fonctions CR. On vérifie sans peine qu'un vecteur réel v de $\mathbb{R}^{2n}=\mathbb{C}^n$ s'écrit d'une seule façon sous la forme $\vec{v}=v+\bar{v}$ où $v=\sum_1^n v_j\,\partial/\partial z_j$ est un vecteur complexe. Si $\vec{v}$ appartient à l'espace tangent complexifié de $\mathbb{R}^{2n}$, alors on a $\vec{v}=v+\bar{w}$ et v n'est égal à w que lorsque $\vec{v}$ est réel. On posera dans ce cas $v=(\vec{v})^{10}$, $w=(\vec{v})^{01}$. Par exemple $[u,\bar{w}]^{01}=\sum_1^n {}_{jk} u_j(\partial_j\bar{w}_k)\partial_k$, $[u,\bar{w}]^{10}=-\sum_1^n {}_{jk}\bar{w}_j(\partial_{\bar{j}}u_k)\partial_k$.

§3 Sous-modules de $C^{\infty}(T)$.

Le fibré tangent complexe T a S comme base. Soit V un $C^{\infty}(S)$-module de sections C^{∞} de T, ses éléments sont des champs vectoriels complexes. On peut associer à V un module de champs vectoriels réels $\mathcal{V} \equiv \{v+\bar{v}, \text{avec } v \in V\}$. Les algèbres de Lie associées à V et à $\mathcal{V}$ ne se correspondent pas par cette identification et la forme de Levi en est responsable comme on verra dans la suite.

On dira que V est analytiquement involutif si tout crochet de champs dans V est encore dans V ($[V,V] \subset V$), et que V est géométriquement involutif si $\mathcal{V}$ est involutif. On voit aisément que V est géométriquement involutif si et seulement si il est analytiquement involutif et l'on a

$$[V,\bar{V}] \subset V+\bar{V}. \tag{3.1}$$

On note par V(z) le sous-espace de T_z formé par tous les vecteurs de V au point z. Puisque dimV(z) est une fonction semi-continue inférieurement, les intérieurs A_j des ensembles où la dimension de V est égale à j ont une réunion qui est dense dans l'ensemble ouvert où V est défini. Si V est géométriquement involutif alors tout point de chaque A_j appartient à une variété qui a V(z) comme espace tangent en chaque point z et qui est donc une variété analytique complexe. $C^{\infty}(T)$ lui-même est analytiquement involutif alors qu'une hypersurface réelle ne contient pas en général une variété analytique complexe. Ceci montre la différence entre les deux types d'involutivité.

Nous avons vu tout à l'heure que V détermine une application de chaque A_j dans la grasmannienne complexe $G(j,n)$.

Définition. On dit que V est un module CR si les applications définies ci--dessus sont des applications CR.

Il est aisé de prouver que V est CR si et seulement si l'on a

$$(3.2) \qquad [V,\bar{T}]^{10} \subset V.$$

En comparant avec (3.1) on peut conclure que les propriétés d'involutivité analytique et géométrique sont équivalentes pour des modules CR.

§4 Formes de Levi sésquilinéaires d'ordre supérieur.

L'équation (3.2) suggère la définition de formes de Levi d'ordre supérieur propres à révéler la présence de modules CR sur une hypersurface.

Pour tout $h \geq 1$ et toute h-ple de champs de vecteurs complexes $v, t_1, \ldots, t_{h-1}$ on pose

$$(3.3) \qquad L_z^h(v,t_1,\ldots,t_{h-1}) = \left[\ldots\left[[v,\bar{t}_1]^{10},\bar{t}_2\right]^{10},\ldots,\bar{t}_{h-1}\right]^{10} F(z).$$

Soit $\underline{N}^{h-1}$ le faisceau des germes de champs vectoriels v tels que, pour tout $j \leq h-1$ et tout $t_1,\ldots,t_{j-1}$ dans $C^\infty(T)$, on ait $L^j(v,t_1,\ldots,t_{j-1}) = 0$, et soit $\tilde{N}_z^{j-1} \subset T_z$ l'espace vectoriel des vecteurs de $\underline{N}^{j-1}$ en z.

Nous considérerons toujours L^h restreint à N^{h-1}.

Par exemple $L^1(v) = vF$ a pour noyau $\underline{N}^1$ = germes des champs $C^\infty(T)$, $L^2(v,t) = \sum_{1}^{n}{}_{jm}(\partial^2 F/\partial z_j \partial \bar{z}_m) v_j \bar{t}_m$ est la forme de Levi usuelle.

Proposition. Soit v un champ vectoriel dans $\underline{N}^{h-1}$ et $t_1,\ldots,t_{h-1}$ des champs complexes tangents, on a

(i) $L^h(v,t_1,\ldots,t_{h-1})$ dépend symétriquement de $t_1,\ldots,t_{h-1}$,

(ii) $L^h_z(v,t_1,\dots,t_{h-1})$ <u>s'annule si l'un des champs</u> $v,t_1,\dots,t_{h-1}$ <u>est nul</u> en z,

(iii) L^h <u>est un opérateur quasi linéaire de degré</u> h <u>portant sur</u> F.

<u>Démonstration</u>. Pour tous les champs $a,b,c\in C^\infty(T)$ l'identité de Jacobi donne $[[a,\bar b]^{10},\bar c]^{10}=[[a,\bar c]^{10},\bar b]^{10}$ d'où (i) suit immédiatement.

Puisque $v\in \underline{N}^{h-1}$, $w \overset{\text{def}}{=} [\dots[v,\bar t_1]^{10},\dots,\bar t_{h-2}]^{10}$ est dans $C^\infty(T)$ et $L^h(v,t_1,\dots,t_{h-1})= =L^2(w,t_{h-1})$, la forme de Levi ordinaire, s'annule en tout point où t_{h-1} s'annule. Pour les champs $t_1,\dots,t_{h-2}$ on n'a qu'à appliquer l'énoncé précédent. On prouvera que $v(z)=0$ entraîne $L^h_z(v,t_1,\dots,t_{h-1})=0$ par récurrence sur h en constatant d'abord que ceci est trivialement vrai pour h=1 ou 2. Si d'ailleurs l'énoncé est vrai jusqu'à h-1 on a

$$L^{h-1}(u,t_2,\dots,t_{h-1})=\sum_1^n{}_j u_j B_j(\bar t_2,\dots,\bar t_{h-1}), \qquad \underline{\text{pour tout}}\ u\in\underline{N}^{h-2}$$

où les B_j sont des fonctions linéaires de $\bar t_2,\dots,\bar t_{h-1}$ à coefficients variables. Puisque v appartient à $\underline{N}^{h-1}$ on a

$$\sum_1^n{}_j v_j B_j(\bar t_2,\dots,\bar t_{h-1})=0 \qquad \underline{\text{sur}}\ S$$

pour tout choix des $t_2,\dots,t_{h-1}$. En appliquant l'opérateur $\bar t_1$ à cette identité, on obtient que $\sum_1^n{}_{jm}\bar t_{1m}(\delta_m v_j)\,B_j(t_2,\dots,t_{h-1})$ s'annule là où v s'annule. Mais alors, puisque

$$[v,\bar t_1]^{10}=-\sum_1^n{}_{jm}\bar t_{1m}(\partial_{\bar m}v_j)\partial_j$$

on peut conclure que $L^{h-1}([v,\bar t_1]^{10},t_2,\dots,t_{h-1})=L^h(v,t_1,\dots,t_{h-1})$ lui aussi s'annule aux mêmes points que v; (ii) est donc prouvé. L'énoncé (iii) est une conséquence immédiate de la définition de L^h. C.Q.F.D.

Les noyaux des formes L^h ont été construits par M.Freeman en [5]. Des formes de Levi d'ordre supérieur se trouvent aussi dans [3] et [4].

§5 Schéma des démonstrations des théorèmes 1,2 et 3.

Les trois démonstrations se ressemblent en leur ligne générale.

Les hypothèses (i) et (ii) du théorème 1 correspondent à l'hypothèse (+) des théorèmes 2 et 3 quand on remplace la forme de Levi ordinaire avec L^h.

Ceci permet de construire un fibré (i.e. module de dimension constante) V qui est CR sur S et est destiné à jouer le rôle du fibré complexe tangent aux génératrices du cylindre à fabriquer.

On vérifie aisément que ce fibré est analytiquement,donc géométriquement involutif. S est donc feuilletée par des variétés complexes de dimension n-k. La propriété remplie par V d'être un fibré CR est une propriété d'immersion et V se prolonge analytiquement à tout voisinage où les fonctions CR de S soient prolongeables. Alors par un examen de la forme de Levi transversale au feuilletage et le théorème classique d'extension on prolonge V au voisinage dans les cas des théorèmes 1 et 2.

Dans le théorème 3 le point O est pseudoplat et il faut s'y prendre autrement. Une simple variation sur le théorème d'extension de Bedford et Fornaess [2] fait alors à la besogne.

Le fibré V étant prolongé il ne reste qu'à résoudre un exercice élémentaire.

Adresse Via D. Silvagni 10, 00152 Roma (Italie)

REFERENCES

[1] Bedford E. et De Bartolomeis P. Levi flat hypersurfaces which are not holomorphically flat. Proc of the A.M.S. (1980)

[2] Bedford E. et Fornaess J.E. Local extension of CR functions from weakly pseudoconvex boundaries. Michigan Math. J. 259-262 (1978)

[3] Bloom T. et Graham I. A geometric characterization of points of type m on real submanifolds of $\mathbb{C}^n$. J. Diff.Geom. 12 (1977) 171-182

[4] D'Angelo J.P. Finite type conditions for real hypersurfaces. J.Diff.Geom. 14 (1979) 59-66

[5] Freeman M. Local biholomorphic straightening of real submanifolds. Ann.of Math.,106 (1977),319-352

[6] Rea C. Levi flat submanifolds and holomorphic extension of foliations. Ann.Sc.Norm.Sup.Pisa,XXVI,(1972) 665-681

[7] Rea C. Cylindrical real hypersurfaces in $\mathbb{C}^n$. à paraître sur Ann.Acc.Naz. dei Lincei.

Produktzerlegung und Äquivalenz von Raumkeimen I
Der allgemeine Fall

K. Spallek
Ruhr-Universität Bochum
Fachbereich Mathematik
Universitätsstr. 150
D 4630 Bochum 1

Einleitung

In [4],[5],[6] werden die C^∞-Äquivalenzklassen irreduzibler komplexanalytischer Mengenkeime vollständig beschrieben. Der Kern des Beweises liegt im Nachweis folgender Faktorisierungssätze, die mit komplex analytischen Methoden für irreduzible komplexanalytische Mengenkeime bewiesen werden:

$$A \times K^\ell \approx B \times K^\ell \curvearrowright A \tilde{\to} B \qquad \text{(vgl. [4];[5], 3.9)}$$

$$A_1 \times \cdots \times A_r \tilde{\to} B_1 \times \cdots \times B_s \qquad r = s,\ A_i \tilde{\to} B_i \ \forall i \quad \text{(O.E.)}$$

(vgl. [5] , 3.8 oder auch [6]).

Dabei seien A, A_i, B, B_j irreduzible komplexanalytische Mengenkeime; K sei hier gleich $\mathbb{C}$; A_i, B_j seien "nicht weiter faktorisierbar", und es bedeute $\xrightarrow{\sim}$ schließlich holomorphe Äquivalenz. Der Beweis der Faktorisierungssätze ist länglich und sehr technisch; der Schritt von ihnen zum Klassifikationssatz dagegen ist im vorliegenden Sonderfall relativ kurz.

Wir möchten diesen Klassifikationssatz so weit als möglich verallgemeinern.

Dazu müssen wir in dieser Arbeit zunächst die obigen Faktorisierungssätze verallgemeinern. Das gelingt mit ganz anderen Methoden als in [4],[5] (nämlich mit solchen aus der differenzierbaren Geometrie) zugleich für alle Differenzierbarkeitsklassen und für sehr allgemein stratifizierbare Raumkeime, speziell für alle sub-analytischen ([9]), semi-analytischen ([1o]), reell-analytischen und komplex-analytischen Raumkeime (1.5; 1.8; 2.1). Die bisher für die Keime getroffenen strengen Voraussetzungen entfallen damit. Wesentliches Hilfsmittel zum Beweis sind Ergebnisse aus [19], aus denen wir hier zunächst einige benötigte Folgerungen (§ 0) für den eigentlichen Beweis in § 1 und § 2 ziehen.

Der nur mit analytischen Mengen vertraute Leser möge diese bei den nachfolgenden Beweisen im Auge haben. Beweisvereinfachungen ergeben sich dadurch allerdings nicht.
In einer nachfolgenden, nunmehr funktionentheoretischen Fortsetzung zu dieser Arbeit leiten wir aus dem vorliegenden Faktorisierungssatz unter Hilfe von [20], [22], [25] einen allgemeinen Klassifikationssatz (für sog. bestimmbare Raumkeime) ab ([21]). Weitere Klassifikationen ähnlicher Art findet man für spezielle Situationen oder andere (wie etwa semianalytische) Fälle etwa in [2], [3], [8], [11], [12], [13], [14], [15], [17], [23].

Die vorliegenden Faktorisierungssätze spielen auch eine Rolle beim Studium von Äquivalenzklassen von Quotientensingularitäten ([8], [13], [15], [16], [23]). Nach [16] faktorisieren sich solche Singularitäten (i.a.) in Faktoren, die selbst wieder Quotientensingularitäten sind. Dabei hängen diese Faktorisierungen eng mit Faktorisierungen der zugehörigen Transformationsgruppen zusammen. Es genügt so das Studium der "p-irreduziblen" Quotientensingularitäten.

0 *Vereinbarungen, Vorbereitungen.*

Sei $K = \mathbb{R}$ oder $= \mathbb{C}$. $A_p \subset K^n$ bezeichne einen Mengenkeim im K^n an der Stelle $p \in K^n$. Mit $q \in A_p$ bezeichnen wir stets eine Punkt eines Repräsentanten A von A_p "in hinreichender Nähe von p" (dem jeweiligen Kontext gemäß). Zur Vereinfachung unterscheiden wir zwischen einem Mengenkeim A_0 im Nullpunkt und einem Repräsentanten A in der Bezeichnungsweise meist nicht. A kann also sowohl einen Mengenkeim wie auch einen Repräsentanten um $0 \in K^n$ bezeichnen. Für $q \in A_p$ bezeichne A_q den Keim eines Repräsentanten von A_p in q. A_q ist i.a. nicht eindeutig definiert; in den folgenden Beweisen stört das nicht, da wir anstelle der Keime in Wirklichkeit stets mit festen Repräsentanten in bestimmten Umgebungen der Aufpunkte und im Rahmen spezieller Situationen arbeiten. Entsprechendes gilt für die Bildung unendlicher Durchschnitte oder Vereinigungen von Keimen und anderem mehr.

C^N - Differenzierbarkeit bedeute im Falle $N = 0$ Stetigkeit, in den Fällen $N = 1,2,\dots\infty$ das Übliche, im Falle $N = \omega$ reelle Analytizität und im Falle $N = \omega^*$ Holomorphie mit $K = \mathbb{C}$. Wir schreiben $1 < 2 < \dots < \infty < \omega < \omega^*$. Einen Mengenkeim A, aufgefaßt als N - differenzierbaren Raumkeim, also zusammen mit der N - differenzierbaren Struktur auf A, bezeichnen wir mit ${}^N\!A$. Wir setzen ${}^N\!A \times {}^N\!B := {}^N(A \times B)$. Für die Einbettungsdimensionen gilt stets (vgl. [17], [18] mit $Tg_0{}^N\!A = C^3({}^N\!A_0)$ für den Tangentialraum):

0.1) $$\text{Einbdim}\ {}^N\!A = \dim Tg_0{}^N\!A,$$

$$\text{Einbdim}\ ({}^N\!A \times {}^N\!B) = \text{Einbdim}\ {}^N\!A + \text{Einbdim}\ {}^N\!B$$

Wir sagen, von ${}^N\!A$ läßt sich ein K^ℓ abspalten, wenn es einen Diffeomorphismus (= Isomorphie) ${}^N\!A \xrightarrow{\sim} {}^N(B \times K_0^\ell)$ gibt. ${}^N\!A$, deutlicher ${}^N\!A_0$, heißt *singulär*, wenn es keine Isomorphie ${}^N\!A_0 \xrightarrow{\sim} {}^N\!K_0^\ell$ gibt, $\ell \in \mathbb{N}$. 0 heißt dann *singulärer Punkt* von ${}^N\!A$. Im anderen Falle heißt ${}^N\!A_0$ *regulär* und *Mannigfaltigkeit(skeim)* und 0 heißt *regulärer Punkt* oder *Mannigfaltigkeitspunkt*. $M({}^N\!A)$ bezeichne die Menge der Mannigfaltigkeitspunkt von ${}^N\!A$ (als Keim oder Repräsentant).

Ein Raumkeim ${}^{\omega^*}\!A$ (bzw. ${}^{\omega}\!A$) heißt *komplexanalytisch* (bzw. *reell-, semi-, sub - analytisch*), wenn $A \subset \mathbb{C}^\ell$ komplexanalytisch (bzw. $A \subset \mathbb{R}^\ell$ reell-, semi- oder sub - analytisch) ist. Ein komplex- (bzw. reell-) analytischer Raumkeim ${}^{\omega^*}\!A$ (bzw. ${}^{\omega}\!A$) heißt analytisch irreduzibel, wenn aus ${}^{\omega^*}\!A = {}^{\omega^*}(A_1 \cup A_2)$ und A_i komplex- (bzw. reell-) analytisch stets folgt $A = A_1$ oder $A = A_2$.

Im folgenden seien alle auftretenden Räume lokalkompakt. Da wir einige Ergänzungen zu [19] benötigen, wiederholen wir zunächst:

Definition 0.2 $Tg_x\,{}^NA \supset Tg_x^i\,{}^NA$ bezeichne die Menge der (Tangential-) Vektoren an ${}^NA_x$, durch die jeweils ein lokal integrables (C^{N-1}-differenzierbares Tangential-) Vektorfeld geht ([19], 1.3).

Aus [19], 1.9 folgt dann sofort:

Lemma 0.3 α) $Tg_x^i\,{}^NA$ ist stets ein Vektorraum.
β) Ein Vektorfeld v an ${}^NA$ ist genau dann lokal integrabel, wenn $v(y) \in Tg_y^i\,{}^NA$ gilt, für jedes $y \in A$ des Definitionsbereiches von v.
γ) Die Menge der lokal integrablen Vektorfelder an ${}^NA$ ist abgeschlossen gegenüber Addition, Multiplikation mit C^{N-1}-Funktionen und Lieproduktbildung (von C^N-Feldern).

Ein Raum ${}^NA \subset \mathbb{R}^n$ heißt *integrabel*, wenn alle (C^{N-1}-differenzierbaren Tangential-) Vektorfelder auf ${}^NA$ lokal integrabel sind ([19], 1.3). Integrable Räume werden in [19] untersucht. Mit 0.3 folgt aus dieser Definition:

Lemma 0.4 ${}^NA = {}^NA_1 \times {}^NA_2$ ist integrabel $\Leftrightarrow$ ${}^NA_1$, ${}^NA_2$ sind integrabel.

Zum Beweis bemerken wir nur, daß die später benötigte triviale Richtung " $\Rightarrow$ " sich direkt aus der Definition ergibt.

Lemma 0.5 α) ${}^NA$ ist eine Mannigfaltigkeit $\Leftrightarrow$ ${}^NA$ ist integrabel und hat lokal konstante Einbettungsdimensionen.
β) ${}^NA$ sei integrabel $\Rightarrow$ Die Menge der Mannigfaltigkeitspunkte von ${}^NA$ ist offen und dicht in A.

Beweis α) " $\Leftarrow$ " O.E. sei $\text{Einbdim}_y {}^NA = n \quad \forall\, y \in A$ und $A \subset \mathbb{R}^n$. Dann liefert jedes Feld an $\mathbb{R}^n$ bei Einschränkung auf A ein Feld an ${}^NA$. Aus [19], 1.9 folgt daher $A \subset_{\text{offen}} \mathbb{R}^n$. β) Die Menge der Punkte von A, in denen ${}^NA$ lokal konstante Einbettungsdimension hat, ist offen und dicht in A. Mit α) folgt jetzt β).

Korollar 0.6 ${}^N A_1 \times \cdots \times {}^N A_r \simeq {}^N \mathbb{R}_o^\ell$ für ein $\ell \in \mathbb{N}$ $\Leftrightarrow$ ${}^N A_i$ ist eine Mannigfaltigkeit $\forall\ i = 1,\ldots,r$.

Beweis " $\Leftarrow$ " Mit ${}^N \mathbb{R}_o^\ell$ ist jedes ${}^N A_i$ integrabel und hat konstante Einbettungsdimension, ist also nach 0.5 eine Mannigfaltigkeit.

Lemma 0.7 $Tg_x^i\, {}^N A = Tg_{x_1}^i\, {}^N A_1 \times Tg_{x_2}^i\, {}^N A_2$,

falls $A_1 \times A_2 = A \ni x = (x_1,x_2)$ und entweder ${}^N A_2$ oder ${}^N A$ integrabel ist, oder die Menge der Mannigfaltigkeitspunkte von ${}^N A$ dicht in A ist.

Beweis Lediglich die Inklusion " $\subset$ " ist kritisch. Sei zunächst ${}^N A_2$ integrabel. Für jedes Feld $v = (v_1,v_2)$ auf ${}^N A = {}^N A_1 \times {}^N A_2$ ist dann (o,v_2) auf ${}^N A$ lokal integrabel. Ist v selbst lokalintegrabel, dann also auch $(v_1,o) = v - (o,v_2)$. Mit $v(x) \in Tg_x^i\, {}^N A$ ist damit $v_1(x) \in Tg_{x_1}^i\, {}^N A_1$, $v_2(x) \in Tg_x^i\, {}^N A_2$. q.e.d.

Der Fall "${}^N A$ integrabel" folgt wegen 0.4 ebenfalls. Die Menge der Mannigfaltigkeitspunkte von ${}^N A$ sei jetzt als dicht vorausgesetzt. Wegen 0.6 gilt Entsprechendes für ${}^N A_1$, ${}^N A_2$. Sei $v = (v_1,v_2)$ ein lokal integrables Feld auf ${}^N A$, (x^ν) eine Folge von Punkten $x^\nu \in A$ mit $x^\nu = (x_1,x_2^\nu) \to x = (x_1,x_2)$ und $x_2^\nu \in A_2$ Mannigfaltigkeitspunkt von ${}^N A_2$. Dann folgt mit dem ersten Teil des Beweises $v_1(x_1,x_2^\nu) \in Tg_{x_1}^i\, {}^N A_1$, im Limes also $v_1(x_1,x_2) \in Tg_{x_1}^i\, {}^N A_1$ (da $Tg_{x_1}^i\, {}^N A_1 \subset \mathbb{R}^{n_1}$ als endlich dimensionaler Untervektorraum abgeschlossen ist). Entsprechend ist $v_2(x_1,x_2) \in Tg_{x_2}^i\, {}^N A_2$. q.e.d.

Lemma 0.7 wird bereits in [19], 1.12 in allgemeinerer Form, aber nur mit einem knappen Beweis-Hinweis ausgesprochen. Da dieser Lücken enthält, haben wir hier für den später benötigten Fall einen anderen Beweis voll durchgeführt.

Für die in 0.7 wichtig gewordene Bedingung sagen wir

Definition 0.8 ${}^N A$ heißt *eckenarm*, wenn die Menge der Mannigfaltigkeitspunkte von ${}^N A$ offen und dicht auf A ist.

Ohnehin dürften höchstens eckenarme Räume A nicht nur im Rahmen der vorliegenden Arbeit relevant sein.

Die Bedeutung lokal-integrabler Felder ergibt sich aus folgendem

<u>*Lemma*</u> 0.9 dim $Tg_x^i\,{}^N A$ ist das Maximum der Zahlen d mit ${}^N A \simeq {}^N B \times K^d$.

Beweis [19], 1.11.

1. C^N - *differenzierbare Produktzerlegung*

Alle auftretenden Mengenkeime seien lokalkompakt.

Der folgende Satz 1.1 geht selbst im komplex analytischen Spezialfall in einem wichtigen Punkt über [5] , 3.9 hinaus:

Satz 1.1 $A \subset K_o^n$ sei ein Mengenkeim, $N \in \{2,3,..,\infty,\omega,\omega^*\}$. Dann gibt es eine größte Zahl $s(A,N) \in \mathbb{N} \cup \{o\}$ und einen Mengenkeim $B \subset K_o^t$ mit der Eigenschaft:

$*)$ $\quad {}^N A \xrightarrow{\sim} {}^N(B \times K_o^s)$.

Ist auch ${}^N(B \times K_o^s) \xrightarrow{\varphi} {}^N(C \times K_o^s)$, so liefert

$**)$ $\quad \pi \circ \varphi \mid B \times \{o\} : B \to C$,

mit $\pi : C \times K_o^s \to C$ als Projektion, einen C^N - Diffeomorphismus.

Beweis: Die Existenz einer maximalen Zerlegung $*)$ ist klar. Zum Beweis von $**)$ sei nun o.E. $\text{Einbdim}\,{}^N B = t$. Wegen 0.1) ist dann auch $\text{Einbdim}\,{}^N C = t$ und o.E. ebenfalls $C \subset K_o^t$. Der Diffeomorphismus $\varphi : {}^N(B \times K_o^s) \xrightarrow{\sim} {}^N(C \times K_o^s)$ wird jetzt von einem Diffeomorphismus $\phi : {}^N K_o^{t+s} \to {}^N K_o^{t+s}$ erzeugt. $v_1,\dots,v_s$ seien l.u. Vektorfelder auf ${}^N(B \times K_o^s)$, die in Richtung der Standardeinheitsvektoren von K^s weisen und konstant sind.

Diese sind im Sinne von [16] lokal integrabel, also auch deren Bilder $v_1^*,\dots,v_s^*$ auf ${}^N(C \times K_o^s)$ unter $d\varphi$. Wegen 0.7; 0.9 und der Maximalität von s haben $v_1^*(o),\dots,v_s^*(o) \in Tg\,{}^N(C \times K_o^s)$ keine von Null verschiedene Komponente in Richtung von $Tg\,{}^N C$. Da $d\varphi(o)$ ein Isomorphismus der Tangentialräume ist, folgt wegen der besonderen Wahl von t aus 0.1):

$d\varphi(o)(Tg\,{}^N B) = d\phi(o)(K_o^t)$; und $d\phi(o)(K_o^t)$ wird durch die Projektion $\pi : K_o^t \times K_o^s \to K_o^t$ isomorph auf K_o^t abgebildet. Es folgt:

$G := \phi(K_o^t \times \{o\}) \subset K_o^t \times K_o^s$ liegt als Graph über K_o^t (: wird durch π C^N - diffeomorph auf K_o^t abgebildet). Damit ist

$$\pi \circ \phi \mid K_o^t \times \{o\} \; : K_o^t \times \{o\} \xrightarrow{\phi} G \xrightarrow{\pi} K_o^t$$

ein C^N - Diffeomorphismus.

Weiterhin ist die Inklusion $\pi\circ\phi(B\times\{o\}) = \pi\circ\varphi(B\times\{o\})\subset C$ klar. Hier gilt sogar die Gleichheit: Da ϕ ein Diffeomorphismus mit $\phi(B\times K_o^s) = C\times K_o^s$ ist, folgt

$$\begin{aligned}\phi(B\times\{o\}) &= \phi((K_o^t\times\{o\})\cap(B\times K_o^s)) = \\ &= \phi(K_o^t\times\{o\})\cap\phi(B\times K_o^s)\\ &= \phi(K_o^t\times\{o\})\cap(C\times K_o^s) = G\cap(C\times K_o^s)\end{aligned}$$

Da $\pi\,|\,G : G\to K_o^t$, also $\pi\,|\,G\cap(C\times K_o^s)\to C$ ein Diffeomorphismus ist, folgt die Behauptung. q.e.d.

Die folgende Aussage enthält selbst im komplexanalytischen Spezialfall eine weitere Verschärfung von [5], 3.9:

Satz 1.2 Sei ${}^NA_1\times\ldots\times{}^NA_r \cong {}^NB\times K^\ell$ und ${}^NB$ eckenarm. Dann gibt es C^N-Diffeomorphismen ${}^NA_i \cong {}^NB_i\times K_o^{n_i}$ mit der Eigenschaft: $\sum_{i=1}^r n_i = \ell$. Es ist sodann ${}^NB_1\times\ldots\times{}^NB_r \cong {}^NB$. Von ${}^NB$ läßt sich schließlich genau dann kein K^1 abspalten, wenn sich von keinem der ${}^NB_i$ ein K^1 abspalten läßt.

Beweis. Nach 0.7 ist

$$\begin{aligned}\dim Tg_o^i\,{}^NB+\ell &= \dim Tg_o^i({}^NB\times K^\ell) = \dim Tg_o^i({}^NA_1\times\ldots\times{}^NA_r)\\ &= \sum_{i=1}^r \dim Tg_o^i\,{}^NA\end{aligned}$$

Es gibt also ganze Zahlen $o\le n_i\le m_i := \dim Tg_o^i\,{}^NA$ mit $\sum_{i=1}^r n_i = \ell$. Ferner ist $n_i = m_i$ für alle $i = 1,\ldots,r$ genau dann, wenn $\dim Tg_o^i\,{}^NB = o$ ist. Aus 1.1 und 0.9 folgt jetzt die Behauptung.

Bemerkung 1.3 α) Die in 1.1 auftretende Zahl s(A,N) hängt echt von A und N ab: So gibt es z.B. $\forall\, k,N\in\mathbb{N}$ k-dimensionale C^N-Mannigfaltigkeitskeime im K_o^ℓ, die keine C^{N+1}-Mannigfaltigkeitskeime sind.

β) Ein komplex - analytischer Mengenkeim ist jedoch ein C^{ω^*} - Mannigfaltigkeitskeim genau dann, wenn er nur ein C^1 - Mannigfaltigkeitskeim ist. Im reell - Analytischen ist das i.a. falsch. Jedoch gilt:
γ) Ein reellanalytischer Mengenkeim ist kein C^{ω} - Mannigfaltigkeitskeim genau dann, wenn er für höchstens endlich viele $N \in \mathbb{N}$ ein C^N - Mannigfaltigkeitskeim ist. Beispiele wie $x^p = y^q$ zeigen, daß diese Aussage i.a. nicht verschärft werden kann. δ) Für einen komplexanalytischen Mengenkeim A gilt: Es gibt ein $N \in \mathbb{N}$ (das von A abhängt) mit der Eigenschaft: ${}^NA \xrightarrow{\sim} {}^NB \times \mathbb{R}^1_o \leftrightarrow {}^{\omega^*}A \xrightarrow{\sim} {}^{\omega^*}C \times \mathbb{C}^1_o$ ([14]) ✓.

Wir interessieren uns für p - irreduzible Zerlegungen von Keimen ${}^NA$:

<u>*Definition*</u> 1.4 α) ${}^NA$ heißt p - (=Produkt) *irreduzibel*, wenn aus ${}^NA \xrightarrow{\sim} {}^NB \times {}^NC$ stets folgt ${}^NB \xrightarrow{\sim} {}^N\{o\}$ oder ${}^NC \xrightarrow{\sim} {}^N\{o\}$. β) ${}^NB_1 \times \ldots \times {}^NB_r$ heißt p-*Zerlegung* von ${}^NA$, wenn es einen Diffeomorphismus ${}^NA \xrightarrow{\sim} {}^NB_1 \times \ldots \times {}^NB_r$ gibt. Die Zerlegung heißt p-*irreduzibel*, wenn alle ${}^NB_i$ p-irreduzibel sind. Die ${}^NB_i$ heißen (*irreduzibele*) *Faktoren* der (irreduziblen) p-Zerlegung.

Beispiel K^ℓ_o p-irreduzibel $\leftrightarrow \ell \in \{o,1\}$

<u>*Satz*</u> 1.5 Jeder Raumkeim besitzt eine p-irreduzible Zerlegung. Jede p-Zerlegung läßt sich sogar zu einer p-irreduziblen "verfeinern".

Der *Beweis* folgt aus der Identität $\dim Tg\, {}^NA = \sum_{i=1}^{s} \dim Tg\, {}^NA_i$ für ${}^NA \xrightarrow{\sim} {}^NA_1 \times \ldots \times {}^NA_s$, wenn man immer weiter zu zerlegen sucht.

Offenbar gilt 1.5 analog auch für nicht reduzierte Räume.
Um genauere Informationen über Zerlegungen zu erhalten, bezeichnen wir einige besondere geometrische Situationen:

<u>*Definition*</u> 1.6 α) Ein Raumkeim ${}^NA$ heißt g-(*geometrisch*) *irreduzibel*, wenn gilt: Die Menge $M({}^NA)$ der Mannigfaltigkeitspunkte von ${}^NA$ (in einer geeigneten Basis von O - Umgebungen) ist zusammenhängend und dicht: $\overline{M({}^NA)} = A$

β) ${}^NB$ heißt g-*irreduzible Komponente* von ${}^NA$, wenn $M({}^NA) \cap B \subset M({}^NA)$ relativ offen und abgeschlossen und in B dicht ist.

γ) ${}^{N}B$ heißt g-*Komponente* von ${}^{N}A$, wenn es g-irreduzible Komponenten ${}^{N}B_i$, $i \in I$ Indexmenge, von ${}^{N}A$ gibt (mit Repräsentanten gemäß α), β) in gemeinsamen festen 0-Umgebungen $\forall\ i \in I$) mit $\bigcup_{i \in I} B_i = B$.

δ) Die zu ${}^{N}B$ *komplementäre Komponente* ${}^{N}B^{C}$ ist die Vereinigung aller der g-irreduziblen Komponenten von ${}^{N}A$, die nicht zu ${}^{N}B$ gehören.

(im Falle $\overline{M({}^{N}A)} = A$ ist $A = B \cup B^{C}$).

ε) ${}^{N}A$ heißt *formal irreduzibel*, falls es eine g-irreduzible Komponente ${}^{N}B$ von ${}^{N}A$ gibt mit $\text{Einbdim}\,{}^{N}A = \text{Einbdim}\,{}^{N}B$. ${}^{N}B$ heißt dann *Bestimmungskomponente* von ${}^{N}A$.

ζ) ${}^{N}A$ heißt g*(geometrisch)-reindimensional*, wenn $M({}^{N}A)$ reindimensional und ${}^{N}A$ Komponente von sich ist.

η) ${}^{N}A$ heißt *kurvenreich*, wenn für einen geeigneten Repräsentanten $A \subset K^{n}$, jeden Punkt $q \in A$ und eine Basis von q-Umgebungen gilt: Jeder Punkt $p \in M({}^{N}A \cap U)$ läßt sich durch eine stetige Kurve $\tau : [o,1] \to A \cap U$ folgender Art mit q verbinden: $\tau(1) = p$, $\tau(o) = q$, $\tau(t) \in M({}^{N}A \cap U)\ \forall\ t \in (o,1]$, $\tau \mid (o,1]$ ist von der Klasse C^{N}.

∂) $\overline{\dim}\,{}^{N}A := \max\{\ell \mid \exists\ p_i \in M({}^{N}A) \text{ mit } p_i \to o, \dim A_{p_i} = \ell\}$

$\underline{\dim}\,{}^{N}A := \min\{\ell \mid \exists\ p_i \in M({}^{N}A) \text{ mit } p_i \to o, \dim A_{p_i} = \ell\}$;

im Falle $N = \omega^{*}$ ist dabei $K = \mathbb{C}$ und die komplexe Dimension zu nehmen.

Die unter 1.6 aufgelisteten geometrischen Eigenschaften sind in vielen konkreten Situationen stets gegeben:

<u>*Bemerkung*</u> 1.7 α) Komplex-, reell-, semi-, sub-analytische, allgemeiner: Whitney-stratifizierbare Mengenkeime sind kurvenreich.

β) Ein komplexanalytischer Raumkeim ist g-irreduzibel genau dann, wenn er analytisch irreduzibel ist. Für reellanalytische Keime gilt das i.a. nicht mehr, jedoch hat man:

γ) Jeder reellanalytische und analytisch irreduzible Keim ist formal irreduzibel. Die Umkehrung gilt i.a. nicht.

δ) Komplexanalytische Mengenkeime sind algebraisch reindimensional genau dann, wenn sie g-reindimensional sind. Ist ein reellanalytischer Keim g-reindimensional, so ist er auch algebraisch reindimensional. Die Umkehrung gilt i.a. nicht.

ε) Raumkeime ${}^{N}A \subset K^{\ell}$ mit 1-codimensionaler irreduzibler und singulärer Komponente sind stets formal irreduzibel.

ζ) "kurvenreich" impliziert "eckenarm".

Beweis: α) Für den semi- (und damit auch reell- und komplex-analytischen) Fall verwende man [10]. Der subanalytische Fall folgt mit [9] entsprechend. Auf einen Beweis für den Whitney-stratifizierten Fall (der übrigens alle anderen Fälle umfassen würde) verzichten wir an dieser Stelle. β) - ζ) sind mehr oder weniger klar.

Zur Vereinfachung der Notation in der folgenden Bemerkung 1.8 bezeichne E eine der folgenden Eigenschaften eines Raumkeimes ${}^NA$:

1) g-irreduzibel, 2) g-reindimensional, 3) formal irreduzibel, 4) kurvenreich, 5) komplexanalytisch, 6) reellanalytisch, 7) semianalytisch, 8) subanalytisch.

<u>*Bemerkung*</u> 1.8 Im Falle ${}^NA \xrightarrow{\varphi} {}^NA_1 \times \ldots \times {}^NA_r$ gilt:

α) Die Zusammenhangskomponenten von $M({}^NA)$ sind (unter φ) die Produkte von Zusammenhangskomponenten der $M({}^NA_i)$.

β) Die g-irreduziblen Komponenten von ${}^NA$ sind die Produkte von g-irreduziblen Komponenten der ${}^NA_i$. Die g-Komponenten von ${}^NA$ sind die Produkte von g-Komponenten der A_i.

γ) ${}^NA$ erfüllt E $\Leftrightarrow$ ${}^NA_i$ erfüllt E $\forall\ i = 1,\ldots,r$

δ) Die Bestimmungskomponenten von ${}^NA$ sind die Produkte von Bestimmungskomponenten der ${}^NA_i$.

Beweis: α), β) Wegen Korollar 0.6 gilt: $p \in M({}^NA) \Leftrightarrow \varphi(p)_i \in M({}^NA_i)$ für jede i-te Komponente von $\varphi(p)$, $i = 1,\ldots,r$.

Daraus folgt α). Durch Abschluß- und Vereinigungsbildung folgt β), und wegen 0,1) auch δ). γ) Wegen α), β), δ) sind die Fälle E = 1), 2) und 3) klar. Der Fall E = 4) ist unproblematisch. Die Fälle E = 5) bis 8) folgen sofort, da endliche Durchschnitte von komplex- (bzw. reell-, semi- oder sub-)analytischen Keimen wieder zur gleichen Kategorie gehören.

Ob unter 1.8) γ) eine analoge Aussage auch für den Whitney-stratifizierten Fall richtig ist, muß hier offen bleiben. Später (2.3) genügt für diesen Fall die Gültigkeit der Richtung "$\leftarrow$" unter 1.8 . γ).

Nun können wir die entscheidende Eindeutigkeit von p-Zerlegungen für alle relevanten Fälle formulieren und beweisen. Wegen 1.7 und 1.8 verallgemeinern wir mit 2.1 den zentralen Satz 3.4 aus [5] nicht nur auf beliebige komplexanalytische, sondern zugleich auf beliebige reell-, semi- oder subanalytische Mengen und darüber hinaus auf geeignet stratifizierbare Mengen

§ 2 *Eindeutigkeit der p-Zerlegung*

Satz 2.1 ${}^NA$ sei kurvenreich. ${}^NA_1 \times \ldots \times {}^NA_r$ und ${}^NB_1 \times \ldots \times {}^NB_s$ seien zwei irreduzible p-Zerlegungen von ${}^NA$. $\rightsquigarrow$

α) $r = s$, und bei geeigneter Numerierung: ${}^NA_i \simeq {}^NB_i \quad \forall\ i = 1,\ldots,r$

β) Ist ${}^NA_1 \times \ldots \times {}^NA_r \cong {}^NC_1 \times {}^NC_2$, so folgt bei geeigneter Numerierung:

$${}^NA_1 \times \ldots \times {}^NA_s \cong {}^NC_1,\ {}^NA_{s+1} \times \ldots \times {}^NA_r \cong {}^NC_2$$

γ) Ist ${}^NA_1 \times \ldots \times {}^NA_r \cong {}^NC_1 \times {}^NC_2$ und ${}^NA_1 \times \ldots \times {}^NA_s \cong {}^NC_1$, so folgt ${}^NA_{s+1} \times \ldots \times {}^NA_r \cong {}^NC_2$.

Beweis β), γ) folgen wegen 1.5 und 1.8.γ) aus α). Zum Beweis von α) führen wir Induktion über Einbdim ${}^NA$. Im Falle Einbdim ${}^NA = 1$ ist ${}^NA$ stets p-irreduzibel, daher ist nichts zu beweisen. Es sei also Einbdim ${}^NA > 1$ und 2.1 für alle Einbettungsdimensionen kleiner als Einbdim ${}^NA$ bewiesen. O.E. sei $r \le s$. Wir setzen ${}^NC_r := {}^NB_r \times \ldots \times {}^NB_s$. N ist im folgenden fest. Zur Vereinfachung lassen wir daher den Index N oft weg. Nach Voraussetzung haben wir einen C^N-Diffeomorphismus

$$\varphi : A_1 \times \ldots \times A_{r-1} \times A_r \cong B_1 \times \ldots \times B_{r-1} \times (B_r \times \ldots \times B_s)$$

Wegen 1.1; 1.2; 1.7,ζ können wir o.E. annehmen, daß alle Keime A_i und B_j singulär sind. Indem wir geeignet numerieren, dürfen wir o.E. annehmen:

$$\ell := \underline{\dim}\,{}^NA_r \le \underline{\dim}\,{}^NA_i \qquad \forall\ i = 1,\ldots,r-1$$

$$L := \sum_{i=r}^{s} \underline{\dim}\,{}^NB_i \le \sum_{i=r}^{s} \underline{\dim}\,{}^NB_{j_i} \quad \text{für jede Wahl } j_r,\ldots,j_s \in \{1,\ldots,s\}$$

paarweise verschiedener Indizes.

Wir gliedern den Beweis in 3 Teile mit Unterfällen und Teilbehauptungen.

Teil A 1. Fall $\ell \le L$

Man wähle einen Mannigfaltigkeitspunkt $p_r \in {}^NA_r$ mit $\dim {}^NA_{r\,p_r} = \underline{\dim}\,{}^NA_r$, setze $p := (o,\ldots,o,p_r)$ und $\varphi(p) = q = (q_1,\ldots,q_s)$.

Behauptung 1 O.E. ist $(q_1,\ldots,q_{r-1}) = o$, und $q_r \in {}^NB_r,\ldots,q_s \in {}^NB_s$ sind Mannigfaltigkeitspunkte. Insbesondere folgt $\ell = L$.

Beweis: Es ist $A_{1o} \times \ldots \times A_{r-1\,o} \times A_{r\,p_r} \Rightarrow B_{1\,q_1} \times \ldots \times B_{s\,q_s}$ und $A_{r\,p_r}$ ein Mannigfaltigkeitskeim. Wegen Satz 1.2 können wie die Induktionsvoraussetzung anwenden. Aufgrund der Wahl von p_r und wegen $\ell \leq L$ erhalten wir mit 1.2; i.5 o.E.:

$B_{r\,q_r}, \ldots, B_{s\,q_s}$ sind C^N - Mannigfaltigkeitskeime,

$B_{1\,q_1}, \ldots, B_{r-1\,q_{r-1}}$ sind keine C^N - Mannigfaltigkeitskeime, aber

p - irreduzibel. Insbesondere spaltet kein ${}^N B_{i\,q_i}$ für $i = 1, \ldots, r-1$ einen K_o^1 ab. Das geht aber nur im Falle $q_1 = o, \ldots, q_{r-1} = o$, wie jetzt gezeigt werden soll:

Wir verbinden $p_r \in A_r$ mit o durch einen Weg $\tau_r : [o,1] \to A_r$ gemäß Def. 1.6, n). Sei $\tau := (o, \ldots, o, \tau_r)$ und σ_i die Komponente des Weges $\sigma := \varphi \circ \tau : [o,1] \to B_1 \times \ldots \times B_s$, die zu B_i gehört. Es ist

$$\sigma(1) = q, \quad \sigma_i(1) = q_i, \quad \sigma(o) = o .$$

$S(B_i)$ bezeichne die Singularitätenmenge von B_i. Sie ist abgeschlossen in B_i. Also ist $\sigma_i^{-1}(S(B_i)) \subset [o,1]$ abgeschlossen, also auch $M := \bigcap_{i=1}^{r-1} \sigma_i^{-1}(S(B_i)) \subset [o,1]$. Ferner ist $M({}^N B_i) = B_i \setminus S(B_i)$ offen in B_i, also ist $M^* := \bigcap_{i=r}^{s} \sigma_i^{-1}(B_i \setminus S(B_i)) \subset [o,1]$ offen.

Für jeden Punkt $\tilde{p} = \tau(t)$, $t \in (o,1)$ und seinen Bildpunkt $\tilde{q} = \sigma(t)$ gelten die gleichen Überlegungen wie oben für p und q. Damit folgt dann $M \subset M^* \subset M$, also $M = M^*$ und deshalb $M = [o,1]$. Das heißt:

Für alle $t \in (o,1)$, $i \in \{1, \ldots, r-1\}$ ist $\sigma_i(t) \in {}^N B_i$ ein singulärer Punkt, und für alle $i \in \{r, \ldots, s\}$ ist $\sigma_i(t)$ ein C^N - Mannigfaltigkeitspunkt. Dann muß ${}^N B_{i\,\sigma_i(t)}$ für $i = 1, \ldots, r-1$ wie oben sogar irreduzibel sein.

Sei etwa $\sigma_1(1) \neq o$. σ_1 ist dann nicht konstant, für ein $t_o \in (o,1)$ ist also $\sigma_1'(t_o) \neq o$. Also ist $\sigma'(t_o) \neq o$, also auch $\tau'(t_o) \neq o$, $\tau_r'(t_o) \neq o$. In der Nähe von $\tau_r(t_o)$ kann τ_r als Kurve eines integrablen Vektorfeldes von (der Mannigfaltigkeit) $A_r \setminus S(A_r)$ aufgefaßt werden. Dann kann τ in der Nähe von $\tau(t_o)$ als Kurve eines integrablen Vektorfeldes auf ${}^N A$ aufgefaßt werden. Diese Situation überträgt sich durch

φ auf die Kurve σ an der Stelle $\sigma(t_o)$. Also ist $\sigma'(t_o) \in Tg^i_{\sigma(t_o)}({}^N B_1 \times \ldots \times {}^N B_s)$, wegen 0.7 also $o \neq \sigma_1'(t_o) \in Tg^{1N} B_{1\sigma_1(t_o)}$. Nach 0.9 müßte sich dann von ${}^N B_{1\sigma(t_o)}$ ein K_o^1 abspalten lassen, und dieser Keim wäre nicht p-irreduzibel. Es folgt: $\sigma_1(t) \equiv o$ und allgemeiner $\sigma_i(t) \equiv o \;\forall\; i = 1,\ldots,r-1$. Die Gleichheit $\ell = L$ folgt jetzt mit Satz 1.2.

2. Fall $\quad \ell \geq L$

Man wähle Mannigfaltigkeitspunkte $q_i \in {}^N B_i \;\forall\; i = r,\ldots,s$ mit $\dim {}^N B_{i\, q_i} = \underline{\dim}\, {}^N B_i$, setze $q := (o,\ldots,o,q_r,\ldots,q_s)$, $\psi := \varphi^{-1}$ und $\psi(q) = p = (p_1,\ldots,p_r)$.

Behauptung 2 O.E. ist $(p_1,\ldots,p_{r-1}) = o$ und $p_r \in {}^N A_r$ ein Mannigfaltigkeitspunkt. Insbesondere folgt $\ell = L$.

Den *Beweis* führe man analog wie oben.

Behauptung 3 O.E. ${}^N A_i \cong {}^N B_i \quad \forall\; i = 1,\ldots,r-1$

Beweis: In den Situationen von Beh. 1 und Beh. 2 treten jeweils Mannigfaltigkeitskeime als p-Faktoren auf. Diese kann man nach Satz 1.1 "wegkürzen". Die Behauptung folgt dann durch Induktion.

Wir haben eigentlich schon mehr gezeigt: $M(p_r) \subset A_r$ bezeichne die Zusammenhangskomponente von $M({}^N A_r)$, in der p_r liegt; dann ist $K(p_r) := \overline{M(p_r)}$ die g-irreduzible Komponente von A_r, die p_r enthält. Nun induzieren die Einschränkungen von φ Isomorphismen

$$0 \times \cdots \times 0 \times M(p_r) \cong 0 \times \ldots \times 0 \times M(q_r) \times \cdots \times M(q_s)$$

$$0 \times \ldots \times 0 \times K(p_r) \cong 0 \times \ldots \times 0 \times K(q_r) \times \cdots \times K(q_s)$$

wegen 1.8, β) also:

$$A_1 \times \cdots \times A_{r-1} \times K(p_r) \cong B_1 \times \cdots \times B_{r-1} \times K(q_r) \times \cdots \times K(q_s)$$

Fassen wir alle jene irreduziblen Komponenten von ${}^N A_r$ und ${}^N C_r := {}^N B_r \times \ldots \times {}^N B_s$ zusammen, für die Entsprechendes gilt, so folgt:

Behauptung 4 Es gibt eine größte g-Komponente A_r^1 von A_r und C_r^1 von C_r derart, daß φ durch Einschränkung C^N - Diffeomorphismen

$$0\times\ldots\times 0\times(M(A_r)\cap A_r^1) \xrightarrow{\sim} 0\times\ldots\times 0\times(M(C_r^1)\cap C_r^1)$$

also auch:

$$0\times\ldots\times 0\times A_r^1 \xrightarrow{\sim} 0\times\ldots\times 0\times C_r^1$$

$$A_1\times\ldots\times A_{r-1}\times A_r^1 \xrightarrow{\sim} B_1\times\ldots\times B_{r-1}\times C_r^1$$

$$A_1\times\ldots\times A_{r-1}\times A_r^2 \xrightarrow{\sim} B_1\times\ldots\times B_{r-1}\times C_r^2,$$

erzeugt, wobei $A_r^2\subset A_r$ (bzw. $C_r^2\subset C_r$) die zu A_r^1 (bzw. C_r^1) komplementäre Komponente von A_r (bzw. C_r) bezeichnet. Es ist $A_r = A_r^1\cup A_r^2, C_r = C_r^1\cup C_r^2$.

Ist $A_r^2=\phi$, speziell etwa ${}^NA$ g-irreduzibel, so sind wir hier mit dem Beweis von Satz 2.1 schon fertig. Der reindimensionale und danach der gemischtdimensionale Fall erfordern weitere Überlegungen. Sei also $A_r^2\neq\phi$.

Teil B 1. Fall Es gebe einen Mannigfaltigkeitspunkt $p_r\in A_r^2$ von ${}^NA_r$ mit folgender Eigenschaft:

*) $(\dim A_{r\,p_r} =) \dim A^2_{r\,p_r} \le \underline{\dim} A_i\ (=\underline{\dim}\, B_i)\ \forall\ i=1,\ldots,r-1$

**) $\le \dim C_{r\,v}$

für jede Wahl eines Mannigfaltigkeitspunktes $v\in C_r^2$ von ${}^NC_r$.

Sei $p := (o,\ldots,o,p_r)$ und $\varphi(p) = q = (q_1,q_2,\ldots,q_s)$. Dann ist $(q_r,\ldots,q_s)\in C_r^2$.

Behauptung 5 Nicht alle ${}^NB_{i\,q_i}$, $i\in\{1,\ldots,r-1\}$, sind singulär.

Beweis: Wären alle ${}^NB_{i\,q_i}$, $i\in\{1,\ldots,r-1\}$, singulär, so müßte jeder Keim ${}^NB_{j\,q_j}$, $j\in\{r,\ldots,s\}$, eine Mannigfaltigkeit sein (1.2; 1.5; Induktion). Wegen **) und Satz 1.2 ließe sich weiter von keinem der ${}^NB_{i\,q_i}$, $i\in\{1,\ldots,r-1\}$, ein K_o^1 abspalten. Wie im Beweis zu Beh. 1 folgt dann $(q_1,\ldots,q_{r-1}) = o$, und es wäre A_r^1 nicht maximal gewählt (vgl. die Feststellungen zu Beh. 4)).

Behauptung 6 Jeder der Keime ${}^NB_{i\,q_i}$, $i\in\{1,\ldots,r-1\}$, ist singulär.

Beweis: O.E. sei etwa ${}^N B_{1q_1}$ ein Mannigfaltigkeitskeim.

Wegen ${}^N A_1 \cong {}^N B_1$ (Beh. 3) und *) ist $\dim B_{1q_1} \geq \dim A_{rp_r}$. Mit Satz 1.2 und der Induktionsvoraussetzung folgt daraus:

$\dim B_{1q_1} = \dim A_{rp_r}$; $r = s$; ${}^N B_{jq_j}$ sind für $j = 2,\dots,r$ p-irreduzibel und singulär.

Wie im Beweis zu Beh. 1 muß daher $q_2 = 0,\dots,q_r = 0$ sein. φ liefert also durch Einschränkung einen C^N - Diffeomorphismus

$$A_{10} \times \dots \times A_{r-1,0} \times A_{rp_r} \xrightarrow{\sim} B_{1q_1} \times B_{20} \times \dots \times B_{r0}$$

Andererseits haben wir durch Einschränkung von φ Isomorphismen:

$$A_1 \times \dots \times A_{r-1} \times A_r^1 \xrightarrow{\sim} B_1 \times \dots \times B_{r-1} \times C_r^1,$$
$$A_1 \times \dots \times A_{r-1} \times A_r^2 \xrightarrow{\sim} B_1 \times \dots \times B_{r-1} \times C_r^2,$$
$$\cup \qquad\qquad\qquad \cup$$
$$A_{10} \times \dots \times A_{r-10} \times A_{rp_r} \xrightarrow{\sim} B_{1q_1} \times \dots \times B_{r-10} \times B_{r0},$$

was nicht möglich ist. q.e.d.

Behauptung 7 Der 1. Fall kann nicht auftreten.

2. Fall Es gebe einen Mannigfaltigkeitspunkt v von ${}^N C_r$, der in C_r^2 liegt und folgende Eigenschaft hat

*) $(\dim C_{rv} =) \dim C^2_{rv} \leq \underline{\dim}\, B_i\ (= \underline{\dim}\, A_i)\quad \forall\ i = 1,\dots,r-1$

**) $\leq \dim {}^N A_{rp_r}$

für jede Wahl eines Mannigfaltigkeitspunktes $p_r \in A_r^2$ von ${}^N A_r$.
Nun folgt wie im Beweis zu Beh. 7:

Behauptung 8 Der 2. Fall kann ebenfalls nicht eintreten.

Ist $C_r^2 = \{0\}$, so sind wir fertig. Dies trifft z.B. mit Teil B zu, wenn ${}^N A$ reindimensional ist: In dem Falle sind nämlich auch ${}^N A_r$ und ${}^N C_r$ reindimensional, und unser Beweis liefert $A_r^1 = A_1$, $C_r^1 = C_r$, also $0 \times \dots \times 0 \times A_r \xrightarrow{\sim} 0 \times \dots \times 0 \times C_r$, womit Satz 2.1 bewiesen wäre. Den gemischtdimensionalen Fall erfassen wir in einem letzten Schritt.
Sei also $C_r^2 \neq \phi$

<u>Teil C</u> O.E. gebe es einen Mannigfaltigkeitspunkt $p_1 \in {}^N A_1$ mit der folgenden Eigenschaft:

*) $\dim A_{1p_1} \leq \underline{\dim} A_i (= \underline{\dim} B_i) \quad \forall\ i = 1,\dots,r-1.$

Sei $p = (p_1, o, \dots, o)$, $\varphi(p) = q = (q_1, \dots, q_s)$

Behauptung 9 Nicht alle ${}^N B_{i q_i}$, $i \in \{1,\dots,r-1\}$, sind singulär.

Beweis: Wären alle ${}^N B_{i q_i}$, $i \in \{1,\dots,r-1\}$, singulär, so müßten alle ${}^N B_{j q_j}$, $j \in \{r,\dots,s\}$, Mannigfaltigkeiten sein (0.6; 1.2; 1.5; Induktion). Es wäre samt einer Umgebung entweder $v := (q_r, \dots, q_s) \in C_r^1$ oder $v \in C_r^2$ (beides gleichzeitig kann nicht eintreten) und damit

$$(A_{1p_1} \times A_{2o} \times \dots \times A_{ro}) \rightleftarrows B_{1q_1} \times \dots \times B_{r-1,q_{r-1}} \times C^1_{rv}$$

oder

$$\text{"} \quad \rightleftarrows B_{1q_1} \dots B_{r-1,q_{r-1}} \times C^2_{rv}$$

Wegen $A_1 \times A_2 \times \dots \times A_{r-1} \times A_r^i \xrightarrow{\sim} B_1 \times \dots \times B_{r-1} \times C_r^i$

und $A_{1p_1} \times A_{2o} \times \dots \times A_{r-1,o} \times A_{r,o} \not\cong A_{1p_1} \times A_{2o} \times \dots \times A_{r-1,o} \times A^i_{r,o}$

für $i = 1,2$ ist das aber nicht möglich.

q.e.d.

Behauptung 10 $r = s$, und ${}^N A_r$ ist zu einem der ${}^N B_j$, ${}^N B_r$ ist zu einem der ${}^N A_i$ isomorph.

Beweis: Für ein bestimmtes $i \in \{1,\dots,r-1\}$ ist ${}^N B_{i q_i}$ nicht singulär. Wegen ${}^N B_i \rightleftarrows {}^N A_i$ (Beh. 3), Satz 1.2 und *) ist dann von keinem der übrigen ${}^N B_{j q_j}$, $j \in \{1,\dots,s\} \setminus \{i\}$, ein K_o^1 abspaltbar, und diese Keime wären alle singulär. Wie im Beweis zu Beh. 1 wäre dann $q_j = o \ \forall\ i \neq j \in \{1,\dots,s\}$, nach Satz 1.2 also $A_{2o} \times \dots \times A_{ro} \xrightarrow{\sim} B_{1o} \times \dots \times \check{B}_{i q_i} \times \dots \times B_{so}$, wobei die mit $\vee$ gekennzeichnete Komponente wegzulassen ist. Beh. 3 und Induktion liefern jetzt $r = s$ und ${}^N A_{ro} \xrightarrow{\sim} {}^N B_{1o}$, ${}^N A_{io} \xrightarrow{\sim} {}^N B_{ro}$ oder ${}^N A_{ro} \xrightarrow{\sim} {}^N B_{ro}$, ${}^N A_{io} \xrightarrow{\sim} {}^N B_{1o}$.

q.e.d.

Damit treten in ${}^N A_1 \times \dots \times {}^N A_r \xrightarrow{\sim} {}^N B_1 \times \dots \times {}^N B_r$ auf jeder Seite dieselben p-Faktoren auf.

Behauptung 11 Auf jeder Seite von ${}^NA_1 \times \ldots \times {}^NA_r \xrightarrow{\sim} {}^NB_1 \times \ldots \times {}^NB_r$ tritt jeder Faktor gleich oft auf.

Beweis: Es ist $\overline{\dim}\, {}^NA_r \leq \underline{\dim}\, {}^NA_i \ \forall\ i = 1,\ldots,r$. Nun sei $o \leq s < r$ irgendeine Zahl mit ${}^NA_{s+1} \not\cong {}^NA_{s+2} \not\cong \ldots \not\cong {}^NA_r$. Man wähle Mannigfaltigkeitspunkte $p_i \in {}^NA_i\ \forall\ i \in \{s+1,\ldots,r\}$ mit

$$\dim A_{i\,p_i} = \underline{\dim}\ A_r^N$$

und setze $p = (o,\ldots,o,p_{s+1},\ldots,p_r)$, $\varphi(p) = q = (q_1,\ldots,q_r)$. Dann sind (wegen der besonderen Wahl von A_r) o.E. genau die Keime ${}^NB_{q_1},\ldots,{}^NB_{q_s}$ singulär, ${}^NB_{q_{s+1}},\ldots,{}^NB_{q_r}$ Mannigfaltigkeiten, und von keinem der ${}^NB_{q_1},\ldots,{}^NB_{q_s}$ läßt sich ein K_o^1 abspalten. Wieder folgt $q_1 = o,\ldots,q_s = o$ (Beweis zu Beh. 1) und

${}^NA_1 \times \ldots \times {}^NA_s \xrightarrow{\sim} {}^NB_1 \times \ldots \times {}^NB_s$, also o.E. ${}^NA_i \cong {}^NB_i\ \forall\ i = 1,\ldots,s$.
Tritt also in ${}^NA_1 \times \ldots \times {}^NA_r$ der Faktor ${}^NA_r$ genau $(r-s)$-mal auf, so tritt der entsprechende Faktor in ${}^NB_1 \times \ldots \times {}^NB_r$ höchstens $(r-s)$-mal auf, aus Symmetriegründen also genau $(r-s)$-mal. Wegen ${}^NA_i \xrightarrow{\sim} {}^NB_i\ \forall\ i = 1,\ldots,s$ folgt die Behauptung. q.e.d.

Damit ist Satz 2.1 bewiesen.

Für später halten wie als Zusatz eine Folgerung fest, die man aus den Beweisteilen B und C sowie mit Satz 1.1 gewinnt:

Zusatz 2.2 Sei ${}^NA_1 \times \ldots \times {}^NA_r \xrightarrow{\sim} {}^NB_1 \times \ldots \times {}^NB_r$, ${}^NA_i$, ${}^NB_i$ p-irreduzibel und singulär $\forall\ i \in \{1,\ldots,r\}$. Ferner sei

$$\overline{\dim}\ {}^NA_r \leq \underline{\dim}\ {}^NA_i \quad \forall\ i \in \{1,\ldots,r-1\}\,.$$

Bei geeigneter Numerierung der B_j gilt dann folgendes:

α) $\varphi(o \times \ldots \times o \times A_r) = o \times \ldots \times o \times B_r$, und $\varphi|\, o \times \ldots \times o \times A_r : {}^NA_r \to {}^NB_r$ ist ein Isomorphismus.

β) Ist $p_r \in {}^NA_r$ ein Mannigfaltigkeitspunkt; $\pi : B_1 \times \ldots \times B_r \to B_1 \times \ldots \times B_{r-1}$ die Projektion, so ist die Komposition

$$A_1 \times \ldots \times A_{r-1} \times \{p_r\} \hookrightarrow A_1 \times \ldots \times A_r \xrightarrow{\varphi} B_1 \times \ldots \times B_r \xrightarrow{\pi} B_1 \times \ldots \times B_{r-1}$$

ein C^N-Diffeomorphismus.

γ) Ist speziell $\overline{\dim}\, {}^N A_i = \underline{\dim}\, {}^N A_i = c \quad \forall\ i \in \{1,\dots,r\}$, so induziert φ durch Einschränkung C^N - Diffeomorphismen

$$0 \times \dots \times 0 \times A_i \times 0 \times \dots \times 0 \xrightarrow{\sim} 0 \times \dots \times 0 \times B_i \times 0 \times \dots \times 0$$

$\forall\ i \in \{1,\dots,r\}$.

Satz 2.1 läßt sich in verschiedener Hinsicht erweitern, z.B. auf allgemeinere Mengenkeime oder auf nicht reduzierte Situationen (vgl. auch [21], 2.2). Uns interessiert der folgende "relative" Fall: K^N sei eine Kategorie kurvenreicher C^N - Raumkeime mit folgenden Eigenschaften:

i) ${}^N A \times {}^N B \in K^N$ für alle ${}^N A,\ {}^N B \in K^N$, ${}^N\{p\} \in K^N$ für alle Punkte p

ii) Für jeden C^N - Raumkeim ${}^N A$ und beliebiges $\ell \in \mathbb{N}$ gilt:

$$ {}^N A \in K^N \iff {}^N(A \times K^\ell) \in K^N$$

$K^N \ni {}^N A$ heiße p - irreduzibel relativ K^N, wenn aus ${}^N A \xrightarrow{\sim} {}^N A_1 \times {}^N A_2$ und ${}^N A_i \in K^N$ stets folgt ${}^N A_1 \xrightarrow{\sim} {}^N\{o\}$ oder ${}^N A_2 \xrightarrow{\sim} {}^N\{o\}$
Offenbar besitzt jeder Keim ${}^N A \in K^N$ eine Zerlegung in Elemente, die p-irreduzibel bzgl. K^n sind. Diese Zerlegung ist auch eindeutig:

Zusatz 2.3 ${}^N A_i,\ {}^N B_j \in K^N$ seien p-irreduzibel relativ K^n, und es sei

$$ {}^N A_1 \times \dots \times {}^N A_r \xrightarrow{\sim} {}^N B_1 \times \dots \times {}^N B_s$$

Dann folgt $r = s$ und ${}^N A_i \xrightarrow{\sim} {}^N B_i$, $\forall i = 1,\dots,r$, bei geeigneter Numerierung.

Beweis: Wie zu 2.1 .

Ein Beispiel für eine solche Kategorie ist die Familie der C^N - Raumkeime, die so Whitney - stratifizierbar sind, daß deren iterierte Singularitätenmengen Unterstratifizierungen bilden.

Manchen Raumkeimen kann man die wichtige p - Irreduzibilität sofort ansehen. Die folgenden Beispiele hierzu beschreiben im analytischen Spezialfall verbreitete Phänomene:

Bemerkung 2.4 In folgenden Fällen z.B. gilt ${}^{N}A \overset{\sim}{\to} {}^{N}(B \times K_o^{\ell})$ mit $\ell \geq o$ und ${}^{N}B$ p-irreduzibel.

α) ${}^{N}A \subset K_o^n$ habe nur integrable (§ 0) g-irreduzible Komponenten, und es gelte $\text{Einbdim}\, {}^{N}A = n$, $\overline{\dim}\, {}^{N}A = n-1$.

β) ${}^{N}A$ sei Komponente von sich, die Singularitätenmenge ${}^{N}(A \setminus M({}^{N}A))$ sei g-irreduzibel.

γ) Im Falle $A \setminus M({}^{N}A) = \{o\}$ ist ${}^{N}A$ stets p-irreduzibel.

δ) Ist ${}^{N}A$ Komponente von sich und $\overline{\dim}\, {}^{N}A \leq 1$, so ist ${}^{N}A$ p-irreduzibel.

Zum *Beweis* ziehe man § 0 und § 1 heran.

Problem 1) Gesucht sind hinreichend allgemeine Kriterien für die Richtigkeit folgender Aussagen: Sei $N' \leq N$

α) Aus ${}^{N'}A \overset{\sim}{\to} {}^{N'}B$ folgt ${}^{N}A \overset{\sim}{\to} {}^{N}B$

β) Ist ${}^{N}A$ p-irreduzibel, so auch ${}^{N'}A$.

Für β) erhält man z.B. mit 2.4 erste Kriterien. Im zweiten Teil [21] geben wir im Rahmen des Benötigten für den analytischen Fall weitere Kriterien.

2) Für welche Raumkeime ${}^{N}A$ ist jede Immersion ${}^{N}A \to {}^{N}A$ ein Diffeomorphismus ? I.a. trifft das schon für semianalytische Keime nicht zu, wie einfache Beispiele zeigen. Dagegen ist das für analytische Abbildungen analytischer Keime stets richtig, wie man algebraisch einsehen kann ([21]). Der Grund sollte allerdings eher rein geometrischer Natur sein, etwa vom Typ "kurvenreich und integrabel oder formalanalytisch ([19])", was auch den differenzierbaren Fall beantworten würde.

Literatur

[1] Campo, N.A' : Le nombre de Lefschetz d'une monodromie. Indag. Math. 35 (1973), 113-118

[2] Becker, J. : C^k and analytic equivalence of complex analytic Varieties, Math. Ann. 225, (1977), 57-67

[3] Bloom, T. : C^1-functions on a complex analytic variety. Duke Math. J. 36 (1969), 283-296

[4] Ephraim, R. : C^∞ and analytic equivalence of singularities. Proc. of Conference of Complex Analysis 1972, Rice Univ. Studies

[5] Ephraim, R. : The cartesian product structure of singularities. Trans. Amer. Math. Soc. 224, (1976), 299-311

[6] Ephraim, R. : Cartesian product structure of singularities. Proc. of Symp. in pure Math., 30 (1977), 21-23

[7] Gibson, C.G.; Wirthmüller, K.; Plessis, A.A. du; Looijenga, E.J.N. : Topological stability of smooth mappings Springer Lecture Notes in Math., 552 (1977)

[8] Gottschling, E. : Invarianten endlicher Gruppen und biholomorphe Abbildungen. Inv. Math. 6 (1969), 315-326

[9] Hironaka, H. : Subanalytic sets. Number Theory, Algebraic Geometry and Commutative Algebra. In honour of Y. Akizuki, Kinokuniya, Tokio (1973), p 453-493

[10] Lojasiewicz, S. : Ensembles semi-analytic, polycopie IHES (1965)

[11] Milnor, J. : Singular points of complex hypersurfaces. Annales of Math. Studies No. 61, New York, Princeton Univ. Press 1968

[12] Mumford, D. : The topology of normal singularities of an algebraic surface and a criterion for simplicity. IHES No. 9 (1961), 5-22

[13] Prill, D. : Local classification of quotients of complex manifolds. Duke math. Journ. 34 (1967), 375-386

[14] Reichard, K. : C^∞-Diffeomorphismen semi- und subanalytischer Mengen. Erscheint in Compositio Mathematica, 1981

[15] Reichard, K. : Lokale Klassifikation von Quotientensingularitäten reeller Mannigfaltigkeiten nach diskreten Gruppen. preprint Bochum

[16] Reichard, K. : Produktzerlegung von Quotientensingularitäten. preprint Bochum

[17] Spallek, K. : Über Singularitäten analytischer Mengen. Math. Ann. 172, (1967), 249-268

[18] Spallek, K. : Differenzierbare Räume. Math. Ann. 180 (1969), 269-296

[19] Spallek, K. : Geometrische Bedingungen für die Integrabilität von Vektorfeldern auf Teilmengen des $\mathbb{R}^n$. manuscripta math. 25 (1978), 147-160

[20] Spallek, K. : L-platte Funktionen auf semianalytischen Mengen. Math. Ann. 227 (1977), 277-286

[21] Spallek, K. : Produktzerlegung und Äquivalenz von Raumkeimen II.

[22] Spallek, K. : Differenzierbare und analytische Struktur auf analytischen und semianalytischen Räumen. preprint

[23] Strub, R. : Vollständige Klassifikation der Singularitäten von Quotienten von unendlich oft reell-differenzierbaren Mannigfaltigkeiten nach eigentlich diskontinuierlichen Gruppen. Dissertation, Mainz (1980)

[24] Thom, R. : Ensembles et morphismes stratifies. Bull. Amer. Math. Soc. 75 (1969), 496-549

[25] Wavrik, J.J. : A theorem on solutions of analytic equations with applications to deformations of complex structures. Math.Am. 216 (1975), 127-142

[26] Whitney, H. : Tangents to an analytic variety. Am. of Math. 81 (1965), 496-549

Abstract

As an application of [19] we generalise Ehpraims product-structure-theorems for irreducible complex-analytic germs in [4],[5],[6] to relative arbitrary germs of spaces with arbitrary classes of differentiability, including all complex analytic, real analytic, semi-analytic or sub-analytic germs. This opens a way to generalise Ephraims C^∞-classification of irreducible complex analytic germs ([21]).

Klassifikation: 32C40, 32K15, 58A35, 58A40, 58C25

Produktzerlegung und Äquivalenz von Raumkeimen II

Der komplexe Fall

Einleitung

In [6],[7],[8] wurden die Diffomorphieklassen irreduzibler komplex-analytischer Mengenkeime bestimmt. Als Anwendung von [17],[19] verallgemeinern wir diese Klassifikation auf nicht notwendig irreduzible Fälle (§ 2, insbesondere 2.8).

Die Frage interessiert, durch welche Daten die (bzw. komplex-) analytische Struktur einer Singularität bereits fest liegt. Ein analytischer Keim, der etwa topologisch "trivial" oder "fast trivial" ist, kann analytisch durchaus sehr singulär sein, wenn nicht zusätzliche algebraisch-analytische Annahmen (Dimension, Normalität u.a.) und/oder Annahmen über die Trivialisierungsabbildungen getroffen werden ([1], [4],[10],[11],[16]). Dahinter steht unteranderem die allgemeinere Frage, wann bestimmte oder gar beliebige (komplex-, bzw. reell-, bzw. semi-, bzw. sub-) analytische Mengenkeime, die C^k-äquivalent sind ($k \in \{0,1,2,\dots,\infty\}$), schon analytisch äquivalent sind. Der reell-analytische Fall folgt mit [17],[21] sofort (1.4). Wenn man außerdem $k < \infty$ und eine gewisse Unabhängigkeit dieser Zahl von konkreten Mengen haben möchte, wird die Situation kompliziert ([2],[3]). Dasselbe trifft für die geometrisch komplizierteren semi-analytischen Mengen zu ([13]), während der sub-analytische Fall merkwürdiger Weise schon aus dem Rahmen fällt: Es gibt C^∞-äquivalente sub-analytische Keime, die nicht analytisch äquivalent sind ([13]). In [14],[15],[20] werden schließlich die C^∞-Äquivalenzklassen gewisser reell-analytischer Quotientensingularitäten mit Bezug auf die zugehörigen Transformationsgruppen beschrieben. Die Situation ist hier ganz anders als im analogen komplex-analytischen Fall, der in [9],[12] unter Holomorphie-Äquivalenz, in [15] unter C^∞-Äquivalenz diskutiert wird.

Der komplex-analytische Fall ist in der Tat etwas verwickelter, da ein komplexer Keim reell-analytisch äquivalent auch konjugiert komplexe und i.a. sogar "gemischt komplexe" Strukturen trägt (vgl. die Beispiele am Ende der Arbeit). Die damit zusammenhängenden Schwierigkeiten bei der Bestimmung von C^∞-Äquivalenzklassen wurden in [6],[7],[8] durch gewisse Faktorisierungssätze für irreduzible komplex-analytische Mengenkeime überwunden. Deren Verallgemeinerung auf nicht notwendig irreduzible komplex- und reell-analytische Mengenkeime in [19] ermöglicht zusammen mit [17],[21] den Beweis allgemeinerer Klassifikationssätze (§ 2). Die Erweiterung erfolgt dabei auf eine optimal größte Klasse von Raumkeimen, wie Beispiele am Ende verdeutlichen.

1. *Differenzierbare und analytische Äquivalenz*

In diesem Abschnitt reduzieren wir die Frage, bis zu welchem Grade differenzierbare Abbildungen komplexer Raumkeime geeignete holomorphe Abbildungen implizieren, auf bestimmte Eigenschaften der Differentiale dieser Abbildungen.

Dazu sei $\mathbf{X} = (X, \mathcal{D}^{\omega^*}/J)$ ein (nicht notwendig reduzierter) komplexer Raumkeim: $\phi \neq \mathbf{X} \subset \mathbb{C}^n_0$ ist ein komplex-analytischer Mengenkeim im Nullpunkt des $\mathbb{C}^n$, $\mathcal{D}^{\omega^*}$ die Algebra der holomorphen Funktionskeime in $0 \in \mathbb{C}^n$ und $J \subset \mathcal{D}^{\omega^*}$ ein Ideal, dessen Nullstellenkeim gerade X ist. Für jedes $N \in \{1,2,\ldots,\infty,\omega,\omega^*\}$ gehört zu $\mathbf{X}$ ein N-differenzierbarer Raumkeim $\mathbf{X}^N := (X, \mathcal{D}^N/J^N)$ mit $J^N := (\bar{J} \cdot \mathcal{D}^{N-1} + J \cdot \mathcal{D}^{N-1}) \cap \mathcal{D}^N$, $\mathcal{D}^N$ = Algebra der C^N-differenzierbaren Funktionskeime in $0 \in \mathbb{C}^n$, sowie ein reduzierter N-differenzierbar Raumkeim ${}^N\mathbf{X}$. $\mathbf{X}^N$ ist auch bei reduziertem X i.a. nicht reduziert. Die Zuordnung $\mathbf{X} \rightsquigarrow \mathbf{X}^N$ und $X \rightsquigarrow {}^N X$ liefert je einen Funktor der entsprechenden Kategorie. Schließlich gehört zu X in natürlicher Weise ein "konjugierter" komplexer Raumkeim $\bar{\mathbf{X}} := (\bar{X}, \mathcal{D}^{\omega^*}/\bar{J})$; z.B. ist $f \in \bar{J}$ genau dann, wenn $g \in J$ für $g(z) := \overline{f(\bar{z})}$ ist. Die Zuordnung $\mathbf{X} \rightsquigarrow \bar{\mathbf{X}}$ liefert einen Funktor in der Kategorie der komplexen Raumkeime. Für die zu den Raumkeimen jeweils gehörenden Tangentialräume gilt:

$$\mathrm{Tg}\,\mathbf{X}^N = \mathrm{Tg}\,\mathbf{X} \qquad \forall\ N \geq 1$$

$$\mathrm{Tg}\,{}^N\mathbf{X} = \mathrm{Tg}\,{}^{\omega^*}\mathbf{X} \qquad \forall\ N \geq N_0 \text{ mit } N_0 \in \mathbb{N} \text{ "hinreichend" groß.}$$

Dabei folgt die erste Identität aufgrund der Definition von $\mathbf{X}^N$ durch einfache Rechnung, die zweite Identität aus [17], 1.9. Insbesondere sind diese Tangentialräume *komplexe* Vektorräume. $\mathbf{Y}$ sei ein weiterer (nicht notwendig reduzierter) komplexer Raumkeim. Eine differenzierbare Abbildung $f: \mathbf{X}^N \to \mathbf{Y}^N$ liefert eine reell-lineare Abbildung (das Differential von f) $df: \mathrm{Tg}\,\mathbf{X}^N \to \mathrm{Tg}\,\mathbf{Y}^N$ komplexer Vektorräume. Diese zerfällt eindeutig in einen holomorphen Anteil hol df und einen antiholomorphen Anteil anhol df. Es ist df = hol df + anhol df. Entsprechendes gilt für alle hinreichend großen $N (\forall\ N \geq N_0 \in \mathbb{N},\ N_0$ fest) im Falle $f: {}^N\mathbf{X} \to {}^N\mathbf{Y}$. Im Unterschied zum ersten Fall ist hier wesentlich, daß N hinreichend groß sein *muß*.

<u>Satz</u> 1.1 Zu $\mathbf{X}$ und $\mathbf{Y}$ gibt es ein $N_0 \in \mathbb{N}$ mit folgenden Eigenschaften: Gegeben seien irgendwelche differenzierbaren Abbildungen mit $N \geq N_0$.

$\alpha)\ f : \mathbf{X}^N \to \mathbf{Y}^N$ bzw. $\beta)\ g : {}^N\mathbf{X} \to {}^N\mathbf{Y}$ ↷

i) Es gibt holomorphe Abbildungen $\alpha)\ F : \mathbf{X} \to \mathbf{Y}$ mit $dF = \mathrm{hol}\, df$ bzw. $\beta)\ G : {}^{\omega^*}\mathbf{X} \to {}^{\omega^*}\mathbf{Y}$ mit $dG = \mathrm{hol}\, dg$

ii) Es gibt holomorphe Abbildungen $\alpha)\ F^* : \mathbf{X} \to \bar{\mathbf{Y}}$ mit $dF^* = \mathrm{anhol}\, df$ bzw. $\beta)\ G^* : {}^{\omega^*}\mathbf{X} \to \overline{{}^{\omega^*}\mathbf{Y}}$ mit $dG^* = \mathrm{anhol}\, dg$

iii) $\alpha)$ Ist $\mathrm{hol}\, df$ injektiv bzw. surjektiv, so ist F eine Immersion bzw. eine Submersion. Ist f ein Diffeomorphismus und $\mathrm{hol}\, df$ bijektiv, so ist F biholomorph. Entsprechendes gilt im Falle $\beta)$ für G.

iv) Unter entsprechenden Voraussetzungen über $\mathrm{anhol}\, df$ und f bzw. $\mathrm{anhol}\, dg$ und g erhält man analoge Aussagen für F^* bzw. G^*.

Beweis: i) $\alpha)$ O.E. sei $\mathbf{X} = (X, \mathcal{D}^{\omega^*}/J) \subset \mathbb{C}^n_o$, $\mathbf{Y} = (Y, \mathcal{D}^{\omega^*}/J^*) \subset \mathbb{C}^m_o$ mit $\mathrm{Tg}\,\mathbf{X} = \mathbb{C}^n$, $\mathrm{Tg}\,\mathbf{Y} = \mathbb{C}^m$. Dann wird $f : \mathbf{X}^N \to \mathbf{Y}^N$ durch eine Abbildung $\varphi : \mathbb{C}^n_o \to \mathbb{C}^m_o$ der Klasse C^N erzeugt mit $d\varphi = df$. $f_1, \ldots, f_r \in J$ bzw. $g_1, \ldots, g_s \in J^*$ seien Erzeugendensysteme der jeweiligen Ideale. Dann gibt es Funktionen d_{ji}, d^*_{ji} der Klasse C^{N-1} mit

$$*)\quad g_j \circ \varphi = \sum_{i=1}^r d_{ji} \cdot f_i + \sum_{i=1}^r d^*_{ji} \cdot \bar{f}_i \quad \forall\ j = 1, \ldots, s$$

$T^N k$ bezeichne die Taylorreihe bis zur Ordnung N im Aufpunkt 0 eines differenzierbaren Abbildungskeim k. $\mathrm{hol}\, P$ ($\mathrm{anhol}\, P$) bezeichne den holomorphen (antiholomorphen) Anteil einer reellen Potenzreihe P im $\mathbb{C}^n$. $m \subset \mathcal{D}^{\omega^*}$ sei das maximale Ideal in 0. ↷

$$\begin{aligned} g_j \circ (\mathrm{hol}\, T^N\varphi) &= \mathrm{hol}(g_j \circ T^N\varphi) \in \mathrm{hol}\, T^N(g_j \circ \varphi) + m^N \\ &\in \sum_i (\mathrm{hol}\, T^N d_{ji}) f_i + m^N \quad \forall\ j = 1, \ldots, s \ ↷ \end{aligned}$$

$$**)\quad g_j \circ (\mathrm{hol}\, T^N\varphi) \in J + m^N \qquad \forall\ j = 1, \ldots, s$$

Nach [20] gibt es bei geeignet großem $N_o \in \mathbb{N}$ (das von den f_i und g_j abhängt) holomorphe ϕ, K_{ji} mit

$$d\phi = d(\mathrm{hol}\, T^N\varphi) \quad (= \mathrm{hol}\, df)$$

$$g_j \circ \phi = \sum_i K_{ji} \cdot f_i \qquad \text{also}$$

$$g_j \circ \phi \in J \qquad \forall\ j = 1, \ldots, r.\ ↷$$

ϕ erzeugt einen holomorphen Morphismus $F : \mathbf{X} \to \mathbf{Y}$ mit $dF = \mathrm{hol}\, df$.

i), β) Der Beweis hierzu ist weniger direkt und funktioniert nur unter Verwendung von [17] : O.E. sei ${}^{\omega^*}\mathbf{X} = (X, \mathcal{D}^{\omega^*}/J) \subset \mathbb{C}^n$, ${}^{\omega^*}\mathbf{Y} = (Y, \mathcal{D}^{\omega^*}/J^*) \subset \mathbb{C}^m$ mit $Tg\,{}^{\omega^*}\mathbf{X} = \mathbb{C}^n$, $Tg\,{}^{\omega^*}\mathbf{Y} = \mathbb{C}^m$. f_i, g_j seien wie oben gewählt. Dann ist

$$g_j \circ \varphi \,|\, X \equiv 0$$

und daher $g_j \circ T^N\varphi$ N-platt auf $\mathbf{X}$ im Sinne von [17]. Nach [17] folgt also zu gegebenem $\ell \in \mathbb{N}$ und hinreichend großem N :

$$g_j \circ (\text{hol } T^N\varphi) \in J + m^\ell,$$

was der Beziehung **) von oben entspricht: Der Rest ergibt sich jetzt wie oben unter α).
Damit ist i) und entsprechend auch ii) bewiesen.

Unter iii) α) ist nur der zweite Satz zu beweisen. Aufgrund der nachfolgenden Bemerkung 1.2 ist mit $\text{hol}\, d f$ auch $\text{hol}(d f)^{-1} = \text{hol}\, d f^{-1}$ bijektiv. Nach dem bisher Bewiesenen gibt es also Morphismen $F : \mathbf{X} \to \mathbf{Y}$, $F^* : \mathbf{Y} \to \mathbf{X}$ mit jeweils bijektivem Differential. $F^* \circ F : \mathbf{X} \to \mathbf{X}$, $F \circ F^* : \mathbf{Y} \to \mathbf{Y}$ sind also Morphismen mit jeweils bijektivem Differential. Insbesondere kommen $F^* \circ F$, $F \circ F^*$ von analytischen Bimorphismen $\mathbb{C}^n_o \to \mathbb{C}^n_o$, $\mathbb{C}^m_o \to \mathbb{C}^m_o$ her. Nach Bem. 1.3 weiter unten sind daher $F \circ F^*$, $F^* \circ F$ biholomorph, also auch F und F^*. Es ist ja z.B. $F \circ (F^* \circ F) = (F \circ F^*) \circ F$, also $(F \circ F^*)^{-1} \circ F = F \circ (F^* \circ F)^{-1}$ also $F \circ [(F^* \circ F)^{-1} \circ F^*] = (F \circ F^*)^{-1} \circ (F \circ F^*) = \text{id}$, $[(F^* \circ F)^{-1} \circ F^*] \circ F = \text{id}$. Damit gilt iii), α) und entsprechend β).
iv) ergibt sich ebenso. q.e.d.

Bemerkung 1.2 Jede reell lineare Abbildung $L : \mathbb{C}^n \to \mathbb{C}^n$ hat in komplexen und konjugiert komplexen Koordinaten die Matrixgestalt

$$M = \begin{pmatrix} A, & B \\ \overline{B}, & \overline{A} \end{pmatrix} \quad \text{mit A,B als komplexen } n \times n\text{-Matrizen}$$

Ist L invertierbar, so hat M^{-1} eine entsprechende Gestalt

$$M^{-1} = \begin{pmatrix} C, & D \\ \overline{D}, & \overline{C} \end{pmatrix}$$

Ist jetzt A (der "holomorphe Anteil" von L) invertierbar, so auch C (der "holomorphe Anteil" von L^{-1}). Entsprechendes gilt für die "antiholomorphen Anteile" B und D.

Beweis: Aus $\det A \neq 0$ und $M \cdot M^{-1} = E$ (jeweilige Einheitsmatrix), folgt $AC + B\bar{D} = E$, $AD + B\bar{C} = 0$, also $AC - B(\bar{A}^{-1} \cdot \bar{B} \cdot C) = E$, $(A - B\bar{A}^{-1}\bar{B}) \cdot C = E$, also $\det C \neq 0$.

<u>*Bemerkung*</u> 1.3 R sei ein noetherscher Ring, $I \subset R$ ein Ideal, $h : R \to R$ ein Isomorphismus mit $h(I) \subset I \curvearrowright h(I) = I$, und h induziert einen Isomorphismus $h^* : R/I \to R/I$.

Beweis: Mit $I_o := I$ und $I_v := h^{-1}(I_{v-1})$ (induktiv) folgt:

$$I_o \subset I_1 \subset I_2 \subset \ldots \subset I_v \subset \ldots$$

Es gibt also ein v_o mit $I_v = I_{v_o}$ $\forall\ v \geq v_o$, also gilt:

$I_{v_o} = h(I_{v_o+1}) = h(I_{v_o}) = I_{v_o-1}$, und induktiv weiter:

$$I_1 = h(I_2) = h(I_1) = I_o$$ q.e.d.

1.3 gilt auch in gewissen nicht-noetherschen (z.B. differenzierbaren) Situationen. Der Beweis muß dann mehr geometrisch geführt werden.

Analog erhält man für reellanalytische (nicht notwendig reduzierte) Raumkeime $\mathbf{X}$ und $\mathbf{Y}$ (wegen [17], und [21]):

<u>*Satz*</u> 1.4 Zu $\mathbf{X}$ und $\mathbf{Y}$ gibt es ein $N^o \in \mathbb{N}$ mit folgenden Eigenschaften: Gegeben seien irgendwelche differenzierbaren Abbildungen

α) $f : \mathbf{X}^N \to \mathbf{Y}^N$ bzw. β) $g : {}^N\mathbf{X} \to {}^N\mathbf{Y}$ mit $N \geq N^o$ $\curvearrowright$

i) Es gibt analytische Abbildungen α) $F : \mathbf{X} \to \mathbf{Y}$ bzw. β) $G : {}^\omega\mathbf{X} \to {}^\omega\mathbf{Y}$ mit $dF = df$ bzw. $dG = dg$.

ii) Ist f bzw. g eine Immersion oder Submersion oder ein Bimorphismus, so gilt Entsprechendes für F bzw. G.

2. *Produktzerlegung und Äquivalenz im analytischen Fall*

Aus [19], 2.1 halten wir zunächst fest:

Satz 2.1 α) Jeder komplex-analytische Keim ${}^{\omega^*}A$ läßt sich bis auf Isomorphie der Faktoren *eindeutig* als Produkt p-irreduzibler (d.h. nicht weiter zerlegbarer) Faktoren ${}^{\omega^*}A_i$ zerlegen, und diese sind komplex-analytisch: ${}^{\omega^*}A \mathrel{\tilde\to} {}^{\omega^*}A_1 \times \dots \times {}^{\omega^*}A_r$. Jede Zerlegung ${}^{\omega^*}A \mathrel{\tilde\to} {}^{\omega^*}B_1 \times \dots \times {}^{\omega^*}B_s$ läßt sich zu einer p-irreduziblen "verfeinern".
β) Eine entsprechende Aussage gilt für die Kategorie der reell- bzw. semi-, bzw. sub-analytischen Mengenkeime.

Schwer beantwortbar in ihrer allgemeinen Form scheint die Frage zu sein, wann für einen p-irreduziblen Keim ${}^{N}A$ auch ${}^{N'}A$ p-irreduzibel ist ($N' < N!$). Eine partielle Antwort enthält:

Satz 2.2. A sei ein reell- (bzw. komplex-) analytischer Keim und ${}^{\infty}A \mathrel{\tilde\to} {}^{\infty}B_1 \times \dots \times {}^{\infty}B_s$, wobei B_i irgendwelche Mengenkeime sind. Dann gibt es reell- (bzw. komplex-) analytische Keime A_i mit ${}^{\omega}A \mathrel{\tilde\to} {}^{\omega}A_1 \times \dots \times {}^{\omega}A_s$. Die ${}^{\infty}B_i$ sind genau dann p-irreduzibel, wenn die ${}^{\omega}A_i$ p-irreduzibel sind.

Beweis. Sei $C_1^{\infty} \times C_2^{\infty} \xrightarrow{\varphi} A^{\infty}$ und ${}^{\infty}C_1$ p-irreduzibel. Wir wählen einen gewöhnlichen Punkt $q_2 \in {}^{\infty}C_2$ und $p = \varphi(0,q_2)$. Dann ist $C_1^{\infty} \times \mathbb{R}^{\ell} \mathrel{\tilde\to} A_p^{\infty}$, mit [16], [19] also $C_1^{\infty} \mathrel{\tilde\to} A_1^{\infty}$, wobei A_1 reell- (bzw. komplex-) analytisch ist. Nun folgt ${}^{\infty}A \mathrel{\tilde\to} {}^{\infty}A_1 \times \dots \times {}^{\infty}A_s$ und mit § 1 die Behauptung.

Für kohärente Keime folgt aus unseren früheren Resultaten zusammen mit § 1 eine entsprechende Aussage bereits aus ${}^{N}A \mathrel{\tilde\to} {}^{N}B_1 \times \dots \times {}^{N}B_s$ bei hinreichend großem $N \in \mathbb{N}$. Der folgende Satz enthält eine weitere Ergänzung, die wir benötigen. Wir verallgemeinern damit zugleich Theorem 5 aus [8] (das dort ohne Beweis behauptet wird).

Als Hilfsmittel benötigen wir Komplexifizierungen reell-analytischer Keime ([5]). Dazu identifizieren wir den $\mathbb{C}^n$ auch mit $\Delta := \{(z,w) \in \mathbb{C}^{2n} \mid z = \bar{w}\}$; einen analytischen Keim $A \subset \mathbb{C}_0^n$ sehen wir also auch als in Δ_0 liegenden reell-analytischen Keim an. Seine

Komplexifizierung $A^c \subset \mathbb{C}_o^{2n}$ ist der kleinste komplex-analytische Keim, der A enthält. A^c ist die Nullstellenmenge all jener holomorphen Funktionskeime auf $\mathbb{C}_o^{2n}$, die auf $A \subset \Delta_o$ verschwinden. Ist ${}^{\omega^*}A$ singularitätenfrei, so wird $A^c = A \times \bar{A}$. Mit $g(z,w) = \sum a_{\nu\mu} z^\nu w^\mu$ und $g(z,\bar{z}) = o$ für alle $z_{\ell+1} = \ldots = z_n = o$ folgt nämlich sofort $a_{\nu\mu} = o \;\forall\; \nu = (\nu_1,\ldots,\nu_\ell,o,\ldots,o), \mu = (\mu_1,\ldots,\mu_\ell,o,\ldots,o)$. Dann wird für jeden irreduziblen komplex-analytischen Keim A ebenfalls $A^c = A \times \bar{A}$. Ist im reduzierten Fall $A = A_1 \cup \ldots \cup A_r$ die Zerlegung in irreduzible Komponenten, so wird damit $A^c = \bigcup_{i=1}^{r} (A_i \times \bar{A}_i)$ die Zerlegung in irreduzible Komponenten. Allgemeiner gehört zu jedem reell-analytischen Keim $B_i \subset \mathbb{R}_o^{n_i} \subset \mathbb{C}_o^{n_i}$ seine Komplexifizierung $B_i^c \subset \mathbb{C}_o^{n_i}$. Es gilt $(B_1 \times \ldots \times B_r)^c = B_1^c \times \ldots \times B_r^c$, und im Falle $n_1 = \ldots = n_r$: $(B_1 \cup \ldots \cup B_r)^c = B_1^c \cup \ldots \cup B_r^c$

Satz 2.3 ${}^{\omega^*}A$ sei ein komplex-analytischer Raumkeim mit einer irreduziblen Komponente, die auch p-irreduzibel ist. Denn ist neben ${}^{\omega^*}A$ auch der zugeordnete reell-analytische Keim ${}^{\omega}A$ p-irreduzibel.

Beweis Der erste Teil der Behauptung ist mit [19], 1.8 klar. Sei nun ${}^{\omega^*}A = {}^{\omega^*}(A_1 \cup \ldots \cup A_r)$ die Zerlegung in irreduzible Komponenten. Für die Komplexifizierung ${}^{\omega^*}A^c$ von ${}^{\omega}A$ gilt nun

$${}^{\omega^*}A^c = {}^{\omega^*}(A_1^c \cup A_2^c \cup \ldots \cup A_r^c)$$

Die Komplexifizierungen A_i^c haben hier die Gestalt $A_i^c = A_i \times \bar{A}_i$, wobei $\bar{A}_i$ der zu A_i konjugiert komplexe Keim ist. A_i und $\bar{A}_i$ sind analytisch irreduzibel. O.E. sei etwa ${}^{\omega^*}A_1$ p-irreduzibel. Dann ist auch ${}^{\omega^*}\bar{A}_1$ p-irreduzibel. Angenommen, es gibt eine p-irreduzible Zerlegung

$$\varphi : {}^{\omega}A \xrightarrow{\sim} {}^{\omega}B_1 \times \ldots \times {}^{\omega}B_s \;, \quad B_i \neq \{o\} \quad \forall\; i = 1,\ldots,s,$$

mit $s \geq 2$. Dann ist $\varphi^c : {}^{\omega^*}A^c \xrightarrow{\sim} {}^{\omega^*}B_1^c \times \ldots \times {}^{\omega^*}B_s^c$, und jeder Keim ${}^{\omega^*}A_i^c$ hat die Gestalt ${}^{\omega^*}A_i^c \xrightarrow{\sim} {}^{\omega^*}B_{1i}^c \times \ldots \times {}^{\omega^*}B_{si}^c$, wobei ${}^{\omega}B_{ji} \subset {}^{\omega}B_j$ eine irreduzible Komponente $\neq \{o\}$ ist (mit [19], 1.8). Insbesondere ist

$${}^{\omega^*}(A_1 \times \bar{A}_1) \xrightarrow{\sim} {}^{\omega^*}(B_{11}^c \times \ldots \times B_{s1}^c) ,$$

wegen 2.1 daher $s = 2$ und o.E.: ${}^{\omega^*}A_1 \xrightarrow{\sim} {}^{\omega^*}B_{11}^c$, ${}^{\omega^*}\bar{A}_1 \xrightarrow{\sim} {}^{\omega^*}B_{21}^c$. Nach [19], 2.2 liefert die Einschränkung von φ^c sogar o.E. Isomorphismen:

$$*) \;\varphi^c : {}^{\omega^*}A_1 \times o \to {}^{\omega^*}B_{11}^c \times o, \quad o \times {}^{\omega^*}\bar{A}_1 \to o \times {}^{\omega^*}B_{21}^c .$$

Ist $A \subset \mathbb{C}_o^n$, $\Delta := \{(z,w) \in \mathbb{C}^n \times \mathbb{C}^n \mid w = \bar{z}\}_o$ und o.E. auch $B_1 \times \ldots \times B_s \subset \mathbb{C}_o^n$, so sind zur Bildung der Komplexifizierung A und $B_1 \times \ldots \times B_s$ als Teilmengen von Δ, und φ als Spur eines reellanalytischen Bimorphismus $\phi : \Delta \to \Delta$ aufzufassen. Dessen Komplexifizierung $\phi^c : \mathbb{C}_o^{2n} \to \mathbb{C}_o^{2n}$ induziert φ^c.

$*$) liefert also durch Einschränkung Isomorphismen

${}^{\omega}(o\times o) = {}^{\omega}((A_1\times o)\cap\Delta) \to {}^{\omega}((B_{11}^c\times o)\cap\Delta) = {}^{\omega}B_{11}\times o$

und damit $B_{11}=o$. Widerspruch ! 2.3 ist bewiesen.

${}^{\omega^*}C$ heiße *Bestimmungskomponente* von ${}^{\omega^*}A$, wenn ${}^{\omega^*}C$ irreduzible analytische Komponente von ${}^{\omega^*}A$ mit $\mathrm{Einbdim}\,{}^{\omega^*}A = \mathrm{Einbdim}\,{}^{\omega^*}C$ ist.

<u>*Satz*</u> 2.4 ${}^{\omega^*}A$, ${}^{\omega^*}B$ seien komplexanalytische Raumkeime, ${}^{\omega^*}A$ habe eine p - irreduzible Bestimmungskomponente. Sind A und B reellanalytisch oder C^∞ - diffeomorph äquivalent, so ist ${}^{\omega^*}A$ zu ${}^{\omega^*}B$ oder zu ${}^{\omega^*}\bar{B}$ isomorph.

Beweis: Wegen 1.4 genügt es, den reellanalytischen Fall zu beweisen: $\varphi: {}^{\omega}A \xrightarrow{\sim} {}^{\omega}B$. φ bildet aus topologischen Gründen irreduzible Komponenten von ${}^{\omega^*}A$ auf solche von ${}^{\omega^*}B$ ab, wobei ([17]) deren komplexanalytische Einbettungsdimensionen nicht verändert werden. Eine Bestimmungskomponente E von ${}^{\omega^*}A$ wird also unter φ auf eine Bestimmungskomponente F von ${}^{\omega^*}B$ abgebildet. In den Komplexifizierungen erhalten wir also eine C^{ω^*} - Isomorphie $\varphi^c : E\times\bar{E} = E^c \xrightarrow{\sim} F^c = F\times\bar{F}$. Ist E noch dazu p - irreduzibel (wie wir aufgrund der Voraussetzungen annehmen dürfen), so ist wegen 2.1 auch F p-irreduzibel. Wegen [19], 2.2 liefert φ^c durch Einschränkung Isomorphismen ${}^{\omega^*}E \to {}^{\omega^*}F$ oder ${}^{\omega^*}E \to {}^{\omega^*}\bar{F}$. Weil E Bestimmungskomponente von A ist, muß entweder hol $d\varphi(o)$ oder anhol $d\varphi(o)$ bijektiv sein. Nach 1.1 ist daher entweder ${}^{\omega^*}A \xrightarrow{\sim} {}^{\omega^*}B$ oder ${}^{\omega^*}A \xrightarrow{\sim} {}^{\omega^*}\bar{B}$.

Wir interessieren uns für komplex - analytische Raumkeime ${}^{\omega^*}A$ der folgenden Art: ${}^{\omega^*}A$ besitzt eine irreduzible Komponente $A_1 \subset A$, die p - irreduzibel ist und $\mathrm{Einbdim}\,{}^{\omega^*}A_1 = \mathrm{Einbdim}\,{}^{\omega^*}A$ erfüllt.

<u>*Definition*</u> 2.5 Wir nennen solche Raumkeime *komplex bestimmt*.

<u>*Beispiele*</u> 2.6 für diese Klasse erhält man zahlreich auf folgende Art: Man wähle irgendeinen irreduziblen und zugleich p-irreduziblen Keim ${}^{\omega^*}A_1 \subset \mathbb{C}^n_o$ mit $\mathrm{Einbdim}\,{}^{\omega^*}A_1 = n$ (z.B. einen irreduziblen singulären Hyperflächenkeim A mit $Tg^i_o A = \{o\}$ ([18]) oder einen Keim mit irreduzibler Singularitätenmenge, etwa einen "genügend krummen" Kurvenkeim ([19], 2.4)) und nehme endlich viele, im übrigen aber beliebige komplex-analytische Keime ${}^{\omega^*}A_i \subset \mathbb{C}^n_o$ als weitere Komponenten hinzu: $A := \bigcup_{i=1}^{r} A_i$.

Definition 2.7 Wir nennen einen Raumkeim ${}^{\omega^*}A$ *komplex bestimmbar*, wenn es eine Isomorphie ${}^{\omega^*}A \xrightarrow{\sim} {}^{\omega^*}A_1 \times \ldots \times {}^{\omega^*}A_s$ mit komplex bestimmten Raumkeimen ${}^{\omega^*}A_j$ gibt.

Insbesondere sind die (von Ephraim betrachteten) irreduziblen Keime alle komplex bestimmbar und viele mehr. Die Diffeomorphie-Klassen komplex bestimmbare Raumkeime können wir nun angeben:

Satz 2.8 ${}^{\omega^*}A$, ${}^{\omega^*}B$ seien komplex bestimmbar; ${}^{\infty}A$, ${}^{\infty}B$ seien diffeomorph. Dann sind die p-irreduziblen Faktoren von ${}^{\omega^*}A$ und ${}^{\omega^*}B$ bis auf Holomorphie oder Antiholomorphie paarweise einander gleich.

Beweis: Sei ${}^{\omega^*}A = {}^{\omega^*}A_1 \times \ldots \times {}^{\omega^*}A_r$, ${}^{\omega^*}B = {}^{\omega^*}B_1 \times \ldots \times {}^{\omega^*}B_s$ und (wegen 1.4) ${}^{\omega}A_1 \times \ldots \times {}^{\omega}A_r \xrightarrow{\sim} {}^{\omega}B_1 \times \ldots \times {}^{\omega}B_s$. Wegen 2.3 sind alle ${}^{\omega}A_i$, ${}^{\omega}B_j$ p-irreduzibel. Also folgt $r = s$ und (bei geeigneter Numerierung) ${}^{\omega}A_i \xrightarrow{\sim} {}^{\omega}B_i$ (2.1). Nach 2.4 wird ${}^{\omega^*}A_i \xrightarrow{\sim} {}^{\omega^*}B_i$ oder ${}^{\omega^*}A_i \xrightarrow{\sim} {}^{\omega^*}\bar{B}_i$. q.e.d.

Beispiele 2.9 für reell-analytisch äquivalente komplexe Raumkeime, die weder komplex noch konjugiert komplex analytisch äquivalent sind, erhält man auf folgende Weise:

Man wähle irreduzible komplex-analytische Keime $A \subset \mathbb{C}^n_o$ mit folgenden Eigenschaften: ${}^{\omega^*}A \not\xrightarrow{\sim} {}^{\omega^*}\bar{A}$, ${}^{\omega^*}A$ p-irreduzibel. Dann ist

$${}^{\omega}(A \times \{o\} \cup \{o\} \times A) \xrightarrow{\sim} {}^{\omega}(\bar{A} \times \{o\} \cup \{o\} \times A),$$

aber es gilt weder

$${}^{\omega^*}(A \times \{o\} \cup \{o\} \times A) \xrightarrow{\sim} {}^{\omega^*}(\bar{A} \times \{o\} \cup \{o\} \times A)$$

noch ${}^{\omega^*}(A \times \{o\} \cup \{o\} \times A) \xrightarrow{\sim} \overline{{}^{\omega^*}(\bar{A} \times \{o\} \cup \{o\} \times A)} = {}^{\omega^*}(A \times \{o\} \cup \{o\} \times \bar{A})$

Hier sind ${}^{\omega^*}(A \times \{o\} \cup \{o\} \times A)$ und ${}^{\omega^*}(\bar{A} \times \{o\} \cup \{o\} \times A)$ zwar p-irreduzibel ([19], 1.8), aber gerade nicht komplex bestimmt. Ähnlich kann man beliebig kompliziertere Beispiele aufbauen.

Wir schließen mit einigen *Fragen*. Verschiedene Eigenschaften von ${}^{N}A$ hängen mit entsprechenden Eigenschaften seiner p-Faktoren zusammen ([19]). Wie sieht das z.B. für algebraische Eigenschaften im komplex-analytischen Fall aus? Wann ist für einen komplex-analytischen Raumkeim ${}^{\omega^*}A$, der p-irreduzibel ist, auch ${}^{\omega}A$ oder ${}^{\omega^*}A^c$ p-irreduzibel? Wann ist für einen reellanalytischen Raumkeim ${}^{\omega}A$, der p-irreduzibel ist, auch ${}^{\omega^*}A^c$ p-irreduzibel?

Literatur

[1] Campo, N.A' : Le nombre de Lefschetz d'une monodromie Indag. Math. 35 (1973), 113-118

[2] Becker, J. : C^k weakly holomorphic functions on an analytic set. Prov. Amer. Math. Soc. 39 (1973), 89-93

[3] Becker, J. : C^k and analytic equivalence of complex analytic varieties. Math. Ann. 225 (1977), 57-67

[4] Bloom, T. : C^1-functions on a complex analytic variety. Duke Math. J. 36 (1969), 283-296

[5] Cartan, H. : Variëtés analytiques réelles et variëtés analytiques complex. Bull. Soc. Math. France 85 (1957), 77-99

[6] Ephraim, R. : C^∞ and analytic equivalence of singularities Proc. of Conf. of Complex Anal. (1972), Rice Univ Studies

[7] Ephraim, R. : The cartesian product structure of singularities Trans. Amer. Math. Soc. 224 (1976), 299-311

[8] Ephraim, R. : Cartesian product structure of singularities Proc. of Sym. in pure Math. 30 (1977), 21-23

[9] Gottschling, E. : Invarianten endlicher Gruppen und biholomorphe Abbildungen. Inv.Math. 6 (1969), 315-326

[10] Milnor, J. : Singular points of complex hypersurfaces. Annales of Math. Studies No. 61, New York Princeton University Press 1968

[11] Mumford, D. : The topology of normal singularities of an algebraic surface and a criterion for simplicity. IHES No. 9 (1961), 5-22

[12] Prill, D. : Local classification of quotients of complex manifolds. Duke Math. Journ. 34 (1967), 375-386

[13] Reichard, K. : C^∞-Diffeomorphismen semi- und subanalytischer Mengen. Erscheint in Composito Mathematica, 1981

[14] Reichard, K. : Lokale Klassifikation von Quotientensingularitäten reeller Mannigfaltigkeiten nach diskreten Gruppen. Preprint Bochum

[15] Reichard, K. : Produktzerlegung von Quotientensingularitäten preprint Bochum

[16] Spallek, K. : Über Singularitäten analytischer Mengen. Math. Ann. 172 (1967), 249-268

[17] Spallek, K. : L-platte Funktionen auf semianalytischen Mengen. Math. Ann. 227 (1977), 266-277

[18] Spallek, K. : Geometrische Bedingungen für die Integrabilität von Vektorfeldern auf Teilmengen im $\mathbb{R}^n$. manuscripta math. 25 (1978), 147-160

[19] Spallek, K. : Produktzerlegung und Äquivalenz von Raumkeimen I

[20] Strub, G. : Vollständige Klassifikation der Singularitäten von Quotienten von unendlich oft reell-differenzierbaren Mannigfaltigkeiten nach eigentlich diskontinuierlichen Gruppen. Dissertation, Mainz 1980

[21] Wavrik, J.J. : A theorem on solutions of analytic equations with applications to deformations of complex structures. Math. Ann. 216 (1975), 127-142

Abstract

As an application of [17],[18],[19] we generalise Ephraims C^∞-classification of irreducible complex-analytic germs in [4],[5],[6].

PSEUDOVARIETES COMPLEXES

par Francesco Succi

Cet exposé contient quelques résultats obtenus dans une recherche en collaboration avec Martin Jurchescu.

Le point de départ était d'élaborer une bonne théorie des variétés complexes bordées ayant pour bord une variété mixte analytique de type (1,n-1), en s'appuyant sur le modèle donné par la situation présentée par les surfaces de Riemann bordées [4].

Puisque les variétés complexes bordées qui offrent le plus grand intérêt sont celles qui admettent un "double", qui est une variété complexe symétrique, nous sommes alors conduits à étudier en premier lieu celles-ci.

1.-Variétés complexes symétriques.

Une variété complexe symétrique est une variété complexe Z munie d'une anti-involution $c : Z \longrightarrow Z$, dite l'anti-involution structurelle de Z.

Un ouvert U d'une telle variété symétrique est dit symétrique si l'on a $c(U) = U$.

Soit $\mathcal{O}_Z$ le faisceau structural de la variété complexe Z. Une fonction holomorphe f, définie sur un ouvert symétrique U de Z, est dite symétrique si $f(c(z)) = \overline{f(z)}$, c'est à dire si $f \circ c = \overline{f}$, où $\overline{f}$ est la fonction complexe qui en chaque point z de U prend la valeur conjuguée de $f(z)$.

Pour chaque ouvert symétrique U de la variété complexe symétrique (Z,c) nous pouvons alors considérer l'algèbre de toutes les fonctions holomorphes symétriques dans U que l'on désignera par $\Gamma_{sym}(U, \mathcal{O}_Z)$. On a alors, pour tout ouvert symétrique U de Z, la décomposition directe :

$$\Gamma(U,\mathcal{O}_Z) = \Gamma_{sym}(U,\mathcal{O}_Z) + i\Gamma_{sym}(U,\mathcal{O}_Z) \quad ,$$

c'est-à-dire que pour U ouvert symétrique de Z toute fonction $f\in\Gamma(U,\mathcal{O}_Z)$ s'écrit d'une manière unique comme $f=f'+if''$ avec $f',f''\in\Gamma_{sym}(U,\mathcal{O}_Z)$, et nous poserons $Re_c(f):=f'$ et $Im_c(f):=f''$.

Nous pouvons maintenant identifier à un seul point tout couple de points de Z du type $z,c(z)$, c'est-à-dire que nous considérons l'espace topologique Z_{sym} ,quotient de l'espace topologique Z par rapport à la relation d'equivalence déterminée par c :

$$z\sim z' \iff z=z' \quad \text{ou} \quad z'=c(z) \quad .$$

Si $p_o : Z \longrightarrow Z_{sym}$ est l'application canonique du quotient, alors elle est continue, ouverte et induit une bijection entre les ouverts symétriques de Z et les ouverts de Z_{sym} . On voit aussi que Z_{sym} est séparé.

A chaque ouvert U_o de Z_{sym} associons maintenant l'algèbre $\Gamma_{sym}(p_o^{-1}(U_o),\mathcal{O}_Z)$ des fonctions holomorphes symétriques dans l'ouvert symétrique $U:=p_o^{-1}(U_o)$ de Z . On définit ainsi sur l'espace topologique Z_{sym} un faisceau de $\mathbb{R}$-algèbres $\mathcal{O}_{Z_{sym}}$, tel donc que $\Gamma(U_o,\mathcal{O}_{Z_{sym}}):=\Gamma_{sym}(p_o^{-1}(U_o),\mathcal{O}_Z)$. Alors $(Z_{sym},\mathcal{O}_{Z_{sym}})$ est un espace $\mathbb{R}$-annelé et on a un morphisme canonique $p : (Z,\mathcal{O}_Z)\longrightarrow(Z_{sym},\mathcal{O}_{Z_{sym}})$, où $(Z,\mathcal{O}_Z)$ est considéré avec la structure $\mathbb{R}$-annelée sous-jacente.

2.-Exemples fondamentaux.

a) $(\mathbb{C}^n,c)$ - Considérons comme variété complexe symétrique l'espace $\mathbb{C}^n$ avec l'anti-involution donnée par l'application de passage au conjugué $c : c(z_1,\ldots,z_n)=(\bar{z}_1,\ldots,\bar{z}_n)$.

b) $(\mathbb{C}\mathbb{P}^n,c)$ - Considérons maintenant la variété complexe symétrique donnée par l'espace projectif complexe $\mathbb{C}\mathbb{P}^n$ avec l'anti-involution c donnée par le passage au conjugué : si $[z_o, z_1, \ldots, z_n] \in \mathbb{C}\mathbb{P}^n$ alors $c([z_o, \ldots, z_n]) = [\bar{z}_o, \ldots, \bar{z}_n]$.

c) Soit Z une variété complexe et soit Z_c la variété conjuguée de Z; alors le double $Z' := Z \sqcup Z_c$ est une variété complexe symétrique.

d) Enfin, le double d'une surface de Riemann bordée [4] est une variété complexe symétrique.

3.-Pséudovariétés complexes et exemples.

Cela dit, donnons la définition suivante :

Définition.-On appelle pséudovariété complexe (PSV) de dimension (complexe) n , tout espace $\mathbb{R}$-annelé $(X, \mathcal{O}_X)$, localement isomorphe à sous-espaces ouverts de $\mathbb{C}^n_{sym} = (\mathbb{C}^n_{sym}, \mathcal{O}_{\mathbb{C}^n_{sym}})$.

Autrement dit, une PSV est un espace $\mathbb{R}$-annelé dont les modèles sont les ouverts de $\mathbb{C}^n_{sym}$.

Si X et Y sont deux PSV, un morphisme de PSV de X dans Y est par définition un morphisme d'espaces $\mathbb{R}$-annelées φ: $X \longrightarrow Y$.

Exemples :

a) $(\mathbb{C}^n_{sym}, \mathcal{O}_{\mathbb{C}^n_{sym}})$ est une PSV de dimension n.

b) D'une façon plus générale, pour toute variété complexe symétrique Z, l'espace $\mathbb{R}$-annelé Z_{sym} est une PSV.

[Ceci résulte du fait que si $s : Z \longrightarrow Z$ est l'anti-involution structurelle de Z, alors pour tout point $a \in Z$ il existe un ouvert symétrique $U \ni a$, un ouvert V symétrique de $\mathbb{C}^n$ et un isomorphisme analytique $U \xrightarrow{\varphi} V$ tel qu'on ait le diagramme commutatif suivant :

$$\begin{array}{ccc} U & \xrightarrow{s} & U \\ \varphi \downarrow & & \downarrow \varphi \\ \mathbb{C}^n \supset V & \xrightarrow{c} & V \subset \mathbb{C}^n \end{array}$$

où $c : \mathbb{C}^n \longrightarrow \mathbb{C}^n$ est le passage au conjugué en $\mathbb{C}^n$.]

c) Pour toute variété complexe Z, $\dim_{\mathbb{C}} Z=n$, l'espace $\mathbb{R}$-annelé sous-jacent à Z est une PSV de dimension n .

En effet, si Z est une variété complexe, on prend la variété complexe $Z'=Z \sqcup Z_c$ avec la symétrie canonique. Alors Z'_{sym} est isomorphe à Z en tant que espace $\mathbb{R}$-annelé.

d) Toute surface de Riemann bordée X, considérée comme espace $\mathbb{R}$-annelé d'une facon évidente est une PSV de dimension 1 . En effet on prend le double $\hat{X}$ et alors X est isomorphe à $(\hat{X})_{sym}$. Par exemple, le cas du demi-disque fermé ou d'un demi-plan fermé .

e) Toute surface de Riemann non orientable X, toujours considérée comme espace $\mathbb{R}$-annelé d'une facon évidente, est une PSV .

Le passage de la variété complexe symétrique Z à l'espace $\mathbb{R}$-annelé Z_{sym} est un foncteur et on a le :

Théorème 1 : Le foncteur canonique

$$\text{Var.compl.sym.} \longrightarrow \text{PSV}$$

défini par $Z \longmapsto Z_{sym}$ est une équivalence de catégories.

En fait il existe même un foncteur canonique PSV $\longrightarrow$ Var.compl.sym. qui est presque l'inverse du premier. On le notera par $X \longmapsto \hat{X}$ et on dira que $\hat{X}$ est le "double" de X ; on a un morphisme canonique $p : \hat{X} \longrightarrow X$.

De ce théorème il résulte par exemple que dans la catégorie PSV il existe le produit direct $X \times X'$.

Quoique du point de vue abstrait les deux catégories Var.compl.sym. et PSV soient identifiables, la considération des PSV est convenable et même indispensable pour développer une théorie des faisceaux sur les variétés complexes symétriques (on pense à une théorie qui tient compte de la structure supplémentaire donnée par la symétrie).

4.-Faisceaux cohérents sur une PSV.

En fait on a le théorème suivant

Théorème 2 : Toute PSV X est un espace de Oka et pour chaque faisceau cohérent sur X , $\mathcal{F} \in \mathrm{Coh}(X)$, le morphisme canonique

$$\mathcal{F} \longrightarrow p_* p^*(\mathcal{F})$$

(où $p : \hat{X} \longrightarrow X$ est le morphisme canonique) se plonge dans une suite exacte de faisceaux sur X :

$$(*) \qquad 0 \longrightarrow \mathcal{F} \longrightarrow p_* p^*(\mathcal{F}) \longrightarrow \mathcal{F} \longrightarrow 0 .$$

(Pour le cas des surfaces de Riemann bordées, ce théorème est dû à M.Jurchescu, cf [4]) .

Donnons maintenant quelques exemples d'applications de ce théorème .

D'abord une définition :

Définition : Une *pseudovariété de Stein* est une pseudovariété complexe X telle que :

S_o) X est séparée et l'ensemble de ses composantes connexes est au plus dénombrable ;

S_1) Pour tout compact $K \subset X$, l'ensemble

$$\hat{K} = \left\{ x \in X \;\middle|\; |f(x)| \leq \sup_K |f| \,, \forall f \in \Gamma(X, \mathcal{O}_X) \right\}$$

est compact ;

S_2) Pour tout point $x_o \in X$ il existe un morphisme $\varphi : X \longrightarrow \mathbb{C}^N_{sym}$ tel que x_o soit un point isolé de la fibre $\varphi^{-1}\varphi(x_o)$.

Les conditions précédentes sont analogues aux conditions utilisées par Grauert [3] pour sa définition des espaces holomorphiquement complets (espaces de Stein) .

Alors les conséquences du théorème sont :

Corollaire 1 : Pour une PSV X les propositions suivantes sont équivalentes :

(i) X est une PSV de Stein ;

(ii) $\hat{X}$ est de Stein en tant que variété complexe ;

(iii) Les théorèmes A et B sont valables pour les $\mathcal{O}_X$-modules cohérents .

Corollaire 2 : Si X est une PSV de Stein, de dimension finie, alors il existe un plongement fermé dans un $\mathbb{C}^N_{sym}$ pour N suffisamment grand .

Corollaire 3 : Si X est une PSV compacte et $\mathcal{F} \in \mathrm{Coh}(X)$ alors $\dim_{\mathbb{R}} H^q(X, \mathcal{F}) < \infty$.

5.-Variétés complexes bordées.

Nous pouvons maintenant appliquer la théorie des PSV pour étudier les variétés complexes bordées avec bord symétrique.

Rappelons qu'une variété mixte de type (m,n) est un espace $\mathbb{C}$-annelé modélé sur les sous-espaces annelés ouverts de $\mathbb{R}^m \times \mathbb{C}^n$, ce dernier étant considéré comme espace annelé avec pour faisceau structural le faisceau des fonctions complexes "morphes" , une fonction $f : U \longrightarrow \mathbb{C}$ étant dite "morphe" si elle est C^∞ dans toutes les variables et holomorphe dans les variables complexes [5] . Remarquons que les variétés mixtes sont des espaces $\mathbb{C}$-annelés réduits; en particulier tout morphisme d'une variété mixte X dans une autre Y est univoquement déterminé par sa composante topologique qu'on appelle alors application morphe de X à Y .

Si au lieu de la condition "C^∞ " on pose la condition " $\mathbb{R}$-analytique" on obtient la notion de variété mixte analytique .

Donnons maintenant la définition suivante :

Définition : Une variété complexe bordée, de dimension n, est un espace $\mathbb{C}$-annelé modélé sur les sous-espaces annelés ouverts de

$$H^n = \{ z \in \mathbb{C}^n \mid \operatorname{Im} z_n \geq 0 \} \quad ,$$

considéré comme espace $\mathbb{C}$-annelé avec $\mathcal{O}_{H^n} := \mathcal{O}_{\mathbb{C}^n} \mid_{H^n}$.

Pour $Z = H^n$, le bord $B(Z)$ est défini par

$$B(Z) = \{ z \in \mathbb{C}^n \mid \operatorname{Im} z_n = 0 \} \subset H^n \, .$$

Pour Z une variété complexe bordée quelconque, on définit $B(Z)$ en utilisant les cartes locales. $B(Z)$ est un ensemble fermé et l'on considérera toujours comme un espace annelé avec $\mathcal{O}_{B(Z)} := \mathcal{O}_Z \mid_{B(Z)}$ comme faisceau structural.

On voit alors sans difficulté que le bord $B(Z)$ d'une variété complexe bordée Z, de dimension n , est une variété mixte analytique de type $(1,n-1)$.

Etant donnée une variété mixte X , une application de conjugaison sur X est une application $c : X \longrightarrow X$ avec la propriété que , pour tout point $a \in X$, il existe un ouvert U de X et une carte structurelle $\phi : U \longrightarrow \mathbb{R}^m \times \mathbb{C}^n$ tels que $a \in U = c(U)$ et que $\phi(c(x)) = \overline{\phi(x)}$ pour tout point $x \in U$, où pour $\phi(x) = (s,z) \in \mathbb{R}^m \times \mathbb{C}^n$ on a posé $\overline{\phi(x)} = (s,\bar{z})$.

Toute application de conjugaison c est une application antimorphe , i.e. une application morphe de X sur X_c la variété mixte conjuguée de X ; en outre on a $c^2 = id$, donc c est une anti-involution de X . Pour les variétés complexes on démontre facilement que tout anti-involution est une application de conjugaison [1] .

Soit Z une variété complexe bordée de dimension n , dont le bord $B(Z)$ soit muni d'une application de conjugaison fixée c : $B(Z) \longrightarrow B(Z)$ (on dira alors que Z est à *bord symétrique*).

On a alors le théorème suivant :

Théorème 3 : Il existe une variété complexe symétrique Z *(avec anti-involution structurelle* s) *et un plongement topologique fermé* $Z \subset \hat{Z}$ *tels que* :

a) $\mathcal{O}_Z = \mathcal{O}_{\hat{Z}}|_Z$,

b) $Z \cup s(Z) = \hat{Z}$, $Z \cap s(Z) = B(Z)$,

c) *l'application de conjugaison* c *de* B(Z) *est induite par* s .

En outre, le couple $(\hat{Z}, s)$ *est uniquement détérminé par les conditions* a),b),et c) *et s'appelle le double de* Z .

On remarque aussi que le double de Z peut être charactérisé par une propriété universelle que voici : Pour toute variété complexe symétrique Z' et toute application holomorphe $\varphi : Z \longrightarrow Z'$ dont la restriction $\varphi|_{B(Z)}$ est symétrique, il existe une et une seule application holomorphe $\psi : \hat{Z} \longrightarrow Z'$ telle que le diagramme

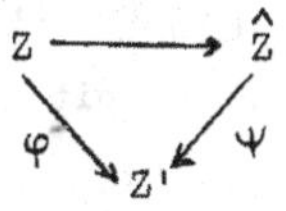

soit commutatif .

Pour toute variété complexe bordée Z , de dimension n , à bord symétrique, on a donc trois opérations : le *bord* B:= B(Z) [B est une variété mixte analytique de type (1,n-1)] , le *double* $X := \hat{Z}$ [X est une variété complexe symétrique] et le *quotient* $S := (\hat{Z})_{sym}$ [S est une pseudovariété complexe de dimension n] .

Ces opérations permettent d'utiliser la théorie des faisceaux cohérents sur S pour obtenir des renseignements relatifs à Z , et nous en conclurons cet exposé par un exemple d'application de ce procédé .

Soit $p : X \longrightarrow S$ le morphisme canonique (d'espaces $\mathbb{R}$-annelés) et soit $C:=p_o(B)$; puisque le bord de Z est symétrique on a $B=p_o^{-1}(C)$. Ona un faisceau de $\mathbb{R}$-algèbres $\mathcal{O}_C$ sur C défini en posant, pour tout ouvert S' de S, $\mathcal{O}_C(S'\cap C):=\Gamma_{sym}(p_o^{-1}(S')\cap B,\mathcal{O}_B)$; muni de ce faisceau C devient un espace $\mathbb{R}$-annelé .

De même , on a un morphisme d'espaces $\mathbb{R}$-annelés $i :$ $C \longrightarrow S$ avec la composante topologique donnée par l'inclusion et avec la composante algébrique définie par les restrictions

$$i_{1,S'}:\Gamma_{sym}(p_o^{-1}(S'),\mathcal{O}_X) \longrightarrow \Gamma_{sym}(p_o^{-1}(S')\cap B,\mathcal{O}_B) .$$

Enfin, on a un morphisme de $\mathcal{O}_S$-modules $\theta:p_*(\mathcal{O}_X)\longrightarrow i_*(\mathcal{O}_C)$ défini en posant , pour tout ouvert S' de S et toute fonction $f\in\Gamma(p_o^{-1}(S'),\mathcal{O}_X)$,

$$\mathcal{O}_{S'}(f):=Im_s(f)\mid_{p_o^{-1}(S')\cap B}=Im_c(f\mid_{p_o^{-1}(S')\cap B}) .$$

Puisque s induit c sur le bord B de Z , on voit que la suite de $\mathcal{O}_S$-modules

$$o \longrightarrow \mathcal{O}_S \longrightarrow p_*(\mathcal{O}_Z) \xrightarrow{\theta} i_*(\mathcal{O}_C) \longrightarrow o$$

est exacte .

Si $H^1(S,\mathcal{O}_S)=0$, il en résulte que l'application

$$\theta_S:\Gamma(p_o^{-1}(S),\mathcal{O}_Z) \longrightarrow \Gamma_{sym}(B,\mathcal{O}_B)$$

est surjective .

D'autre part ,

$$H^1(X,\mathcal{O}_X) = H^1(S,\mathcal{O}_S) \oplus H^1(S,\mathcal{O}_S)$$

d'après le théorème 2 .

Ainsi on a le

Théorème 4 : Soit Z une variété complexe bordée , à bord B symétrique (avec application de conjugaison c) , et telle que $H^1(\hat{Z},\mathcal{O}_{\hat{Z}})=0$. Alors, pour toute fonction $h\in\Gamma_{sym}(B,\mathcal{O}_B)$, il existe une fonction $f\in\Gamma(Z,\mathcal{O}_Z)$ telle que $\mathrm{Im}_c(f\,|_B)=h$.

La condition $H^1(\hat{Z},\mathcal{O}_{\hat{Z}})=0$ est vérifié par exemple lorsque $\hat{Z}$ est une variété de Stein, en particulier pour Z une surface de Riemann bordée non-compacte (dans ce cas le théorème 4 est dû à M.Jurchescu, cf. [4]) . Remarquons aussi que pour Z le disque unité dans **C** , le double $\hat{Z}$ est la droite projective qui satisfait la condition $H^1(\hat{Z},\mathcal{O}_{\hat{Z}})=0$. Ainsi le théorème de l'existence d'une solution du problème de Dirichlet pour le disque unité avec une fonction bord ℝ-analytique est contenu dans le théorème 4 .

Bibliographie

(1) Andreotti,A. and P.Holm: Quasianalytic and parametric spaces. Nordic Summer School/NAVF - Symposium in Math., Oslo, August 5-25, 1976; p.13-97.

(2) Cartan,H. et J.-P.Serre: Un théorème de finitude concernant les variétés analytiques compactes. C.R.Acad.Sci.Paris, 237(1953), p.128-130.

(3) Grauert,H.: Charakterisierung der holomorph vollständigen komplexen Räume. Math.Ann.,129(1955),p.233-259.

(4) Jurchescu,M.: Coherent sheaves on borded Riemann surfaces. Trans.Am.Math.Soc.,144(1969), p.557-563.

(5) - : Variétés mixtes. L.N.in Math.("Springer-Verlag"), vol.743: Roumanian-Finnish Seminar on Complex Analysis. Proceedings, 1976.

COMMUTATIVE BANACH ALGEBRAS AND ANALYTIC FUNCTIONS OF COUNTABLE-MANY VARIABLES

Toma V. Tonev (Sofia, Bulgaria)

1. Let Γ be a subgroup of additive group of real rational numbers and $G = \hat{\Gamma}$ be the group of all (not only continuous) characters of Γ (i.e. of all homomorphisms of Γ to the unit circle). The big complex plane $\mathbb{C}_G$ we call the infinite cone over G : $[0,\infty)\times G / \{0\} \times G$. Every element $p \in \Gamma_+ = \Gamma \cap [0,\infty)$ generates a continuous function $\tilde{\chi}_p$ on $\mathbb{C}_G$, namely: $\tilde{\chi}_p(\lambda, g) = \lambda^p g(p)$, if $\lambda \neq 0$, $p \neq 0$, $\tilde{\chi}_p(\{0\}\times G/\{0\}\times G) = 0$ for $p \neq 0$ and $\tilde{\chi}_0 \equiv 1$.

Let B be a uniform comutative Banach algebra with unit and $Sp\,B$ be its maximal ideal space. Let $\{f_{p(j)}\}_{j=1}^{\infty}$, $\|f_{p(j)}\| \leq 1$, is a multiplicative subsemigroup of B, isomorphic to the additive semigroup R_+ of nonnegative real rational numbers. We define the mapping $\tau : Sp\,B \to \mathbb{C}_G$ in the following way: $\tau(\varphi) = (\lambda, g)$, where $\lambda = |\varphi(f_1)| \neq 0$, $g(p) = \varphi(f_p)/|\varphi(f_p)| = \varphi(f_p).\lambda^{-p}$, $g(-p) = \overline{\tilde{\chi}_p(g)}$, for $p \in R_+$. If $\lambda = |\varphi(f_1)| = 0$, we define $\tau(\varphi)$ as the "origin" $* = \{0\}\times G/\{0\}\times G$ of $\mathbb{C}_G$. It easy to check that τ is continuous.

Definition. The joint spectrum $\sigma = \sigma(\{f_{p(j)}\})$ of a multiplicative semigroup $\{f_{p(j)}\}_{j=1}^{\infty} \subset B$ we call the compact set $\tau(Sp\,B) \subset \bar{\Delta}_G = \{(\lambda, g) \in \mathbb{C}_G \mid \lambda \leq 1\}$.

An easy verification shows that a point (λ, g) belongs to σ iff all the functions $\{\tilde{\chi}_{p(j)}(\lambda, g) - f_{p(j)}\}$ belong together to a maximal ideal of B. Moreover, if $(\lambda, g) \in \sigma$ and $\varphi \in \tau^{-1}(\lambda, g)$ then $\varphi(f_p) = \tilde{\chi}_p(\lambda, g)$ for any $p \in R_+$.

In the opposite case $((\lambda,g)\notin\sigma)$ there will exist such $g_{j_n}\in B$, that $\sum_{n=1}^{K_m}(\tilde{\chi}_{p(j_n)}(\lambda,g)-f_{p(j_n)})g_{j_n}$ tend to 1 if $K_m\to\infty$. Hence for any ε, $0<\varepsilon<1$, there will exist such a finite set $(j_1,\dots,j_s)$ of natural numbers, that $\|\sum_{n=1}^{s}(\tilde{\chi}_{p(j_n)}(\lambda,g)-f_{p(j_n)})g_{j_n}-1\|<\varepsilon$. But then the element $\sum_{n=1}^{s}(\tilde{\chi}_{p(j_n)}(\lambda,g)-f_{p(j_n)})g_{j_n}$ will have an inverse element in B, say d. If denote the elements $g_{j_n}\cdot d$ by h_{j_n}, we have that there will exist elements h_{j_n} in B, with $\sum_{n=1}^{s}(\tilde{\chi}_{p(j_n)}(\lambda,g)-f_{p(j_n)})h_{j_n}=1$, i.e. $(\tilde{\chi}_{p(j_1)}(\lambda,g),\dots,\tilde{\chi}_{p(j_s)}(\lambda,g))\notin\sigma(f_{p(j_1)},\dots,f_{p(j_s)})$.
Both remarks above imply the following

<u>Proposition 1</u>. The joint spectrum σ of a multiplicative subsemigroup $\{f_{p(j)}\}_{j=1}^{\infty}$ of a uniform commutative Banach algebra with unit consists of these points $(\lambda,g)\in\mathbb{C}_G$, such that $(\tilde{\chi}_{p(j_1)}(\lambda,g),\dots,\tilde{\chi}_{p(j_n)}(\lambda,g))\in\sigma(f_{p(j_1)},\dots,f_{p(j)})$ for any finite n-tuple of indeces $j_1,\dots,j_n$.

Here $\sigma(g_1,\dots,g_n)$ is the usual joint spectrum of n-tuples $(g_1,\dots,g_n)\subset B^n$, i.e. the image of the mapping $(\hat{g}_1,\dots,\hat{g}_n): sp\,B\to\mathbb{C}^n$.

Let the commutative Banach algebra B is linearly generated by its elements $f_1,\dots,f_n$, what means that the linear combinations of these elements with complex coeffitients are dence in B. It is well known (cf. [1]) that the joint spectrum $\sigma(f_1,\dots,f_n)$ of generators $f_1,\dots,f_n$ of B is a polynomially convex set in $\mathbb{C}^n$. We generalize this result for algebras, linearly generated by semigroups of its elements. If K is a bounded set in $\mathbb{C}_G$, the <u>generalized-polynomial hull</u> $\hat{K}$ of K we call as usual the set of these $(\lambda,g)\in\mathbb{C}_G$, for which $|P(\lambda,g)|\le\sup_K|P(\lambda,g)|$ for any generalized-polynomial P, i.e. for any linear combination of functions $\tilde{\chi}_p$. A bounded set $K\subset\mathbb{C}_G$ is called generalized-polynomially convex, iff $\hat{K} = K$.

Theorem 1. Let B be a commutative Banach algebra, linearly generated by a multiplicative subsemigroup $\{f_{p(j)}\}_{j=1}^{\infty}$ of B , isomorphic to R_+ , $\|f_{p(j)}\| \leq 1$. Then the joint spectrum $\sigma = \sigma(\{f_{p(j)}\})$ of semigroup $\{f_{p(j)}\}$ is generalized-polynomially convex and the mapping $\tau : \operatorname{sp} B \to \sigma$ is a homeomorphism.

Proof: The mapping τ is one-to-one, because any point $\varphi \in \operatorname{sp} B$ is defined uniquely by its values $\{\varphi(f_{p(j)})\}$. Hence τ is a homeomorphism. Let (λ_0, g_0) belongs to the generalized-polynomial hull $\hat{\sigma}$ of the joint spectrum σ . Then $|\sum c_j \tilde{\chi}_{p(j)}(\lambda_0, g_0)| \leq$

$$\leq \sup\{|\sum c_j \tilde{\chi}_{p(j)}(\lambda, g)| \mid (\lambda, g) \in \sigma\} = \sup\{|\sum c_j \varphi(f_{p(j)})| \mid \varphi \in \operatorname{sp} B\}$$

$$= \sup\{|\varphi(\sum c_j f_{p(j)})| \mid \varphi \in \operatorname{sp} B\} \leq \|\sum c_j f_{p(j)}\|.$$

Consequently the mapping $\sum c_j f_{p(j)} \mapsto \sum c_j \tilde{\chi}_{p(j)}(\lambda_0, g_0)$ induces a well defined linear multiplicative functional ϕ on B . Let us compute $\tau(\phi)$: $|\phi(f_1)| = |\tilde{\chi}_1(\lambda_0, g_0)| = \lambda_0$; for $p(j) \in R_+$ we have $\phi(f_{p(j)}) = \tilde{\chi}_{p(j)}(\lambda_0, g_0) = \lambda_0^{p(j)} g_0(p(j))$ and if $\lambda \neq 0$ $\phi(f_{p(j)})/|\phi(f_{p(j)})| = g_0(p(j))$, i.e. $\tau(\phi) = (\lambda_0, g_0)$. Hence $(\lambda_0, g_0) \in \sigma(\{f_{p(j)}\})$ and $\hat{\sigma} = \sigma$, Q.E.D.

By $P_G(K)$ we denote the closure of all generalized-polynomials in sup-norm on K .

Corollary 1. A uniform algebra B has the elements of a semigroup $\{f_{p(j)}\}_{j=1}^{\infty} \subset B$, isomorphic to R_+ for its linear generators, iff B is isometrically isomorphic to the algebra $P_G(K)$, for some generalized-polynomially convex compact K in $\mathbb{C}_G$.

The proof is based on Theorem 1 and on the following result: if K is a compact subset of $\mathbb{C}_G$, then the spectrum of algebra $P_G(K)$ coincides with the generalized-polynomial hull $\hat{K}$ of K [2].

It is easy to see, that both Theorem 1 and Corollary 1 hold for the case of arbitrary subgroups Γ of rational numbers. If Γ is such a subgroup of R, that $q>p$ and $p\in\Gamma_+\subset R_+$ imply $q\in\Gamma_+$, we can say more about the compact $\hat{K}\subset\mathbb{C}_G$. Namely:

<u>Theorem 2.</u> (Grigorjan, Tonev [3]) Let the uniform algebra B is antisymmetric and linearly generated by a semigroup $\{f_{p(j)}\}$ of its elements, isomorphic to Γ_+. If for some $p\in\Gamma$, $|f_p|_{\partial B}\equiv$ const, then $sp\,B$ is homeomorphic to the big disc $\bar{\Delta}_G=\{(\lambda,g)\in\mathbb{C}_G \,|\, \lambda\leq 1\}$. The Shilov boundary ∂B can be equipped with the structure of a compact abelian group, homeomorphic and isomorphic to the group $G=\hat{R}$.

Actually this theorem holds for more wide classes of groups and the antisymmetric condition above can be replaced by the condition $\partial B\neq sp\,B$.

2. The joint spectrum of n-tuples are connected with the holomorphic functional calculus of these n-tuples. According to it, if F is a function of n variables, analytic in a neighbourhood of the joint spectrum $\sigma(f_1,\dots,f_n)$ of elements $f_1,\dots,f_n\in B$, then there exists an element $g\in B$, unique in a way, for which $\hat{g}=F\circ(\hat{f}_1,\dots,\hat{f}_n)$. Let $\{f_{p(j)}\}_{j=1}^{\infty}$ be a multiplicative semigroup in B and F be a function, generalized-analytic in the proximity of $\sigma=\sigma(\{f_{p(j)}\})$ (i.e. F can be approximated by generalized polynomials on compact subsets of some open set containing σ). There arise the question if there exist elements $g\in B$, for which $\hat{g}(\varphi)=F\circ\tau(\varphi)$, $\varphi\in sp\,B$, i.e. if there exists a generalized-holomorphic calculus of multiplicative subsemigroups of B? There are known only partial answers to that question. Applying the holomorphic functional calculus of n Banach algebra elements, one can see that besides the generalized polynomials, on a semigroup $\{f_{p(j)}\}\subset B$ act also the functions $F(\lambda,g)=f_0(\tilde{\chi}_{p(j_1)}(\lambda,g),\dots,\tilde{\chi}_{p(j_n)}(\lambda,g))$

where f is analytic in the proximity of $\sigma(f_{p(j_1)}, \dots, f_{p(j_k)})$. Because $\tilde{X}_p(\tau(\varphi)) = \hat{f}_p(\varphi)$, this class of functions coincides with the class of functions of type $G(\lambda, g) = g(\tilde{X}_{\frac{1}{s}}(\lambda, g))$, where s is the smallest common multiple of denomoiators of indeces $p(j_1)$, $\dots, p(j_k)$ and g is an analytic function in the proximity of $\tilde{X}_{\frac{1}{s}}(\sigma)$.

Let $\{g_j\}_{j=1}^{\infty} \subset B$, $\|g_j\| \leq 1$ and the joint spectrum σ' of sequence $\{g_j\}$ be the image of $sp\, B$ through the continuous mapping $\gamma : sp\, B \to \Delta^{\infty} : \varphi \mapsto (\hat{g}_1(\varphi), \hat{g}_2(\varphi), \dots)$. Let F be an analytic function of countable-dimensional argument on a neighbourhood of the compact σ' , i.e. F is locally uniform limit of polynomials on $\mathbb{C}^{\infty}$. If η be the continuous injective mapping $\eta : \Delta_G \to \Delta^{\infty}$, defined in the following way: $\eta(\lambda, g) = (\tilde{X}_{p(1)}(\lambda, g), \tilde{X}_{p(2)}(\lambda, g), \dots)$, then $\sigma'(\{f_{p(j)}\}) = \eta(\sigma(\{f_{p(j)}\}))$.

Theorem 3. (D. Stankov) Let $\{f_{p(j)}\}$ be a multiplicative semi-group of B and F be an analytic function of countable-dimensional argument on an open set, containing $\sigma'(\{f_{p(j)}\})$. If Φ is the generalized-analytic function $\Phi(\lambda, g) = F(\eta(\lambda, g))$ in the proximity of $\sigma(\{f_{p(j)}\})$, then there exists an element $g \in B$, for which $\hat{g}(\varphi) = \Phi(\tau(\varphi))$ for all $\varphi \in sp\, B$.

Sketch of the proof: Let F is defined on V : $\Delta^{\infty} \supset V \supset \sigma'$, and $U = \eta^{-1}(V) \supset \sigma$. According to a theorem of Rickart [5], there exist a neighbourhood $V_0 \subset V$ of σ' and a projection $\pi : \mathbb{C}^{\infty} \to \mathbb{C}(z_{t_1}, z_{t_2}, \dots, z_{t_m}) \cong \mathbb{C}^m$, $t_k \in Z_+$, such that F can be extended to the open set $V_0^{\pi} = \pi^{-1}(\pi(V_0))$ of $\mathbb{C}^{\infty}$ as: F_{π}: $F_{\pi}(\pi^{-1}(\pi(Z))) = F(Z)$ for any $Z \in V_0$. Because π is an open continuous mapping, it is easy to see that $\pi(\sigma')$ is compact. Actually $\pi(\sigma')$ coincides with the joint spectrum of elements $f_{p(t_1)}$, $\dots, f_{p(t_m)}$, and $F_{\pi} \circ \pi^{-1}$ can be approximated by polynomials of

$z_{t_1}, \ldots, z_{t_m}$ on the open set $\pi(V_0) \supset \pi(\sigma')$. Then, according to the holomorphic functional calculus of several Banach algebra elements, there exists such a $g \in B$, that $\hat{g}(\varphi) = F_\pi \circ \pi^{-1}(\hat{f}_{p(t_1)}(\varphi), \ldots, \hat{f}_{p(t_m)}(\varphi))$ for any $\varphi \in sp B$. Now on $sp B$ we have: $\phi(\tau(\varphi)) = F \circ \eta \circ \tau(\varphi) =$
$= F(\tilde{\chi}_{p(1)}(\tau(\varphi)), \tilde{\chi}_{p(2)}(\tau(\varphi)), \ldots) = F_\pi \circ \pi^{-1}(\tilde{\chi}_{p(t_1)}(\tau(\varphi)), \ldots, \tilde{\chi}_{p(t_m)}(\tau(\varphi)))$
$= F_\pi \circ \pi^{-1}(\hat{f}_{p(t_1)}(\varphi), \ldots, \hat{f}_{p(t_m)}(\varphi)) = \hat{g}(\varphi)$, Q.E.D.

Unfortunately not any generalized-analytic function F near σ is of the type, considered in Theorem 3. It seems that we need a new and more general holomorphic functional calculus for to be able to answer completely to the stated above question.

Recently D. Stankov found some sufficient conditions for the joint spectrum $\sigma(\{f_{p(j)}\})$ under which all the functions, generalized-analytic near σ will operate on the semigroup $\{f_{p(j)}\}$. For $g \in G$ by j_g we denote the injective mapping from $\mathbb{C}$ to $\mathbb{C}_G$ trough $g \in G$ with dence image (cf. [1]).

<u>Theorem 4.</u> (D. Stankov [4]) Let B be a commutative Banach algebra and $\{f_{p(j)}\}_1^\infty$, $\|f_{p(j)}\| \leq 1$ is a semigroup of nonnegative rational numbers R_+. Let the joint spectrum $\sigma = \sigma(\{f_{p(j)}\})$ has finite many connected components, any of which is contained in **some** $j_{g_\nu}(\mathbb{C})$, $g_\nu \in G$. If $\{K_1, \ldots, K_m\}$ is the decomposition of σ into connected components, let there exists an index $p(j_0)$, such that $\tilde{\chi}_{p(j_0)}(K_\nu) \cap \tilde{\chi}_{p(j_0)}(K_\mu) = \phi$ for $\nu \neq \mu$. Then if F is a generalized-analytic function in the proximity of σ, there exists a unique $f \in B$, for which $\hat{f}(\varphi) = F(\tau(\varphi))$ on $sp B$.

Note, that under the conditions of this **theorem**, the point $\{*\}$ necessarily belongs to the unbounded connected components of $\mathbb{C}_G \setminus K_\nu$ for every $\nu = 1, 2, \ldots, m$. The finiteness assumption for the decomposition of σ into connected components can be dropped (cf. [4]).

R E F E R E N C E S

1. T. W. Gamelin, Uniform algebras, Prentice-Hall, N.J., 1969.

2. T. Tonev, Algebras of generalized-analytic functions, Banach Center Publ., v. 8, Warsaw (to appear).

3. S. Grigorjan, T. Tonev, A characterization of the algebra of generalized-analytic functions, Comt. Rend. de l'Acad.bulg. des Sci., v. 33, Nr 1, 1980, 25-25 (Russian).

4. D. Stankov, Holomorphic functional calculus of some semigroups of elements of a commutative Banach algebra, Proc. of Shumen Pedagogical Inst.(to appear).

5. C. Rickart, Analytic functions of an infinite number of complex variables, Duke Math. J. 36 (1969), 581-597.

Institute of Mathematics
Bulgarian Academy of Sciences
1090 Sofia, P.O.Box 373, Bulgaria

POTENTIALS ON LOCALLY COMPACT NON-ABELIAN GROUPS

Martha Bănulescu

The present paper studies the locally compact non-abelian groups from the point of view of potential theory. The first three chapters develop some aspects of potential theory on a locally compact topological group on which there exists a potential kernel k, such that k is absolutely continuous with respect to the left invariant Haar measure; the last chapter gives a large class of such groups, namely all Lie groups.

For the first three chapters we give the following notations and preliminaries: G will be a fixed locally compact group, where the composition is written multiplicatively and the neutral element is denoted by e; λ_s will be a fixed left invariant Haar measure on G, Δ will be the modular function of G, λ_d will be the right invariant Haar measure on G, such that $d\lambda_d(x)=\frac{1}{\Delta(x)}\, d\lambda_s(x)$ for every $x\in G$. $L^1(\lambda_s)$ (resp. $L^1_{loc}(\lambda_s)$) will be the integrable (resp. locally integrable) functions on G. For a subset A of G, I_A will be the characteristic function of A. For a numerical function f on G and for a G, f_a, ${}_af$, $\check{f}$ and f^* will be the functions on G defined by: $f_a(x)=:f(xa)$, $({}_af)_{(x)}=:f(ax)$, $\check{f}(x)=:f(x^{-1})$, $f^*(x)=:\Delta(x^{-1})\, f(x^{-1})$. $\mathcal{B}$ will be the set of all Borel subsets of G, $\mathcal{F}$ will be the numerical Borel functions on G, C(G) will be the real continuous functions on G, K(G) will be the functions from C(G) which have compact support, M(G) will be the Radon measures on G, $M_b(G)$ will be the bounded Radon measures on G; ε_a is the Dirac measure for $a\in G$. For a measure μ on G, $\check{\mu}$ will be the measure defined by $\check{\mu}(f)=:\mu(\check{f})$ for $f\in K^+(G)$. One can easily verify that if $\mu=f\cdot\lambda_s$, then $\check{\mu}=f^*\cdot\lambda_s$. For $\mu,\nu\in M^+(G)$, we define the measure $\mu*\nu$ (when it exists) by: $(\mu*\nu)_{(f)}=:\int f(xy)\, d\mu(x)\, d\nu(y)$ for $f\in K(G)$. For $\mu\in M^+(G)$ and $g:G\to\bar{R}^+$, we define the function $\mu*g$ (when it exists) by:

(1) $(\mu*g)_{(x)}=:\int g(y^{-1}x)\, d\mu(y)$ for $x\in G$.

One can easily verify that if $\mu,\nu\in M^+(G)$, $\nu=f\cdot\lambda_s$, where $f\in L^1_{loc}(\lambda_s)$, and if $\mu*\nu$ exists, then we have $\mu*\nu=(\mu*f)\cdot\lambda_s$. For two positive numerical functions f and g defined on G,we denote by $f*g$ (when it exists)the function $(f*g)_{(x)}=:\int f(y)\ g(y^{-1}x)\ d\lambda_s(y)=\int\frac{1}{\Delta(z)}f(xz^{-1})\ g(z)\ d\lambda_s(z)$,for $x\in G$.If we take in (1) $\mu=f\cdot\lambda_s$,then we have $f*g=\mu*g$ on G.One can easily verify that if $\mu\in M^+(G)$ and $f\in K^+(G)$,then $\mu*f\in C^+(G)$,hence we can consider the map $N_\mu:K^+(G)\dashrightarrow C^+(G)$,defined by $N_\mu(f)=:\mu*f$ for $f\in K^+(G)$.

Remark:If $f\in L^1(\lambda_s)$,the proposition 26,cap.IV, §24 of [3] tells us that $f*g\in C(G)$ for every bounded $g\in\mathcal{F}^+$;it follows that $f*h$ is lower semicontinuous on G,for every $h\in\mathcal{F}^+$.

I ABSOLUTELY CONTINUOUS POTENTIAL KERNELS ON G

The main result of this chapter is theorem 1.3.8.,which gives some necessary conditions for a potential kernel k on G,to be an absolutely continuous measure with respect to λ_s and which gives some properties of its density.

1.1.Continuous kernels on G

1.1.1.Definition.A linear and positive map $N:K(G)\longrightarrow C(G)$ will be called, as in [1],a continuous kernel on G.For every $x\in G$ we consider the linear and positive map $N_x:K(G)\longrightarrow R$, $N_x(f)=:N(f)_{(x)}$ for $f\in K(G)$;hence N_x is a Radon measure on G and thus we can define for every $f\in\mathcal{F}^+$ and for every $x\in G$ $N(f)_{(x)}=:N_x(f)$.

1.1.2.Example.For every $\mu\in M^+(G)$,the map N_μ,defined above,is a continuous kernel on G,which satisfies:

(2) $[N_\mu(f_a)]_{(e)}=[{}_a(N\ (f))]_{(e)}$ for every $a\in G$ and $f\in\mathcal{F}^+$

(2') $N_\mu(f_a)=(N_\mu(f))_a$ for every $a\in G$ and $f\in\mathcal{F}^+$.

Conversely,we have:

1.1.3.Proposition.If N is a continuous kernel which satisfies (2) or (2'),then there exists a unique measure $\mu\in M^+(G)$,such that:

(3) $N(f)=\mu*f$ for every $f\in K(G)$.

Proof.For the uniqueness,let $\mu\in M^+(G)$ which satisfies (3).Then we have:

(4) $\mu(f)=N(\check{f})_{(e)}$ for every $f\in K(G)$.

Indeed, $N(\check{f})_{(e)}=(\mu*\check{f})_{(e)}=\int\check{f}(y^{-1})\ d\mu(y)=\int f(y)\ d\mu(y)=\mu(f)$.For the existence,we remark that the function $f\longmapsto N(\check{f})_{(e)}$ is a linear and positive map on K(G),hence it is a measure $\mu\in M^+(G)$,which satisfies (4).Using

(4) and one of the relations (2) or (2'), we obtain (3).

1.2. Potential Kernel on G

1.2.1. Definition. A family $(\mu_t)_{t>0} \subseteq M_b(G)$ with the properties:

(i) $\mu_t(G) \leq 1$ for every $t>0$;

(ii) $\mu_t * \mu_s = \mu_{t+s}$ for every $t,s>0$;

(iii) $\lim_{t\to 0} \mu_t(f)=f(e)$ for every $f\in K(G)$

is called a (vaguely continuous) convolution semigroup on G.

1.2.2. Let $(\mu_t)_{t>0}$ be a convolution semigroup on G; for every $\lambda>0$ we define the measure $k_\lambda \in M^+(G)$ by $k_\lambda(f)=:\int_0^\infty e^{-\lambda t}\mu_t(f)\,dt$, for $f\in K(G)$. We have:

(5) $k_\lambda(G) \leq \frac{1}{\lambda}$, hence $k_\lambda \in M_b^+(G)$;

(6) $k_\lambda - k_\mu = (\mu-\lambda)(k_\lambda * k_\mu)$ for every $\lambda,\mu>0$.

The family $(k_\lambda)_{\lambda>0}$ is called a resolvent of measures associated with the convolution semigroup $(\mu_t)_{t>0}$. For a fixed $f\in K^+(G)$, the map on $(0,\infty)$ with values in R^+ $\lambda \longmapsto k_\lambda(f)$ is decreasing, hence we can denote $k(f)=:\sup_{\lambda>0} k_\lambda(f)=\lim_{\lambda\to 0} k_\lambda(f)=\int_0^\infty \mu_t(f)\,dt \leq +\infty$. If $k(f)<+\infty$, for every $f\in K^+(G)$, the convolution semigroup $(\mu_t)_{t>0}$ is called transient and the measure k is called the potential kernel associated with $(\mu_t)_{t>0}$. We have

(6') $k - k_\lambda = \lambda(k * k_\lambda)$ for every $\lambda>0$;

(7) $k * k_\lambda = k_\lambda * k$ for every $\lambda>0$.

1.2.3. Definition. A measure $k\in M^+(G)$ is called a potential kernel if there exists a transient convolution semigroup $(\mu_t)_{t>0}$, such that k is its potential kernel. We write $k=\int_0^\infty \mu_t\,dt$.

1.2.4. Let $k=\int_0^\infty \mu_t dt$ be a potential kernel on G and let $(k_\lambda)_{\lambda>0}$ be the resolvent of measures associated with $(\mu_t)_{t>0}$. Then we can define, by the previous section, the continuous kernels $N_0=:N_k$ and $N_\lambda=:N_{k_\lambda}$, for every $\lambda>0$. We remark that we have:

(8) $N_\lambda - N_\mu = (\mu-\lambda)\, N_\lambda N_\mu$ and $N_\lambda N_\mu = N_\mu N_\lambda$ for every $\lambda,\mu \geq 0$;

(9) $\lim_{\lambda\to\infty} \lambda N_\lambda(f)_{(x)}=f(x)$ for every $f\in K^+(G)$ and $x\in G$;

(10) $N_0(f)=\lim_{\lambda\to 0} N_\lambda(f)=\sup_{\lambda>0} N_\lambda(f)$ for every $f\in \mathcal{F}^+$.

The family $\mathcal{N}=\{N_\lambda\}_{\lambda>0}$ is called the resolvent of operators associated with the potential kernel $k\in M^+(G)$.

1.2.5.Proposition. A potential kernel k satisfies the complete maximum principle, i.e. if for $f,g\in K^+(G)$ and $a\in R^+$ we have:

(11) $(k * f)_{(x)} \leq (k * g)_{(x)} + a$,for every $x\in \mathrm{supp}(f)$,

then (11) holds for every $x\in G$. (For proof, see [1], cor.16.27)

1.3.Basic Kernels with respect to a Measure.

1.3.1.Definition. A continuous kernel $N:K(G)\longrightarrow C(G)$ is called basic if there exists a σ-finite measure μ on $(G,\mathcal{B})$, such that $N(I_A)\equiv 0$ for every $A\in\mathcal{B}$ with $\mu(A)=0$. This is equivalent with the fact that all the measures $\{N_x\}_{x\in G}$ are absolutely continuous with respect to μ. The measure μ can be chosen to be finite.

1.3.2.Proposition. For $\mu\in M^+(G)$ the following assertions are equivalent:

a) $A\in\mathcal{B}$, $\mu(A)=0 \Rightarrow \mu(xA)=0$;

b) $\lambda_s \ll \mu$ (λ_s is absolutely continuous with respect to μ) and $\mu\ll\lambda_s$.

For the proof, one can see [6], §1, nr.9, prop.11.

1.3.3.Remark. The previous proposition is still true if we replace λ_s by λ_d and a) by:

a') $A\in\mathcal{B}$, $\mu(A)=0 \Rightarrow \mu(Ax)=0$ for every $x\in G$.

1.3.4.Remark. It is known from the measure theory that for every $A\in\mathcal{B}$:

a) $\lambda_s(A)=0$ iff $\lambda_s(A^{-1})=0$, and $\lambda_d(A)=0$ iff $\lambda_d(A^{-1})=0$;

b) $\lambda_d(A)=0$ iff $\lambda_s(A)=0$

c) if $\mu\in M^+(G)$, then we have $\mu\ll\lambda_s$ iff $\mu\ll\lambda_d$.

1.3.5.Proposition. Let us consider a potential kernel $k=\int_0^\infty \mu_t dt$, such that the continuous kernel $\check{N}_0=:N_{\check{k}}$ is basic with respect to a measure $\mu\in M^+(G)$, for which there exists $k*\mu$. If we define the measure μ_0 by $\mu_0(f)=:\mu(\check{N}_0(f))=(k*\mu)_{(f)}$ for $f\in K(G)$, then we have:

a) μ_0 satisfies a') of 1.3.3.

b) $\mu_0\ll\lambda_d$ (and hence $\mu_0\ll\lambda_s$, too, by 1.3.4.)

c) N_0 is basic with respect to λ_s.

d) $k\ll\lambda_s$.

Proof: Let us first make the following remarks, which can be easily verified: (12) $k=\int_0^\infty \mu_t dt$ is a potential kernel iff $\check{k}=\int_0^\infty \check{\mu}_t dt$ is a potential kernel

(13) $A\in\mathcal{B}$,$\check{N}_o(I_A)=0$ μ-a.e. $\Rightarrow \check{N}_o(I_A)\equiv 0$

(13') $A\in\mathcal{B}$,$N_o(I_A)=0$ μ-a.e. $\Rightarrow N_o(I_A)=0$,if $N_o=:N_k$ is basic with respect to a measure $\mu\in M^+(G)$,such that there exists $\check{k}*\mu$.

For a),let $A\subseteq G$ be a compact set,such that $\mu_o(A)=0$,let $x_o\in G$ and let us show that $\mu_o(Ax_o)=0$.Indeed,we have $0=\mu_o(A)=\mu(\check{N}_o(I_A))$,hence $\check{N}_o(I_A)=0$ μ-a.e.;by (13) we obtain $\check{N}_o(I_A)\equiv 0$;it follows that:

(14) $k(Ax^{-1})=\int I_{Ax^{-1}}(y)dk(y)=\int I_A(yx)dk(y)=\check{N}_o(I_A)_{(x)}=0$,for every $x\in G$,

$\mu_o(Ax_o)=\int(\int I_{Ax_o}(yz)dk(y)\ )d\mu(z)=\int(\int I_{Ax_oz^{-1}}(y)\ )d\mu(z)=\int k(Ax_oz^{-1})d\mu(z)=0$

b) follows from prop.1.3.2. For c),let $A\in\mathcal{B}$,such that $\lambda_s(A)=0$;by b) we have $0=\mu_o(A)=\mu(\check{N}_o(I_A))$,hence $\check{N}_o(I_A)=0$ μ-a.e. and by (13) we obtain $\check{N}_o(I_A)\equiv 0$.For d),let $A\in\mathcal{B}$,such that $\lambda_s(A)=0$;by c) we have $\check{N}_o(I_A)\equiv 0$;by the proof of a),taking in (14) $x=e$,we obtain $k(A)=0$.

<u>1.3.6.Corollary.</u>If $k\in M^+(G)$ is a potential kernel and if $\mu\in M^+(G)$ is such that there exists $\check{k}*\mu$ and $N_o=:N_k$ is basic with respect to μ, then k is absolutely continuous with respect to λ_s .

<u>Proof</u>:By prop.1.3.5.we have $\check{k}\ll\lambda_s$.Let us take $A\in\mathcal{B}$ such that $\lambda_s(A)=0$; by a) of 1.3.4.we have $\lambda_s(A^{-1})=0$,hence $\check{k}(A^{-1})=0$.i.e. $k(A)=0$.

<u>1.3.7.Corollary.</u>With the assumptions of the previous cor.,and with the notations of the previous section,let us consider the sets:

(15) $\mathcal{S}(\mathcal{N})=:\{s\in\mathcal{F}^+\mid \alpha N_\alpha(s)\leq s$ for every $\alpha>0\}$

(15') $\mathcal{S}'(\mathcal{N})=:\{s\in\mathcal{S}(\mathcal{N})\mid (\alpha N_\alpha(s))\uparrow s$ when $\alpha\to+\infty\}$

(16) $\mathcal{E}(\mathcal{N})=:\{s\in\mathcal{S}'(\mathcal{N})\mid s$ is finite $\mathcal{N}$-a.e.$\}$ (we recall that $A\in\mathcal{B}$ is $\mathcal{N}$-negligible if $N_\alpha(I_A)\equiv 0$ for every $\alpha>0$) and <u>let us assume</u> that N_o is proper,i.e. there exists $(f_n)_{n\in N}\subseteq\mathcal{F}^+$,such that $(f_n)_n\uparrow 1$ and $N_o(f_n)$ is bounded for every n N.<u>Then $\mathcal{E}(\mathcal{N})$ is a standard H-cone</u> (for this definition and for the proof,see [4],prop.2.3.).

<u>1.3.8.Theorem.</u>Let us consider a potential kernel $k=\int_0^\infty\mu_t dt$,let $\mathcal{N}=\{N_\alpha\}_{\alpha>0}$ be the resolvent of operators associated with k by 2.2.4.and let $\mu\in M^+(G)$ be a measure such that there exists $\check{k}*\mu$ and the kernel $N_o=:N_k$ is basic with respect to μ .<u>Then we have</u>:

a) k (and hence every k_α) is absolutely continuous with respect to

λ_s, hence there exist $f\in L^1_{loc}(\lambda_s)$ and $f_\alpha\in L^1_{loc}(\lambda_s)$ for every $\alpha>0$, such that $k=f\cdot\lambda_s$ and $k_\alpha=f_\alpha\cdot\lambda_s$. Moreover, $f_\alpha\in L^1(\lambda_s)$ for every $\alpha>0$;

b) using the notation (16), one can assume that $f\in\mathcal{E}(\mathcal{N})$ and $f_\alpha\in\mathcal{E}(\mathcal{N}_\alpha)$, where $\mathcal{N}_\alpha =: \{N_{\alpha+\beta}\}_{\beta>0}$, for every $\alpha>0$;

c) f and f_α $(\alpha>0)$ are lower semicontinuous functions on G; let us denote $G_f =: \{x\in G \mid f(x)>0\}$;

d) G_f is an open set and a subsemigroup of G;

e) <u>If</u> $e\in G_f$, there exists an open symmetrical neighbourhood V of e in G, such that $G_f\supseteq V$, hence $G_f\supset G_0 =: \bigcup_{n\in N} V^n$; G_0 is a subsemigroup of G, which is an open and closed set at the same time;

f) <u>If</u> G is connected and <u>if</u> $e\in G_f$, then $G=\bigcup_{n\in N} V^n$, hence the function f is strictly positive on G.

<u>**Proof**</u>: a) follows from the cor.1.3.6.; by the relation (5) we have for every $\alpha>0$ $\alpha\int f_\alpha\, d\lambda_s\le 1$, hence $f_\alpha\in L^1(\lambda_s)$. For b), we replace the functions f and f_α by the functions $f' =: \lim_{\alpha\to\infty}\alpha N_\alpha(f)=\sup_{\alpha>0}\alpha N_\alpha(f)=\lim_{n\to\infty} n\, N_n(f)$, $f'_\alpha =: \lim_{\beta\to\infty}\beta N_{\alpha+\beta}(f_\alpha)=\sup_{\beta>0}\beta N_{\alpha+\beta}(f_\alpha)=\lim_{n\to\infty} n\, N_{\alpha+n}(f_\alpha)$. c) By the remark of the preliminaries, the functions $N_\alpha(f)$ and $N_{\alpha+\beta}(f_\alpha)$ are lower semicontinuous on G, for every $\alpha,\beta>0$. Using: (17) $f=f_\alpha+\alpha(f_\alpha * f)$ on G, we conclude that f is lower semicontinuous on G. d) We consider a fixed $x\notin G_f$, i.e. $f(x)=0$; using (17) we obtain $0=(f*f_\alpha)_{(x)}=\int f_\alpha(y^{-1}x)\, f(y)\, d\lambda_s(y)$, and therefore:

(18) $f_\alpha(y^{-1}x)=0$ or $f(y)=0$ for every $y\in G$.

Remarking that $y\,(y^{-1}x)=x$, we can state: (18') $x=ab\notin G_f \Rightarrow f(a)=0$ or $f_\alpha(b)=$

(18") $x\notin G_f$ and $a\in G_f \Rightarrow f_\alpha(a^{-1}x)=0$ for every $\alpha>0$.

From this last relation and the fact that $(f_\alpha)_\alpha\uparrow f$, when $\alpha\longrightarrow 0$, we have:

(19) $x\notin G_f$ and $a\in G_f \Rightarrow f(a^{-1}x)=0$,

hence G_f is a semigroup; it is open because f is lower semi-continuous.

e) G_0 is obvious a semigroup; it is closed, because V is symmetrical.

<u>1.3.9. Remark.</u> The assumption on the potential kernel k of the previous theorem can be omitted in the case when G is a $\mathcal{K}_\sigma$-set, i.e. $G=\bigcup_{n\in N} K_n$, with K_n compact subsets in G.

II H-CONES OF FUNCTIONS ON THE TOPOLOGICAL GROUP G

In the first section of this chapter we give two remarkable examples of standard H-cones of functions on the locally compact group G; in the second section we study the natural and fine topologies on G, with respect to the two H-cones of functions on G.

2.1. H-Cones of Functions on the Locally Compact Group G

2.1.1. Definition. A set $\mathcal{S}$ of positive, numerical functions on a set X is called a H-cone of functions on X if $\mathcal{S}$ endowed with the pointwise algebraic operations and order relation is a H-cone (see [4], section 1), with the convention $0\cdot\infty=0$, such that:

F1) if $(s_i)_{i\in I}$ is an increasing net in $\mathcal{S}$, such that there exists its l.u.b. in $\mathcal{S}$, denoted by s, then for every $x\in X$ we have $s(x)=\sup_{i\in I} s_i(x)$;

F2) For every $s,t\in\mathcal{S}$ and for every $x\in X$ we have $(s\wedge t)_{(x)}=\inf(s(x),t(x))$, where $s\wedge t$ is the l.u.b. of s and t in $\mathcal{S}$.

F3) The set $\mathcal{S}$ separates X and contains the positive constant functions.

2.1.2. Definition. If the H-cone of functions $\mathcal{S}$ on the set X is standard (see [4], section 2), then it is called a standard H-cone of functions on the set X. The set of universally continuous elements of $\mathcal{S}$ will be denoted by $\mathcal{S}_0$.

Throughout this chapter we assume that the assumptions of theorem 1.3.8. are fulfild and that N_0 is proper; f and f_α $(\alpha>0)$, given by a) of th.1.3.8., will be fixed. For every $g\in K^+(G)$ and $x\in G$ we have:

(20) $N_0(g)_{(x)}=(k*g)_{(x)}=\int \frac{f(xz^{-1})}{\Delta(z)} g(z)\, d\lambda_s(z)$

(21) $N_\alpha(g)_{(x)}=(k_\alpha * g)_{(x)}=\int \frac{f_\alpha(xz^{-1})}{\Delta(z)} g(z)\, d\lambda_s(z)$. We also can define:

(20') $\check{N}_0(g)_{(x)}=:(k*g)_{(x)}=\frac{1}{\Delta(x)}\int f(zx^{-1})\, g(z)\, d\lambda_s(z)$,

(21') $\check{N}_\alpha(g)_{(x)}=:(k_\alpha * g)_{(x)}=\frac{1}{\Delta(x)}\int f_\alpha(zx^{-1})\, g(z)\, d\lambda_s(z)$.

One can easily see that $\check{\mathcal{N}}=:\{\check{N}_\alpha\}_{\alpha>0}$ is the resolvent of operators associated with the potential kernel $k=\int_0^\infty \mu_t\, dt$, and that the resolvents

$\mathcal{N}$ and $\check{\mathcal{N}}$ are in duality, i.e. for every $g,h\in K(G)$ we have:

(22) $\int g(x)\, N_\alpha(h)_{(x)} d\lambda_s(x) = \int \check{N}_\alpha(g)_{(x)} h(x)\, d\lambda_s(x)$, for every $\alpha\geq 0$,

hence the dual of $\mathcal{E}(\mathcal{N})$, which is also a standard H-cone, coincides with $\mathcal{E}(\check{\mathcal{N}})$ (see for proofs [4]). From (5) we obtain that the resolvents $\mathcal{N}$ and $\check{\mathcal{N}}$ are sub-Markovian. One can easily verify that:

(23) $(N_\alpha g)_a = N_\alpha(g_a)$ for every $\alpha\geq 0, g\in\mathcal{F}^+$, and $a\in G$

(23') $(\check{N}_\alpha g)_a = \check{N}_\alpha(g_a)$ for every $\alpha\geq 0, g\in\mathcal{F}^+$, and $a\in G$

(24) $s\in\mathcal{S}(\mathcal{N})$, lower semicontinuous on G $\Longrightarrow$ $s\in\mathcal{S}'(\mathcal{N})$.

__2.1.3. Remark.__ Because $f\geq 0, f\neq 0$, for every $x\in G$ we have: $0 < N_0(1)_{(x)} = \check{N}_0(1)_{(x)} = \int f(y)\, d\lambda_s(y) \leq +\infty$, hence $N_0(1)$ and $\check{N}_0(1)$ are strictely positive constants and therefore $s\in\mathcal{E}(\mathcal{N})$ is a weak unit in $\mathcal{E}(\mathcal{N})$ iff $s(x)>0$ for every $x\in G$. We have $1\in\mathcal{S}(\mathcal{N})$ because $\mathcal{N}$ is sub-Markovian; by (24) we obtain $1\in\mathcal{E}(\mathcal{N})$, hence 1 is a weak unit in $\mathcal{E}(\mathcal{N})$; analoguously, 1 is a weak unit in $\mathcal{E}(\check{\mathcal{N}})$, too.

__2.1.4. Remark.__ Because $\mathcal{E}(\mathcal{N})$ is a standard H-cone, we can assert that:

a) if $(s_i)_{i\in I}\subseteq\mathcal{E}(\mathcal{N})$ is an increasing family, dominated in $\mathcal{E}(\mathcal{N})$, then there exists $(i_n)_{n\in N}\subseteq I$, such that $\bigvee_{i\in I} s_i = \bigvee_{n\in N} s_{i_n}$

b) if $(s_i)_{i\in I}\subseteq\mathcal{E}(\mathcal{N})$, then there exists $(i_n)_{n\in N}\subseteq I$, such that:

$$\bigwedge_{i\in I} s_i = \bigwedge_{n\in N} s_{i_n}$$

__2.1.5. Lemma.__ If $g\in\mathcal{F}_b$, with compact support K, then: a) $N_0(g)\in C_b(G)$; b) $N_0(g)\in(\mathcal{E}(\mathcal{N}))_0$.

__Proof:__ a) follows by the remark of preliminary (because $N_0(g)=f*g$) and by the fact that the potential kernel k is a shift-bounded measure on G (see [1], 13.10), i.e. $N_0(h)=k*h\in C_b(G)$ if $h\in K(G)$. For b), by Hunt's theorem we have $N_0(g)\in\mathcal{E}(\mathcal{N})$; let $(s_n)_{n\in N}\subseteq\mathcal{E}(\mathcal{N})$, such that $(s_n)_{n\in N}\uparrow N_0(G)$, let $\varepsilon\in R_{>0}$ and let $s\in\mathcal{E}(\mathcal{N})$ be an arbitrary weak unit (hence $s(x)>0$ for every $x\in G$). We denote $\alpha =: \inf_{x\in K} s(x) > 0$. By Dini's theorem, there exists $n_\varepsilon\in N$, such that for every $n\geq n_\varepsilon$ we have $s_n+\varepsilon s\geq s_n+\varepsilon\alpha\geq N_0(g)$ on K; by 1.2.5. it follows that $s_n+\varepsilon s\geq N_0(g)$ on G. Thus we proved b).

__2.1.6. Notation.__ $\mathcal{A} =: \{h\in\mathcal{F}^+ \mid h$ is bounded, with compact support$\}$

2.1.7.Lemma. Let us consider $g\in\mathcal{F}_b^+$, such that $N_o(g)$ is bounded; by Hunt's theorem, $N_o(g)\in\mathcal{E}(\mathcal{N})$. Moreover, there exists an increasing sequence $(g_n)_{n\in N}$ of functions from $\mathcal{A}$, such that $(N_o(g_n))_n\uparrow N_o(g)$.

Proof: Let $(K_i)_{i\in I}$ be an increasing net of compact subsets of G, such that $\bigcup_{i\in I} K_i=G$ and let $g_i=:g\cdot I_{K_i}$. Using the previous lemma, remark 2.1.4. and the fact that $(N_o(g_i))_{i\in I}$ is increasing to $N_o(g)$, we obtain the existence of the desired sequence.

2.1.8.Proposition. Let $s\in\mathcal{E}(\mathcal{N})$; then there exists a sequence $(g_n)_{n\in N}\subseteq\mathcal{A}$, such that $(N_o(g_n))_n\uparrow s$.

Proof: By Hunt's theorem, there exists $(h_n)_{n\in N}\subseteq\mathcal{F}_b^+$, with $N_o(h_n)$ bounded for every $n\in N$, such that $(N_o(h_n))_{n\in N}\uparrow s$. The existence of the desired sequence follows by the previous lemma, applied to every h_n.

2.1.9.Corollary. a) Every $s\in\mathcal{E}(\mathcal{N})$ is lower semicontinuous on G;
b) Every $s\in(\mathcal{E}(\mathcal{N}))_o$ is continuous and bounded on G.

2.1.10.Proposition. a) $s\in\mathcal{E}(\mathcal{N}), a\in G \implies s_a\in\mathcal{E}(\mathcal{N})$;
b) $s\in\mathcal{E}(\check{\mathcal{N}}), a\in G \implies s_a\in\mathcal{E}(\check{\mathcal{N}})$.

Proof: a) Let us consider, by previous proposition, a sequence $(g_n)_{n\in N}\subseteq\mathcal{A}$, such that $(N_o(g_n))_{n\in N}\uparrow s$, hence: (25) $\{(N_o(g_n))_a\}_{n\in N}\uparrow s_a$; by (23) and by Hunt's theorem, we have $(N_o(g_n))_a\in\mathcal{E}(\mathcal{N})$ for every $n\in N$. By assumption, $s<+\infty$ $\mathcal{N}$-a.e., hence $s_a<+\infty$ $\mathcal{N}$-a.e.; using (25),(24) we obtain $s_a\in\mathcal{E}(\mathcal{N})$; b) can be proved analoguously.

2.1.11.Theorem. $\mathcal{E}(\mathcal{N})$ (resp. $\mathcal{E}(\check{\mathcal{N}})$) is a standard H-cone of functions on G (see definition 2.1.2).

Proof: We already know that $\mathcal{E}(\mathcal{N})$ is a standard H-cone; for F3) we know that $\mathcal{E}(\mathcal{N})$ contains the positive constant functions (see 2.1.3); we now prove that $\mathcal{E}(\mathcal{N})$ separates the points of G: by 2.1.8, it is sufficient to consider $x\in G$, such that $(N_og)_{(x)}=(N_og)_{(e)}$ for every $g\in\mathcal{A}$, and to show that $x=e$. Indeed, N_α being bounded kernels, by the relation $N_o(g)=N_\alpha(g)+\alpha N_\alpha N_o(g)$ for every $\alpha>0$, we obtain:

(26) $(N_\alpha g)_{(x)}=(N_\alpha g)_{(e)}$ for every $g\in K(G)$ and $\alpha>0$,

hence, by (9) we obtain $g(x)=g(e)$ for every $g\in K(G)$; but this last relation is possible iff $x=e$, hence we have F3); F2) follows by 2.1.8. and (24); F2) follows by 2.1.9. and (24).

2.1.12. Lemma. If $U\neq\emptyset$ is an open subset of G, such that $\overline{U}$ is compact, then $N_o(I_U)\neq 0$.

Proof: Indeed, otherwise we would obtain $f\equiv 0$, which is false.

2.2. The Natural and the Fine Topologies on G, with respect to $\mathcal{E}(\mathcal{N})$ and $\mathcal{E}(\check{\mathcal{N}})$.

Because we have just proved that $\mathcal{E}(\mathcal{N})$ and $\mathcal{E}(\mathcal{N})$ are H-cones of functions on G, we can define on G (see [4], section 3) the natural topology with respect to $\mathcal{E}(\mathcal{N})$ (resp. to $\mathcal{E}(\check{\mathcal{N}})$), as being the coarsest topology on G for which the universally continuous elements of $\mathcal{E}(\mathcal{N})$ (resp. of $\mathcal{E}(\check{\mathcal{N}})$) are continuous functions on G and it will be denoted by $\mathcal{T}_n$ (resp. $\check{\mathcal{T}}_n$), and the fine topology on G with respect to $\mathcal{E}(\mathcal{N})$ (resp. to $\mathcal{E}(\check{\mathcal{N}})$), as being the coarsest topology on G for which all the elements of $\mathcal{E}(\mathcal{N})$ (resp. of $\mathcal{E}(\check{\mathcal{N}})$) are continuous functions on G and it will be denoted by $\mathcal{T}_f$ (resp. $\check{\mathcal{T}}_f$); G endowed with $\mathcal{T}_n$ (resp. $\check{\mathcal{T}}_n$) is a metrisable and separable topological space and G endowed with $\mathcal{T}_f$ (resp. $\check{\mathcal{T}}_f$) is a completely regular topological space (see [4]). Obviously $\mathcal{T}_n\subseteq\mathcal{T}_f$ and $\check{\mathcal{T}}_n\subseteq\check{\mathcal{T}}_f$. By 2.1.5. and 2.1.8. it follows that $\mathcal{T}_n\subseteq\mathcal{T}_G$, where $\mathcal{T}_G$ is the initial given topology of the topological group G.

2.2.1. Lemma. a) $V\in\mathcal{T}_f$ (resp. $V\in\check{\mathcal{T}}_f$), $a\in G \Rightarrow V\cdot a\in\mathcal{T}_f$ (resp. $V\cdot a\in\check{\mathcal{T}}_f$);

b) $V\in\mathcal{T}_n$ (resp. $V\in\check{\mathcal{T}}_n$), $a\in G \Rightarrow V\cdot a\in\mathcal{T}_n$ (resp. $V\cdot a\in\check{\mathcal{T}}_n$).

Proof: a) If we consider the map $f:G\longrightarrow G, f(x)=:xa^{-1}$, by prop. 2.1.10 $s\circ f=s_{a^{-1}}\in\mathcal{E}(\mathcal{N})$ (resp. $\mathcal{E}(\check{\mathcal{N}})$) is finely continuous for every $s\in\mathcal{E}(\mathcal{N})$ (resp. for every $s\in\mathcal{E}(\check{\mathcal{N}})$), hence (see [8], ch. I, §2, n°3, prop. 4) it follows that f is finely continuous, which gives a); b) follows from a).

2.2.2. Lemma. If $V\in\mathcal{T}_G, V\neq\emptyset$, there exists $x\in V$ such that V is a $\mathcal{T}_n$-neighbourhood of x.

Proof: Let us consider $U\in\mathcal{T}_G, U\neq\emptyset$, such that U is $\mathcal{T}_G$-compact and $\overline{U}\subseteq V$, and the compactification $\overline{G}$ of G with respect to the family of bounded $\mathcal{T}_G$-continuous functions on G $\mathcal{F}_o=:(\mathcal{E}(\mathcal{N}))_o\cup K(G)$ (see [11], th. 1.1).

We also consider the following cone of continuous functions on the compact space $\overline{G}$ (the extension by continuity on $\overline{G}$ of a function $s \in \mathcal{F}_o$ will also be denoted by s) : $\mathcal{S}_U =: \{s+\alpha - N_o(I_U) \mid s \in (\mathcal{E}(\mathcal{N}))_o, \alpha \in R^+\}$. By lemma 2.1.12 we have $N_o(I_U) \neq 0$, hence there exists the Šilov boundary $\partial\overline{G} \neq \emptyset$ of G, with respect to $\mathcal{S}_U$. Because N_o satisfies the complete maximum principle and because every function from $\mathcal{E}(\mathcal{N})$ is N_o-dominating on G (see [10]), it follows that the closure $\overline{U}$ of U in $\overline{G}$ is a closed boundary set with respect to $\mathcal{S}_U$, i.e.: $f \in \mathcal{S}_U, f \geq 0$ on $\overline{U} \Rightarrow f \geq 0$ on $\overline{G}$; thus we have $\partial\overline{G} \subseteq \overline{U}$. We now assert that:

(27) there exists $s \in \mathcal{S}_U$, such that $s \geq 0$ on $C_V =: \overline{G} - V$, and $s(x) < 0$ for some $x \in U$.

Indeed, otherwise, for every $s \in \mathcal{S}_U$ we would have:

(28) $s \geq 0$ on $C_V \Rightarrow s \geq 0$ on $U \Rightarrow s \geq 0$ on $\overline{G}$,

and therefore we would obtain that C_V is a closed boundary set with respect to $\mathcal{S}_U$, hence $\emptyset \neq \partial\overline{G} \subseteq \overline{U} \cap C_V = \emptyset$, which is a contradiction. Thus we proved (27). Let us now consider $s = u + \alpha - N_o(I_U) \in \mathcal{S}_U$, given by (27); if we take $t =: N_o(I_U) - (u+\alpha) \wedge N_o(I_U)$, which is $\mathcal{T}_n$-continuous, we have $t \geq 0$ on G, $t = 0$ on C_V and $t(x) > 0$, hence $x \in \{y \in G \mid t(y) > 0\} \subseteq V$; therefore V is a $\mathcal{T}_n$-neighbourhood of x.

__2.2.3. Theorem.__ a) $\mathcal{T}_n = \mathcal{T}_G$; b) $\check{\mathcal{T}}_n = \mathcal{T}_G$.

__Proof__: a) Let $V \in \mathcal{T}_G$ and let $x \in V$ arbitrary; we show that V is a $\mathcal{T}_n$-neighbourhood of x: Indeed, by the continuity of multiplication in the point $(e,x) \in G \times G$, there exists $U', W' \in \mathcal{T}_G$, such that $e \in U' \cap W'$ and $U' \cdot (W' \cdot x) \subseteq V$. Let now consider $W \in \mathcal{T}_G$, such that $e \in W$, $W^{-1} = W$ and $W \subseteq U' \cap W'$. By lemma 2.2.2, there exists $y \in W \cdot x$, such that $W \cdot x$ is a $\mathcal{T}_n$-neighbourhood of y, hence, by 2.2.1. we can take $\Gamma \in \mathcal{T}_n$ such that $e \in \Gamma$ and $\Gamma \cdot y \subseteq W \cdot x$; We have: $x \in \Gamma x \subseteq (Wxy^{-1})x = W(xy^{-1})x \subseteq W \cdot W \cdot x \subseteq V$; using again 2.2.1, it follows that V is a $\mathcal{T}_n$-neighbourhood of x; b) can be proved analoguous.

__2.2.4. Corollary.__ a) $\mathcal{T}_n = \check{\mathcal{T}}_n = \mathcal{T}_G$; b) $(G, \mathcal{T}_G)$ is a locally compact topological group, whose topology is metrisable and separable.

III POTENTIALS ON THE TOPOLOGICAL GROUP G

In this chapter we again assume that G is a locally compact non-abelian group, which satisfies the assumptions of th.1.3.8. All the notations of the first two chapters are valid in this chapter, too. If f and f_α ($\alpha>0$) are the functions given by th.1.3.8. and if we define for every $x,y\in G$ and $\alpha>0$ $G(x,y)=:\frac{f(xy^{-1})}{\Delta(y)}$, $G_\alpha(x,y)=:\frac{f_\alpha(xy^{-1})}{\Delta(y)}$, then the function $G(.,.)$ defined on $G\times G$ has the following properties, with respect to the pair of resolvents $\mathcal{N}$ and $\check{\mathcal{N}}$ (analoguously for $G_\alpha(.,.)$, with respect to $\mathcal{N}_\alpha$ and $\check{\mathcal{N}}_\alpha$):

(29) for every fixed $y\in G$, the map $G_y: x\longmapsto G(x,y)$ belongs to $\mathcal{E}(\mathcal{N})$;

(30) for every fixed $x\in G$, the map $G^x: y\longmapsto G(x,y)$ belongs to $\mathcal{E}(\check{\mathcal{N}})$;

(31) the function $G(.,.)$ gives the duality between N_0 and $\check{N}_0$, i.e.:
$N_0(g)_{(x)}=:\int G(x,y)\ g(y)\ d\lambda_s(y)$, and $\check{N}_0(g)_{(x)}=\int G(y,x)\ g(y)\ d\lambda_s(y)$;

(32) if $G(x,y)=G(x,y')$ for every $x\in G$, then $y=y'$.

Proof: We have (31) by (20) and (20'); (29) follows by prop.2.1.10, remarking that $G_y(x)=\frac{1}{\Delta(y)}\ f_{y^{-1}}(x)$ for every $x\in G$. For (30) let us consider a fixed $x\in G$ and an arbitrary $y\in G$; we know that $f\in\mathcal{E}(\mathcal{N})$ and that $f_\alpha * f=f*f_\alpha$ for every $\alpha>0$ (see (7)), hence:

(i) $\{\alpha N_\alpha(f)\}_{\alpha>0}\uparrow f$, i.e. $\{\alpha(f*f_\alpha)\}_{\alpha>0}\uparrow f$, when $\alpha\longrightarrow+\infty$, i.e.:

(i') $\lim_{\alpha\to\infty}\alpha\int f_\alpha(u^{-1}t)\ f(u)\ d\lambda_s(u)=f(t)$ for every $t\in G$.

Taking $t=xy^{-1}$ in (i'), we obtain:

(ii) $\lim_{\alpha\to\infty}\alpha\int f_\alpha(u^{-1}xy^{-1})\cdot f(u)\frac{1}{\Delta(u^{-1}x)}\ \frac{\Delta(x)}{\Delta(u)}\ d\lambda_s(u)=f(xy^{-1})$, i.e.:

(ii') $\frac{1}{\Delta(y)}\lim_{\alpha\to\infty}\alpha\int f_\alpha(zy^{-1})\ \frac{f(xz^{-1})}{\Delta(z)}\ d\lambda_s(z)=\frac{f(xy^{-1})}{\Delta(y)}$;

hence, using (20'), we obtain (30); (32) will follow from cor.3.2.7, taking $\mu=\varepsilon_y$ and $\nu=\varepsilon_{y'}$.

3.1. Potentials on G.

3.1.1. Definition. Let $\mu\in M^+(G)$ and let us define the positive numerical functions P^μ and $\check{P}^\mu$ on G, by $P^\mu(x)=:\int G(x,y)d\mu(y)=(f*\mu)_{(x)}\leq+\infty$

$\check{P}^\mu(x)=:\int G(y,x)d\mu(y)=(\check{f}*\mu)_{(x)}\leq+\infty$.

The function P^μ (resp. $\check{P}^\mu$) is called the k-potential (resp. $\check{k}$-potential) associated with the measure μ if it is finite $\mathcal{N}$-a.e. (resp. $\check{\mathcal{N}}$-a.e.).
The set of all k-potentials (resp. $\check{k}$-potentials) will be denoted by $\mathcal{P}(k)$ (resp. $\mathcal{P}(\check{k})$). Obviously $\mathcal{P}(k)$ and $\mathcal{P}(\check{k})$ are convex cones.

3.1.2. Proposition. a) For every $\mu\in M^+(G)$ we have $P^\mu\in\mathcal{S}'(\mathcal{N})$; b) $\mathcal{P}(k)\subseteq\mathcal{E}(\mathcal{N})$

Proof: We must show that for every $x\in G$ we have:

(33) $\{\alpha N_\alpha(P^\mu)_{(x)}\}_{\alpha>0}\uparrow P^\mu(x)$, when $\alpha\longrightarrow+\infty$.

Indeed, let us fix $x\in G$ and let $z\in G$ be arbitrary; $f_{z^{-1}}\in\mathcal{E}(\mathcal{N})$, hence we have:

$\{\alpha\int\frac{1}{\Delta(y)}f_\alpha(xy^{-1})\, f_{z^{-1}}(y)\, d\lambda_s(y)\}_{\alpha>0}\uparrow f_{z^{-1}}(x)$, for every $z\in G$;

$\{\alpha\int\frac{f_\alpha(xy^{-1})}{\Delta(y)}(\int\frac{f(yz^{-1})}{\Delta(z)}d\mu(z))\,d\lambda_s(y)\}_{\alpha>0}\uparrow\int\frac{f(xz^{-1})}{\Delta(z)}\,d\mu(z)$, i.e.(33).

3.1.3. Lemma. If $s\in\mathcal{E}(\mathcal{N})$, then $s\in L^1_{loc}(\lambda_s)$.

Proof: In [12], th.4, it is shown that the isomorphism between $\mathcal{E}(\mathcal{N})$ and $(\mathcal{E}(\check{\mathcal{N}}))^*$ is given by the map θ, defined for every $s\in\mathcal{E}(\mathcal{N})$ by:

(34) $\langle s,\check{N}_0 g\rangle=\theta(s)_{(\check{N}_0 g)}=:\int s(x)\, g(x)\, d\lambda_s(x)$, for every $g\in\mathcal{A}$;

(35) $\langle s,s'\rangle=\sup_{n\in N}\langle s,\check{N}_0 g_n\rangle$, where $\{\check{N}_0 g_n\}_{n\in N}\uparrow s'$, with $g_n\in\mathcal{A}$ for every $n\in N$.

We also know that $\check{N}_0(g)\in(\mathcal{E}(\check{\mathcal{N}}))_0$ if $g\in\mathcal{A}$, hence, by [4], cor. of prop.2.6, we obtain that $\int s\cdot g\, d\lambda_s=\theta(s)_{(\check{N}_0 g)}<+\infty$ if $s\in\mathcal{E}(\mathcal{N})$ and $g\in\mathcal{A}$, i.e. $s\in L^1_{loc}(\lambda_s)$.

3.1.4. Proposition. If $s\in\mathcal{S}'(\mathcal{N})$, the following assertions are equivalent:

i) $s<+\infty$ λ_s-a.e.; ii) $s<+\infty$ $\mathcal{N}$-a.e.; iii) $s\in L^1_{loc}(\lambda_s)$.

Proof: i) $\Rightarrow$ ii) follows by (21); ii) $\Rightarrow$ iii) follows by 3.1.3; iii) $\Rightarrow$ i) is true because the topology of G is metrisable, by 2.2.4.

3.1.5. Corollary. If we denote $D^+(k)=:\{\mu\in M^+(G)\mid \text{there exists } k*\mu\}$, and if we consider $\mu\in M^+(G)$, then the following assertions are equivalent:

a) $P^\mu<+\infty$ λ_s-a.e. (resp. $\check{P}^\mu<+\infty$ λ_s-a.e.);

b) $P^\mu<+\infty$ $\mathcal{N}$-a.e. (resp. $\check{P}^\mu<+\infty$ $\check{\mathcal{N}}$-a.e.);

c) $P^\mu\in\mathcal{P}(k)$ (resp. $\check{P}^\mu\in\mathcal{P}(\check{k})$);

d) $P^\mu\in L^1_{loc}(\lambda_s)$ (resp. $\check{P}^\mu\in L^1_{loc}(\lambda_s)$);

e) $\mu\in D^+(k)$ (resp. $\mu\in D^+(\check{k})$).

Proof: We have only to remark that d) $\Leftrightarrow$ e); indeed, for every $g\in K^+(G)$ we have $(k*\mu)_{(g)}=\int P^\mu(z)\, g(z)\, d\lambda_s(z)$.

We also remark that: (36) $(k*\mu)=(f*\mu)\cdot\lambda_s=P^\mu\cdot\lambda_s$, if $\mu\in D^+(k)$.

3.1.6. Remark. If $h\in\mathcal{A}$, then $N_0(h)\in\mathcal{P}(k)$. Indeed, for every $x\in G$ we have:

$$N_0(h)_{(x)}=\int G(x,y)\, h(y)\, d\lambda_s(y)=P^{h\cdot\lambda_s}(x).$$

3.1.7. Lemma. If K is a compact subset of G, there exists $p\in\mathcal{P}(k)\cap C_b^+(G)\cap(\mathcal{E}(\mathcal{N}))_0$, such that $p(x)>0$ for every $x\in K$.

Proof: For every $x\in K$ there exists $s\in\mathcal{E}(\mathcal{N})$ such that $s(x)>0$, hence by 2.1.8. we can consider a function $g_x\in\mathcal{A}$, such that $(N_0g_x)(x)>0$, and a neighbourhood V_x of x, such that $N_0(g_x)>0$ on V_x; let us take $K\subseteq\bigcup_{i\in I}V_{x_i}$, $g=:\sum_{i\in I}g_{x_i}$; then $p=N_0(g)$ satisfies our requirements.

3.1.8. Proposition. If $(p_n)_{n\in N}\subseteq\mathcal{P}(k)$ is such that $s=:\sum_{n\in N}p_n\in\mathcal{E}(\mathcal{N})$, then $s\in\mathcal{P}(k)$.

Proof: For every $n\in N$ we can write $p_n(x)=P^{\sigma_n}(x)=\int G(x,y)\,d\sigma_n(y)$, where $\sigma_n\in D^+(k)$; hence $s(x)=\int G(x,y)\,d\sigma(y)=P^{\sigma}(x)$ for every $x\in G$, where $\sigma=\sum_{n\in N}\sigma_n$; s being excessive, by 3.1.5. it follows that $s\in\mathcal{P}(k)$.

3.1.9. Proposition. In our context there exists a bounded $p\in\mathcal{P}(k)$, which is strictely positive on G.

Proof: By 2.2.4, we can write $G=\bigcup_{n\in N}K_n$, where K_n are compact sets; using 3.1.7, we can choose for every $n\in N$ a potential p_n, strictely positive on K_n, such that $0\leq p_n\leq\frac{1}{2}$ on G; by previous prop., $p=:\sum_{n\in N}p_n$ satisfies 3.1.9.

3.2. The Riesz Decomposition Theorem

3.2.1. Definition. A function $h\in\mathcal{E}(\mathcal{N})$ is called k-invariant if it satisfies one of the following equivalent relations: i) $\mu_t*h=h$ for every $t>0$; ii) $\alpha\, k_\alpha*h=h$ (i.e. $\alpha N_\alpha(h)=h$) for every $\alpha>0$. We denote the set of all k-invariante functions on G by $\mathcal{H}(k)$.

3.2.2. Definition. A measure $\xi\in M^+(G)$ is called k-excessive if it satisfies one of the following equivalent relations:

i) $\{\mu_t*\xi\}_{t>0}\uparrow\xi$ vaguely, when $t\to 0$;

ii) $\{\alpha k_\alpha*\xi\}_{\alpha>0}\uparrow\xi$ vaguely, when $\alpha\to+\infty$.

One can easily prove that:

3.2.3. Proposition. For every $\xi\in M^+(G)$, we have the following equivalences:

i) ξ is a k-excessive measure;

ii) there exists $s\in\mathcal{E}(\mathcal{N})$, such that $\xi=s\cdot\lambda_s$.

3.2.4. Corollary. If $\eta\in M^+(G)$, then η is k-invariant (i.e. $\mu_t*\eta=\eta$ for every $t>0$) iff there exists $h\in\mathcal{H}(k)$, such that $\eta=h\cdot\lambda_s$.

3.2.5. Theorem. If $s\in\mathcal{E}(\mathcal{N})$, then there exist $p=P^{\sigma}\in\mathcal{P}(k)$ and $h\in\mathcal{H}(k)$, such that $s=p+h$. In such a decomposition σ and h are uniquely determinated and are given by the formulas:

i) $\sigma = \lim_{t\to 0} \frac{1}{t} (s\cdot\lambda_s - \mu_t * (s\cdot\lambda_s))$,vaguely;

ii) $h = \lim_{t\to\infty} \mu_t * s$ λ_s-a.e.

Remark.The proof of this theorem follows the line of C.Berg's proof of the theorem 16.7. of [1].hence we don't repeat the ideas of proof here.

3.2.6.Remark.a) By the uniqueness of the previous theorem,for every $q\in\mathcal{P}(k)$ we have $\lim_{t\to\infty} \mu_t * q = 0$; b) every $h\in\mathcal{H}(k)$ has the following property:

(37) $q\in\mathcal{P}(k)$, $q\prec h$ (i.e.$(\exists)q'\in\mathcal{E}(\mathcal{N})$ such that $q+q'=h$) $\Longrightarrow$ $q=0$.

Indeed, $\mu_t*q + \mu_t*q' = \mu_t*h = h = q+q'$, $\lim_{t\to\infty}\mu_t*q + \lim_{t\to\infty}\mu_t*q' = q + q'$,hence $\lim_{t\to\infty} \mu_t*q = q$,and by a),we obtain $q=0$.

3.2.7.Corollary.If $P^{\mu}=P^{\nu}\in\mathcal{P}(k)$,then $\mu=\nu$.

3.2.8.Corollary.If $s\leq q$,where $s\in\mathcal{E}(\mathcal{N})$ and $q\in\mathcal{P}(k)$,then $s\in\mathcal{P}(k)$.

Proof:By previous theorem,$s=p+h$,where $p\in\mathcal{P}(k)$ and $h\in\mathcal{H}(k)$;using 3.2.6.a), we have $h=\lim_{t\to\infty} \mu_t*s \leq \lim_{t\to\infty} \mu_t*q = 0$, $s=p\in\mathcal{P}(k)$.

3.2.9.Corollary.If $s\prec t$,where $s\in\mathcal{E}(\mathcal{N})$ and $t\in(\mathcal{E}(\mathcal{N}))_0$,then $s\in\mathcal{P}(k)$.

Proof:Let us consider,by 3.1.9,a potential $p\in\mathcal{P}(k)$,which is strictely positive on G,hence p is a weak unit for $\mathcal{E}(\mathcal{N})$.In [13],page 1015,it is shown that there exists $\alpha\in R^+$,such that $t\leq\alpha p$,hence $s\leq\alpha p\in\mathcal{P}(k)$; by 3.2.8,we obtain $s\in\mathcal{P}(k)$.

3.3.Carrier of Potentials

We proved that $\mathcal{E}(\mathcal{N})$ and $\mathcal{E}(\check{\mathcal{N}})$ are standard H-cones of functions on our locally compact group G;we have already remarked in section 3.1. that we have the following isomorphisms of H-cones:

$$s\in\mathcal{E}(\mathcal{N}) \longleftrightarrow \tilde{s}\in(\mathcal{E}(\check{\mathcal{N}}))^*,\quad \tilde{s}(s^*)=:\langle s,s^*\rangle \quad \text{for } s^*\in\mathcal{E}(\check{\mathcal{N}});$$

$$s^*\in\mathcal{E}(\check{\mathcal{N}}) \longleftrightarrow \tilde{s^*}\in(\mathcal{E}(\mathcal{N}))^*,\quad \tilde{s^*}(s)=:\langle s,s^*\rangle \quad \text{for } s\in\mathcal{E}(\mathcal{N}).$$

3.3.1.Remark.By [10],prop.4.2.8, $\mathcal{E}(\mathcal{N})$ and $\mathcal{E}(\check{\mathcal{N}})$,endowed with their natural topologies are completely metrisable and separable spaces,hence the two sets endowed with their fine topologies are Baire spaces (see [10],th.4.3.4.).

3.3.2.We consider now the following (natural) compact convex sets: $K_1 =: \{s^*\in\mathcal{E}(\check{\mathcal{N}}) \mid \langle 1,s^*\rangle\leq 1\}$, $K_1^* =: \{s\in\mathcal{E}(\mathcal{N}) \mid \langle s,1\rangle\leq 1\}$,which are Choquet simplexes in $\mathcal{E}(\check{\mathcal{N}})$,resp. $\mathcal{E}(\mathcal{N})$ (see [10],section 4.2.) and

Lusin subsets of the Luzin spaces $\mathcal{E}(\check{\mathcal{N}})$ and $\mathcal{E}(\mathcal{N})$ resp.(see [10], section 5.2.)Further,let:

$X_1=:\text{ext}(K_1)-\{0\}, Y_1=:\overline{X}_1 \subseteq K_1, \quad (\mathcal{E}(\mathcal{N}))_1=:\{\overline{s}=:\tilde{s}|_{X_1} \mid s\in\mathcal{E}(\mathcal{N})\}$;

$X_1^*=:\text{ext}(K_1^*)-\{0\}, Y_1^*=:\overline{X_1^*} \subseteq K_1^*, \quad (\mathcal{E}(\check{\mathcal{N}}))_1=:\{\tilde{s}^*|_{X_1} \mid s^*\in\mathcal{E}(\check{\mathcal{N}})\}$.

By [10] ,th.4.2.12,we have: a) X_1 is a $\mathcal{G}_\delta$-subset of K_1 (endowed with the natural topology),hence X_1 is a Lusin space; b) $(\mathcal{E}(\mathcal{N}))_1$ is a H-cone of functions on X_1,isomorphic with $\mathcal{E}(\mathcal{N})$;we can assert analoguously for X_1^* and $(\mathcal{E}(\check{\mathcal{N}}))_1$.

<u>3.3.3</u>. By [10],section 4.3,we have two embedings:

$G \hookrightarrow X_1 \subseteq (\mathcal{E}(\mathcal{N}))^*, \quad x\mapsto\tilde{x}, \ \tilde{x}(s)=:s(x)$ for $s\in\mathcal{E}(\mathcal{N})$

$G \hookrightarrow X_1^* \subseteq (\mathcal{E}(\check{\mathcal{N}}))^*, \quad x\mapsto\tilde{\tilde{x}}, \ \tilde{\tilde{x}}(s^*)=:s^*(x)$ for $s^*\in\mathcal{E}(\check{\mathcal{N}})$,

and the restriction of the natural (resp.fine) topology of $(\mathcal{E}(\mathcal{N}))^*$ to G coincides with $\mathcal{T}_n$ (resp. $\mathcal{T}_f$) defined in section 2.2.

Because $G=\bigcup_{n\in N}K_n\in\mathcal{K}_\sigma$ in K_1(endowed with the natural topology),which is a Lusin space,it follows that $(G,\mathcal{T}_G)$ is a Lusin space,hence every Borel subset of G is a Lusin set (and therefore a Suslin set).One can easily verify that for every $s\in\mathcal{E}(\mathcal{N})$ and for $x\in G$ arbitrary,we have: $\langle s,x\rangle=\langle s,G^x\rangle$,hence $\tilde{x}=G^x=\check{P}^{\varepsilon_x}\in\mathcal{P}(\check{k})\cap X_1$ for every $x\in G$.Analoguously $\tilde{\tilde{x}}=G_x=P^{\varepsilon_x}\in\mathcal{P}(k)\cap X_1^*$ for every $x\in G$;thus,the embedings $G\subseteq X_1$and $G\subseteq X_1^*$ are given respectively by the mappings $x\mapsto G^x$ and $x\mapsto G_x$.

<u>3.3.4</u>.For $A\subseteq G$ and $s\in\mathcal{E}(\mathcal{N})$ (resp.$s^*\in\mathcal{E}(\check{\mathcal{N}})$)we define the function $B_s^A\in\mathcal{E}(\mathcal{N})$ (resp.$B_{s^*}^A\in\mathcal{E}(\check{\mathcal{N}})$) by $B_s^A=:\wedge\{t\in\mathcal{E}(\mathcal{N}) \mid t\geq s \text{ on } A\}$ (resp. $B_{s^*}^A=:\wedge\{t^*\in\mathcal{E}(\check{\mathcal{N}}) \mid t^*\geq s^* \text{ on } A\}$).If A is fine open in G,we have $B_s^A=R_s^A=:\inf\{t\in\mathcal{E}(\mathcal{N}) \mid t\geq s \text{ on } A\}$,the map $s\mapsto(B^A)_{(s)}=:B_s^A$ is a balayage on $\mathcal{E}(\mathcal{N})$,and we have $B_s^A=s$ on A.For every $A\subseteq X_1$ and $s\in\mathcal{E}(\mathcal{N})$ we also can define the function ${}^1B_{\overline{s}}^A=:\wedge\{\overline{t} \mid t\in\mathcal{E}(\mathcal{N}), \overline{t}\geq\overline{s} \text{ on } A\}\in(\mathcal{E}(\mathcal{N}))_1$.

<u>3.3.5</u>.A set $A\subseteq G$ is called <u>thin at a point $x\in G$</u> (with respect to $\mathcal{E}(\mathcal{N})$), if there exists $s\in\mathcal{E}(\mathcal{N})$ such that $B_s^A(x)<s(x)$.The set of all $x\in G$ such that A is not thin at x is called <u>the base of A</u> and is denoted by b(A). A set $A\subseteq G$ is called <u>basic</u> if A=b(A).For every $A\subseteq G$,the set $A\cup b(A)$ is the fine closure of A.A subset A of G is called <u>totally thin</u> if $b(A)=\emptyset$,

or, equivalently, if A is thin at any point $x \in G$. Countable unions of totally thin sets are called <u>semi-polar sets</u>. A subset A of G is called <u>negligible</u> (with respect to $\mathcal{E}(\mathcal{N})$), if any compact subset of A is semi-polar. Analoguously, we can define these notions for subsets of X_1, with respect to $(\mathcal{E}(\mathcal{N}))_1$.

<u>3.3.6. Remark.</u> If $s^* \in (\mathcal{E}(\mathcal{N}))_o$, we have by 3.2.9. that $s^* = \check{P}^{\mu} \in \mathcal{P}(\check{k})$, where $\mu \in M^+(G)$. One can easily verify that $\langle s, s^* \rangle = \int s \, d\mu$ for every $s \in \mathcal{E}(\mathcal{N})$ and we say that <u>s^* is representable by means of the measure μ</u> and the H-integral s^* is then called <u>H-measure</u>; because of this property, <u>G is nearly saturated with respect to $\mathcal{E}(\mathcal{N})$</u>; analoguously G is nearly saturated with respect to $\mathcal{E}(\check{\mathcal{N}})$. By [10], prop.5.4.1. it follows that every negligible (with respect to $\mathcal{E}(\mathcal{N})$) Borel subset of G is semi-polar. In [10], prop.5.4.2. it is shown that for every $A \subseteq G$ the following two assertions are equivalent: a) A is negligible with respect to $\mathcal{E}(\mathcal{N})$; b) $\mu(K)=0$ for every $s^* \in (\mathcal{E}(\check{\mathcal{N}}))_o$ (where μ is uniquely given by $s^* = \check{P}^{\mu}$) and for every compact subset K of A.

If we take in b) $s^* =: \check{N}_o(I_K)$, then the correspondent measure for it is $\mu = I_K \cdot \lambda_s$. Hence, if $A \subseteq G$ is negligible with respect to $\mathcal{E}(\mathcal{N})$ (resp. $\mathcal{E}(\check{\mathcal{N}})$), then $\lambda_s(K)=0$ for every compact $K \subseteq A$.

<u>3.3.7. Remark.</u> If Γ is a Borel subset of X_1, which is fine open, then it is not negligible with respect to $(\mathcal{E}(\mathcal{N}))_1$ (because a negligible set is of the first category in the fine topology (see [5], cor.6.3.3) and because K_1 endowed with the fine topology is a Baire space (see [10], theorem 4.3.4)).

<u>3.3.8.</u> By [10], th.5.3.8, $X_1 - G$ (resp. $X_1^* - G$) is negligible with respect to $(\mathcal{E}(\mathcal{N}))_1$ (resp. to $(\mathcal{E}(\check{\mathcal{N}}))_1$). Using the previous remark, one can easily verify that $({}^1B_s^A)|_G = B_s^{A \cap G}$ for every $s \in \mathcal{E}(\mathcal{N})$ and every fine open $A \subseteq X_1$;

(38) $G_y = \frac{1}{\Delta(y)}(G_e)_{y^{-1}}$, $B_{G_y}^{Ay} = \frac{1}{\Delta(y)}(B_{G_e}^A)_{y^{-1}}$ for every $A \subseteq G$ and for every $y \in G$.

<u>3.3.9.</u> For every $s \in \mathcal{E}(\mathcal{N})$ we define, as in [10], section 5.5.:

$\mathrm{carr}^1(s) =: \{x \in X_1 \mid {}^1B_s^{X_1 - V} \neq s$, for every (natural) neighbourhood V of x in $X_1\}$

$\mathrm{carr}(s) =: \{x \in G \mid B_s^{G-V} \neq s$ for every neighbourhood V of x in $G\}$. We remark

that $(\mathrm{carr}^1 s)\cap G=\mathrm{carr}(s)$.The following properties are immediately :
a) $\mathrm{carr}(0)=\emptyset$;$\mathrm{carr}(s)$ is closed in G for every $s\in\mathcal{S}(\mathcal{N})$; b) $\mathrm{carr}(s_1+s_2)=\mathrm{carr}(s_1)\cup\mathrm{carr}(s_2)$ for every $s_1,s_2\in\mathcal{S}(\ )$; c)for every $y\in G$ $G_y=P^{\varepsilon_y}\in\mathcal{P}(k)$ is an extreme element in $\mathcal{S}(\mathcal{N})$ and,hence,by the cor.of th.2.3.of [9], we have that either $\mathrm{carr}(G_y)=\emptyset$,or $\mathrm{carr}(G_y)$ contains a single point.

The main aim of this section is to show that $\mathrm{carr}(G_y)=\{y\}$ for every $y\in G$.For a balayage B on $\mathcal{S}(\mathcal{N})$ (see [10],section 2.2),<u>the base of B</u> is the set $b(B)=:\{x\in G \mid (Bs)_{(x)}=s(x) \text{ for every } s\in\mathcal{S}(\mathcal{N})\}$.

<u>3.3.11.Lemma</u>.If $0\neq p\in(\mathcal{S}(\mathcal{N}))_o$,let $B:\mathcal{S}(\mathcal{N})\dashrightarrow\mathcal{S}(\mathcal{N})$ be the map defined for every $s\in\mathcal{S}(\mathcal{N})$ by $Bs=:\bigvee\{q\in(\mathcal{S}(\mathcal{N}))_o \mid q\prec\alpha p \text{ for some } \alpha\in\mathbb{R}^+;q\leq s\}$

<u>Then</u>:a) B is a balayage on $\mathcal{S}(\mathcal{N})$, $Bp=p$, $b(B)\neq\emptyset$;

b) $\mathrm{carr}(p)=b(B)$ (hence $\mathrm{carr}(p)\neq\emptyset$);

c)if M is a closed subset of G such that $B_p^M=p$,then $\mathrm{carr}(p)\subseteq M$.

(for proof,see [10],prop.4.3.13)

<u>3.3.12.Lemma</u>.If $g\in\mathcal{A}$,then $\mathrm{carr}(N_o g)\subseteq\mathrm{supp}(g)$.

<u>Proof</u>:By 1.2.5.we have $B_{N_o(g)}^{\mathrm{supp}(g)}=N_o(g)$,hence by c) of the previous lemma we obtain 3.3.12.

<u>3.3.13.Remark</u>.For every open set $V\subseteq G,V\neq\emptyset$,the set $V_o=:\{y\in G\mid \mathrm{carr}(G_y)=\emptyset$ is negligible with respect to $\mathcal{S}(\mathcal{N})$ (see [10],section 5.5),hence it is λ_s-negligible.In the proof of the following lemma,we shall use the notation $V'=:V-V_o$; V' is not λ_s-negligible.

<u>3.3.14.Lemma.</u>If $U\neq\emptyset$ is an open set in G,then there exists $y\in U$,such that $\mathrm{carr}(G_y)\neq\emptyset$ and $\mathrm{carr}(G_y)\in U$.

<u>Proof</u>:Let $V\neq\emptyset$ be an open relatively compact subset of G,such that $\overline{V}\subseteq U$, and let us prove that <u>there exists $y_o\in V'$</u> (the notation of 3.3.13),<u>such that $\mathrm{carr}(G_{y_o})\in V$</u>. Indeed,if we assume the contrary,$X_1-\overline{V}$ is a neighbourhood of $\mathrm{carr}(G_y)$ for every $y\in V'$,hence $1_{B_{G_y}^{X_1-(X_1-V)}}\neq G_y$; thus we would have <u>(39)</u> $B_{G_y}^V\neq G_y$ for every $y\in V'$. Let us consider now $0\neq g\in K^+(G)$,such that $\mathrm{supp}(g)\subseteq V$ and let us denote $\sigma:=g\cdot\lambda_s\in M^+(G)$; we have $N_o(g)=$ $=\int G_y\,d\sigma(y)$.By previous lemma,$B_{N_o(g)}^V=N_o(g)$,hence,for every $x\in G$ we can write $\int B_{G_y}^V(x)\,d\sigma(y)=\int G_y(x)\,d\sigma(y)$;if we denote $W=:\{y\in G\mid g(y)>0\}\subseteq V$, because W' is not λ_s-negligible,there exists $y_o\in W'\subseteq V'$such that:

$\int B^{V}_{G_{y_o}} d\lambda_s(x)=\int G_{y_o}(x)\, d\lambda_s(x)$, hence $B^{V}_{G_{y_o}}=G_{y_o}$ λ_s-a.e.; $B^{V}_{G_{y_o}}$ and G_{y_o} being excessive functions, from the last relation it follows that $B^{V}_{G_{y_o}}=G_{y_o}$ on G, where $y_o\in V'$, which contradicts (39).

<u>3.3.15. Lemma.</u> $Carr(G_e)=\{e\}$.

<u>Proof:</u> Let V be an arbitrary neighbourhood of e; by the continuity of the map $(x,y)\longmapsto xy^{-1}$ in the point $(e,e)\in G\times G$, there exist $r_1,r_2\in R_{>0}$, such that $B(e,r_1)(B(e,r_2))^{-1}\subseteq V$, where $B(e,r_1)$ is the discus of center e and radius r_1, taken with respect to a metric d, which is compatible with the topology of G. If we take $r=:\min(r_1,r_2)$, by previous lemma, there exists $z\in B(e,r)$, such that $\{t_o\}=carr(G_z)\subseteq B(e,r)$; then $t_oz^{-1}\in V$ and using the continuity of the map $t\longrightarrow tz^{-1}$ in the point t_o, there exists $\varepsilon>0$, such that $B(t_o,\varepsilon)\, z^{-1}\subseteq V$. Because $\{t_o\}=carr(G_z)$, we have $B^{G-B(t_o,\varepsilon)}_{G_z}\neq G_z$, hence $\Delta(z)\, (B^{(G-B(t_o,\varepsilon))z^{-1}z}_{G_z})_z\neq\Delta(z)(G_z)_z$, and using (38), written in an equivalent form, we obtain $B^{G-V}_{G_e}\leq B^{G-B(t_o,\varepsilon)z^{-1}}_{G_e}\neq G_e$, hence $B^{G-V}_{G_e}\neq G_e$; V being arbitrary, it follows that $e\in carr(G_e)$.

<u>3.3.16. Proposition.</u> For every $y\in G$ we have $carr(G_y)=\{y\}$.

<u>Proof:</u> Let us consider an arbitrary $y\in G$ and arbitrary neighbourhood V of y; by the continuity of the map $t\longrightarrow ty$ in the point $e\in G$, there exists a neighbourhood V' of e, such that $V'\cdot y\subseteq V$; by previous lemma, we have $B^{G-V'}_{G_e}\neq G_e$, hence $\frac{1}{\Delta(y)}(B^{G-V'}_{G_e})_{y^{-1}}\neq\frac{1}{\Delta(y)}(G_e)_{y^{-1}}$, hence, using again (38), we obtain $B^{G-V}_{G_y}\leq B^{G-V'y}_{G_y}\neq G_y$; V being arbitrary, it follows that $y\in carr(G_y)$.

<u>3.3.17. Lemma.</u> a) For every $P^{\mu}\in\mathcal{P}(k)$ we have $supp(\mu)\subseteq carr(P^{\mu})$; in particular, if $carr(P^{\mu})=\emptyset$, it follows that $P^{\mu}\equiv 0$; b) the map $p\longmapsto carr(p)$ <u>is an abstract carrier on $\mathcal{P}(k)$</u>, with values in the set of closed subsets of G (for the definition of an abstract carrier, see [5], section 8.1); c) if $(g,p)\longrightarrow g\cdot p$ is the map defined on $K^+(G)\times\mathcal{P}(k)$ with values in $\mathcal{P}(k)$, like in [5], section 8.1, than $carr(g\cdot p)\subseteq supp(g)\cap carr(p)$ for every $g\in K^+(G)$ and for every $p\in\mathcal{P}(k)$; the same is true for a bounded $g\in\mathcal{F}^+$.

<u>Proof:</u> a) can be proved following the proof of th.3.3.b) of [9]; for b), follow the proof of th.2.3. of [9]; c) is proved in [5], prop.8.1.6.

<u>3.3.18. Proposition.</u> If $p=:P^{\mu}\in\mathcal{P}(k)$, then we have $carr(P^{\mu})=supp(\mu)$.

<u>Proof:</u> By 3.2.7, we have $P^{\mu}\equiv 0$ iff $\mu=0$, hence we can assume that $p\neq 0$.

We have already the inclusion „$\supseteq$". For the other inclusion, we denote $F=:\operatorname{supp}(\mu)$; it is sufficient to prove that $I_F \cdot p=p$ (hence, by previous lemma, $\operatorname{carr}(p) \subseteq \operatorname{supp}(I_F)=F$.). Indeed, let K be a compact set, such that $K \cap F=\emptyset$ and let us denote $q=:I_K \cdot p=P^{\nu} \in \mathcal{P}(k)$. We have $\operatorname{supp}(\nu) \subseteq \operatorname{carr}(q) \subseteq K$, hence $\nu(G-K)=0$; on the other hand, $q \prec p$, hence $\nu \leq \mu$, hence $\nu(G-F)=0$, i.e. $q=0$; K being arbitrary, it follows that $I_{G-F}p=0$, hence $I_F \cdot p=p$.

<u>3.3.19. Corollary.</u> If $p \in \mathcal{P}(k)$ is such that $\operatorname{carr}(p)=\{y\}$, then there exists $\alpha \in R^+$, such that $p=\alpha G_y$. In particular, the following property of proportionality is true: (P) if $p_1, p_2 \in \mathcal{P}(k)$ are such that $\operatorname{carr}(p_1)=\operatorname{carr}(p_2)=\{y\}$, and if $p_1 \neq 0$, then there exists $\alpha \in R^+$, such that $p_2=\alpha p_1$.

<u>Proof:</u> It is sufficient to recall that $G_y=P^{\varepsilon_y}$.

IV POTENTIAL THEORY ON LIE GROUPS

The present chapter gives a large class of non-abelian topological groups, to which we can apply the results developed in the first three chapters, namely the class of Lie groups.

4.0. Notations and preliminaries

Throughout this chapter, G will be a fixed Lie group, of dimension n; e will be the neutral element of G; L(G) will be the Lie algebra of G, $F=:C^{\infty}(G)$ will be the real algebra of all (real) differentiable functions on G, $C^{\infty}(p)$ will be the set of differentiable functions at $p \in G$, $\mathcal{D}^1=\mathcal{D}^1(G)$ will be the F-module of all vector fields on G, $G_p=:\mathcal{D}^1(p)$ will be tangent space to G at $p \in G$, $\mathcal{D}_1=\mathcal{D}_1(G)$ will be the dual of the F-module $\mathcal{D}^1(G)$, i.e. the set of all differential 1-forms on G, $\mathcal{D}_1(p)$ will be the dual of the real vector space $G_p=\mathcal{D}^1(p)$, $\mathcal{D}^r_s=\mathcal{D}^r_s(G)$ will be the set of tensor fields of type $(r,s) \in N \times N$ on G, where $\mathcal{D}^0_0=F$; $\mathcal{D}=\mathcal{D}(G)=\bigoplus_{r,s\geq 0}^{\infty} \mathcal{D}^r_s$, endowed with the tensorial product $\otimes$, will be the mixed tensor algebra over G; K(G), M(G) and $\mathcal{F}$ will be the same sets like in the previous chapters. For every $a \in G$, R_a will be the diffeomorphism of G, defined by $R_a(g)=:ga$ for g G; if f is a function on G, we have $f_a=f \circ R_a$.

<u>4.0.1.</u> If $\phi: G \to G$ is a diffeomorphism of G and if $p \in G$, then $d_p\phi: G_p \to G_{\phi(p)}$, $(d_p\phi)_{(A)}(h)=:A(h \circ \phi)$ for every $A \in G_p$ and $h \in C^{\infty}(\phi(p))$, will be the dif-

ferential of ϕ at p; the differential of ϕ will be the mapping $d\phi : \mathcal{D}^1(G) \dashrightarrow \mathcal{D}^1(G)$, defined by $[(d\phi)_{(X)}]_q =: (d_{\phi^{-1}(q)}\phi)_{(X_{\phi^{-1}(q)})} \in G_p$ for $X \in \mathcal{D}^1(G)$ and $q \in G$. It is convenient to write $X^{\phi} =: (d\phi)_{(X)}$. We say that $X \in \mathcal{D}^1(G)$ is <u>ϕ-invariant</u>, if $X^{\phi} = X$; X is called <u>right invariant</u> if it is R_a-invariant for every $a \in G$. If $\omega \in \mathcal{D}_r = \mathcal{D}^0_r(G)$, then $\phi^*\omega \in \mathcal{D}_r$ will be the r-form which satisfies: $(\phi^*\omega)_{(X_1, X_2, \ldots, X_r)} = \omega(X_1^{\phi}, \ldots, X_r^{\phi}) \circ \phi$ for every $X_1, X_2, \ldots, X_r \in \mathcal{D}^1(G)$; $\omega \in \mathcal{D}_r(G)$ is called <u>ϕ-invariant</u> if $\phi^*\omega = \omega$; it is called <u>right invariant</u> if it is R_a-invariant for every $a \in G$.

<u>4.0.2.</u> <u>A Riemannian structure on G</u> is a tensor field $g \in \mathcal{D}^0_2$ which satisfies: a) $g(X,Y) = g(Y,X)$ for every $X, Y \in \mathcal{D}^1(G)$; b) for each $p \in G$, g_p is a non-degenerate and positive definite bilinear form on $G_p \times G_p$. By 4.0.1, g is called <u>right invariant</u> if $g(X^{R_a}, Y^{R_a}) \circ R_a = g(X,Y)$ for every $X, Y \in \mathcal{D}^1(G)$. Let $\{x_1, \ldots, x_n\}$ be a system of coordinates, valid on an open neighbourhood U of $p \in G$; then there exist vector fields X_i $(1 \le i \le n)$ and 1-forms ω_j $(1 \le j \le n)$ on G, and an open neighbourhood N of p, $p \in N \subseteq U$, such that we have on N: $X_i = \frac{\partial}{\partial x_i}$, $\omega_j = dx_j$, $\omega_j(X_i) = \delta_{ij}$. Then the Riemannian structure g can be written: $g = \sum_{j,k}^{n} g_{jk} dx_j\, dx_k$, where $(g_{jk})_{j,k} \in \mathcal{M}(n, F)$ is a patratic matrix with elements from F and $dx_j\, dx_k = \frac{1}{2}(dx_j \otimes dx_k + dx_k \otimes dx_j)$.

<u>4.1. The construction of a right invariant Riemannian structure on G</u>

<u>4.1.1. Lemma.</u> a) Given a tangent vector $u \in G_e$, there exists exactely one right invariant vector field $\overline{u}$ on G, such that $(\overline{u})_e = u$, and this u is analytic; b) the mapping $u \mapsto \overline{u}$ of G_e into $\mathcal{D}^1(G)$ is linear and injective.

<u>Proof</u>: a) For every $x \in G$, we define $(\overline{u})_x =: (d_e R_x)_{(u)} \in G_x$. It is obvious that $\overline{u} \in \mathcal{D}^1(G)$ is right invariant and analytic. For the uniqueness, let us consider $Y, Z \in \mathcal{D}^1(G)$, which are right invariant, such that $Y_e = Z_e$ and let us show that Y=Z. Indeed, for an arbitrary $x \in G$, by the right invariance of Y and Z, we have: $Y_x = (d_e R_x)_{(Y_e)} = (d_e R_x)_{(Z_e)} = Z_x$, hence Y=Z. b) It is known that G_e (i.e. L(G)) is the real vector space spanned by the n linearly independent vectors $e_i =: (\frac{\partial}{\partial x_i})_e : f \mapsto (\frac{\partial f}{\partial x_i})_e$ for $f \in C^{\infty}(G)$, $1 \le i \le n$. If we consider, by a), the right invariant vector fields $\overline{e}_1, \overline{e}_2, \ldots, \overline{e}_n$, they are linearly independent, i.e. for every $x \in G$ we have: $\sum_{i=1}^{n} \lambda_i (\overline{e}_i)_x = 0,\ \lambda_i \in R \Rightarrow \lambda_i = 0\ (1 \le i \le n)$

Hence, we have proved that <u>there exist n right invariant vector fields on G, which are linear independent</u>, namely $\bar{e}_1, \bar{e}_2, \ldots, \bar{e}_n$, where e_i are given above.

<u>4.1.2.</u> We consider now the 1-forms $\{\omega_i\}_{1\le i\le n} \subseteq \mathcal{D}_1(G)$, such that $\omega_i(\bar{e}_j) = \delta_{ij}$ on G (obviously, they are right invariant 1-forms) and let $(c_{ij})_{i,j}$ be a symmetrical and positive definite matrix of real numbers. Then: $g =: \sum_{i,j=1}^{n} c_{ij}\ \omega_i\ \omega_j \in \mathcal{D}_2^0(G)$ is a Riemannian structure on G, which is right invariant. If we take $c_{ij} =: \delta_{ij}$, $1\le i.j\le n$, and if we write in a local coordinate system $(\omega_i)_x = \sum_{j=1}^{n} A_{ij}(x_1,\ldots,x_n)\, dx_j$ for every $1\le i\le n$, then the Riemannian structure g can be written (locally):

(40) $g_x = \sum_{i=1}^{n} (\omega_i)_x\ (\omega_i)_x = \sum_{j,k=1}^{n} g_{jk}(x_1,\ldots,x_n)\, dx_j dx_k$, where:

$g_{jk}(x_1,\ldots,x_n) =: \sum_{i=1}^{n} A_{ij}(x_1,\ldots,x_n)\ A_{ik}(x_1,\ldots,x_n)$, for $1\le j,k\le n$. For every $x\in G$ the matrix $(g_{ij}(x))_{i,j}$ is a nondegenerate, symmetrical and positive definite patratic matrix of real numbers; let us denote by $(g^{ij}(x))_{i,j}$ its inverse and let $\delta(x) =: \det(g_{ij}(x))_{i,j} \in C^{\infty}(G)$; then we can define: $dv =: \sqrt{\delta}\, dx_1 \wedge dx_2 \wedge \ldots \wedge dx_n \in \mathcal{D}_n(G)$, which is called <u>the volume element</u> of G ($dv = \omega_1 \wedge \omega_2 \wedge \quad \wedge \omega_n$); it is a right invariant n-form and it does not depend on the choise of the local coordinate system.

<u>4.2. The Laplacian Operator</u>

Throughout this chapter $g \in \mathcal{D}_2^0(G)$ will be the right invariant Riemann structure on G, constructed in the previous section and written locally by (40). For $f\in C^2(G)$ we can define locally $\Delta(f) =: \frac{1}{\sqrt{\delta}} \sum_{i=1}^{n} \frac{\partial}{\partial x_i}(\sqrt{\delta} \sum_{j=1}^{n} g^{ij} \frac{\partial f}{\partial x_j})$; more generaly, we can define, by previous formula, $\Delta(f)$ for $f\in L^1_{loc}(dv)$, in the sense of distributions. One can easily verify that:

(41) $\Delta(f_a) = (\Delta f)_a$ for every $f\in L^1_{loc}(dv)$ and for every $a\in G$.

<u>4.3. The cone of superharmonic functions on G</u>

<u>4.3.1. Definition.</u> If L is the operator $Lu =: \Delta u - u$, then an element of the set $\mathcal{S} = \mathcal{S}(G) =: \{ s\in L^1_{loc}(dv) \mid s \text{ is lower semi-continuous on } G, s > -\infty, Lu \le 0\}$, will be called <u>a superharmonic function on G</u>; a function $t \in -\mathcal{S}(G)$ will be called <u>a subharmonic function on G</u> and a function $h \in -\mathcal{S}(G) \cap \mathcal{S}(G)$ will be called <u>a harmonic function on G</u>. Analoguously, we can define these

notions for every open subset U of G (which is a submanifold of G); a function $p \in \mathcal{S}^+(U)$ will be called <u>a potential on U</u> if for every $h \in \mathcal{H}(U)$ such that $h \leq p$, we have $h \leq 0$. The set of all potentials on U will be denoted by $\mathcal{P}(U)$. If $p \in \mathcal{S}(G)$ and if U is the greatest open subset of G such that $p_{|U} \in \mathcal{H}(U)$, then the set G-U is called the harmonic carrier of p and is denoted by carr(p). We say that the axiom of proportionality is fulfilled, if: (P) $p_1, p_2 \in \mathcal{P}(G)$, $\mathrm{carr}(p_1) = \mathrm{carr}(p_2) = \{x\} \Rightarrow p_1 = \alpha p_2$, where $\alpha \in R_{>0}$. In [17], ch.I, th.2, it is proved that G endowed with the sheaf of harmonic functions is a harmonic Brelot space, which satisfies axiom (P). $\mathcal{S}$ is a standard H-cone of functions on G, which is, by (41), right invariant, i.e.:

(42) $s \in \mathcal{S}(G)$, $a \in G \Longrightarrow s_a \in \mathcal{S}(G)$.

<u>4.3.2.Remark</u>. If $p \in \mathcal{P}(G)$, G with countable base, there exist $\{p_n\}_{n\in N} \subseteq \mathcal{P}(G) \cap C_b^2(G)$ such that $\mathrm{carr}(p_n)$ is compact for every $n \in N$ and such that $\{p_n\}_{n\in N} \uparrow p$ (see [18], ch.2, and [5], th.2.3.1).

<u>4.3.3</u>. If $p \in \mathcal{P}(G)$, $s \in \mathcal{S}(G)$ and if $p+s \geq 0$ on G, then $s \geq 0$ (see [5], prop.2.2.1).

<u>4.4. The Green Function on G</u>

<u>4.4.1</u>. If L is the operator introduced at 4.3.1, then, by [16], there exists a unique Green function $G(.,.)$ on GxG, for the equation $Lu=0$, i.e. a function with the following properties: a) $G(x,y) = G(y,x)$ for every $x, y \in G$;

b) $G(.,.) \in C^\infty(GxG - \Delta_G)$, where Δ_G is the diagonal of GxG;

c) if $G_y =: G(.,y)$, then $G_y \in \mathcal{P}(G)$, with $\mathrm{carr}(G_y) = \{y\}$, i.e. for every $y \in G$ we have:

(43) $L\, G(.,y) = -\varepsilon_y$ (in the sense of distributions).

In [15], it is given the following local expresion of the Green :

(44) $G_y(x) = \frac{1}{(n-2)s_n} r^{2-n}(x,y)\,\alpha(x,y) + o(r^{2-n})$, where α is continuous, $\alpha(z,z)=1$, s_n is the area of the unit sphere of the n-dimensional Euclidian space and $r(x,y)$ represents the distance between the points x and y, taken with respect to our right invariant metrique g on G.

<u>4.4.2.Definition</u>. For every $\mu \in M^+(G)$, we put $P^\mu(x) =: \int G(x,y)\, d\mu(y)$, for every $x \in G$. In [18], ch.4, § 3, it is shown that for every $\mu \in M^+(G)$ such that $P^\mu \in L^1_{loc}(dv)$ we have $P^\mu \in \mathcal{P}(G)$. Obviously, $G_y = G(.,y) = P^{\varepsilon_y} \in \mathcal{P}(G)$.

<u>4.4.3.Lemma</u>. For every $a, x, y \in G$, we have $G(xa, ya) = G(x,y)$.

Proof: Using (42), the function $p: x \longmapsto G(xa,ya) = (G_{ya})_a(x)$ is an element of $\mathcal{P}(G)$ and $carr(p)=\{y\}$, hence, by (P), there exists $\lambda \in R_{>0}$, such that: $G(xa,ya)=\lambda G(x,y)$ for every $x,y \in G$; using (44) in a neighbourhood of a point $x_0 \in G$, we obtain $\lambda = 1$.

4.5. The construction of a potential kernel on the Lie group G.

4.5.1. Definition. For every $f \in \mathcal{F}^+$, let us consider the function $N_0(f) \in \mathcal{F}^+$, given by $N_0(f)_{(x)} =: \int G(x,y)\, f(y)\, dv(y) = P^{f\cdot dv}(x)$, for $x \in G$. The mapping $N_0 : \mathcal{F}^+ \longrightarrow \mathcal{F}^+$ is obviously linear, hence it is a positive kernel on G. Using 4.4.3. and the fact that dv is a right invariant measure on G, we have:

(45) $N_0(f_a) = (N_0 f)_a$ for every $f \in \mathcal{F}^+$ and $a \in G$.

4.5.2. Remark. For every $f \in \mathcal{F}^+$ we can write $N_0(f) = h * f$, where $h =: G_e \in L^1_{loc}(dv)$, hence, by the remark before ch. I, we have: $f \in \mathcal{A} \Longrightarrow N_0(f) \in C(G)$. ($\mathcal{A}$ was introduced in ch. II).

4.5.3. Lemma. The kernel N_0 is bounded; in particular, we have $G_e \in L^1(dv)$.

Proof: Let us take $f \in \mathcal{F}^+$, such that $0 \le f \le 1$ and let us prove that $0 \le N_0(f) \le 1$; we can assume that f has compact support; by (43) we have $L\, N_0(f) = -f$, hence $L(1-N_0(f)) = -1+f \leqslant 0$; thus $1-N_0(f) \in \mathcal{S}(G)$, $N_0(f) \in \mathcal{P}(G)$ and by 4.3.3. we obtain $1-N_0(f) \geqslant 0$.

4.5.4. By [15], prop. 5, N_0 satisfies the complete domination principle, hence, by Hunt's theorem ([10], page 127), there exists a unique sub-Markovian resolvent $\mathcal{N} = \{N_\alpha\}_{\alpha>0}$ on G, such that N_0 is its initial kernel and such that $\mathcal{N}$ is continuous, i.e. for every bounded $f \in \mathcal{F}^+$ we have $\lim_{\alpha\to\infty} \alpha N_\alpha(f) = f$; by Hille-Yoshida theorem, there exists a sub-Markovian semigroup $\{P_t\}_{t>0}$ of positive kernels on G, which is continuous (i.e. $\lim_{t\to 0} P_t(f) = f$ for every bounded $f \in \mathcal{F}^+$) and which satisfies:

(46) $N_\alpha(f) = \int_0^\infty e^{-\alpha t} P_t(f)\, dt$, for every $f \in \mathcal{F}^+$ and $\alpha > 0$.

For every $f \in \mathcal{F}^+$, $\alpha > 0$ and $t > 0$, we also have: $N_\alpha(f_a) = (N_\alpha f)_a$, $P_t(f_a) = (P_t f)_a$, hence, by 1.1.3. it follows that there exist, uniquely, the positive measures k, $\{k_\alpha\}_{\alpha>0}$ and $\{\mu_t\}_{t>0}$ on G, such that $N_0(f) = k * f$, $N_\alpha(f) = k_\alpha * f$, $P_t(f) = \mu_t * f$ for every $f \in \mathcal{F}^+$, $\alpha > 0$, $t > 0$, where the convolution is taken with respect to the left invariant Haar measure λ_s on G. Using (46), we have $k_\alpha(f) = \int_0^\infty e^{-\alpha t} \mu_t(f)\, dt$, for every $f \in \mathcal{F}^+$ and $\alpha > 0$. One can easily verify that $(\mu_t)_{t>0}$ is a transient convolution semigroup on G, that $(k_\alpha)_{\alpha>0}$ is the resolvent of measures associated with $(\mu_t)_{t>0}$ and that $k(f)=\sup_{\alpha>0} k_\alpha(f)=\lim_{\alpha\to 0} k_\alpha(f)=\int_0^\infty \mu_t(f)\, dt$ for every $f\in K^+(G)$, hence k is a potential kernel on G, such that the associated positive kernel N_o is basic with respect to the measure $\mu:=dv$. In order to apply the theory developped in the first three chapters, the following condition also must be fulfilled: $(\check{k}*\mu)_{(f)}=\int f(x^{-1}y)\, dk(x)\, dv(y)=\int N_o(f)\, dv<+\infty$, for every $f\in K^+(G)$. Indeed, it is satisfied, because N_o satisfies the domination principle and because $G_e\in L^1(dv)$.

__4.5.5.Remark.__ If $\mathcal{E}(\mathcal{N})$ is the H-cone of functions associated with k by (16), then we have $\mathcal{E}(\mathcal{N})=\mathcal{S}(G)$. Indeed, the inclusion „$\subseteq$" is obvious by 2.1.8. For the other inclusion, by the aproximation of superharmonic functions with sequences of potentials (G_e is a strictely positive potential on G), it is sufficient to take $P^\mu\in\mathcal{P}(G)$ and to prove that $P^\mu\in\mathcal{E}(\mathcal{N})$; let us consider, by 4.3.2, a sequence $\{p_n\}_{n\in N}\subseteq\mathcal{P}(G)\cap C_b^2(G)$, such that: $\{p_n\}_{n\in N}\uparrow P^\mu$ and let us denote $f_n=:L(-p_n)\in C_b^+(G)$ for every $n\in N$. Because p_n and $N_o(p_n)$ are potentials such that $L(N_o(f_n)-p_n)=0$, we have $p_n=N_o(f_n)\in\mathcal{E}(\mathcal{N})$, for every n N, hence $P^\mu\in\mathcal{E}(\mathcal{N})$.

BIBLIOGRAPHY

[1] C.Berg, G.Forst: Potential Theory on Locally Compact Abelian Groups, Springer Verlag, 1975.

[2] E.Herwitt, A.Ross: Abstract Harmonic Analysis, vol.I, Springer Verlag, 1963.

[3] N.Dinculeanu: Integrarea pe spaţii local compacte, Bucureşti 1965.

[4] N.Boboc, Gh.Bucur, A.Cornea: H-cones and Potential Theory, Ann.Instit. Fourier de l'Université de Grenoble, t.XXV, fasc.2 et 3, 1975.

[5] C.Constantinescu, A.Cornea: Potential Theory on Harmonic Spaces, Springer Verlag 1972.

[6] N.Bourbaki: Elements de mathématique, Intégration, ch.7, Hermann, Paris.

[7] E.M.Alfsen: Compact Convex Sets and Boundary Integrals, Springer Verlag, 1971.

[8] N.Bourbaki:Topologie Générale,ch.1,2,Hermann,Paris,1965.

[9] N.Boboc,Gh.Bucur,A.Cornea:Carrier Theory and Negligible Sets on a Standard H-cone of Functions,Revue Roumaine de math.pures et appl.,t.XXV n°2,1980.

[10] N.Boboc,Gh,Bucur,A,Cornea:Order and Convexity in Potential Theory: H-Cones,Lecture Notes in Maths.853.1981.

[11] N.Boboc,Gh.Sireţki:Sur la compactification d'un espace topologique, Bull.Math.de la Soc.Sci.Math.Phys.de la R.P.R.,t.5(53),n°3-4,1961.

[12] N.Boboc,A.Cornea:Cônes convexes ordonnés.H-cônes et adjoints de H-cones,C.R.Acad.Sc.Paris,t.270,p.596-599 (2 mars 1970).

[13] N.Boboc,Gh.Bucur,A.Cornea:Natural Topologies on H-Cones.Weak Completeness,Rev.Roum.de Math.Pures et Appl.,t.XXIV,n°7,1979.

[14] S.Helgason :Differential Geometry,Lie Groups and Symmetric Spaces, Academic Press,1978.

[15] N.Boboc:Certains principles dans la theorie du potentiel sur variétés riemanniennes,Revue Roum.de math.pures et appl.,t.VI,3,494-500,1961.

[16] N.Boboc,N.Radu:Sur l'existence de la fonction de Green pour les équations du type elliptique,definies sur des variétés différentiable, C.R.Acad.Sci.Paris 246(1958),p.3204-3207.

[17] N.Boboc,P.Mustaţă:Espaces harmoniques associés aux opérateurs différentiels linéares du second ordre de type elliptique,Lecture Notes in Math.68,1968.

[18] M.Brelot:Eléments de la théorie classique du potentiel,4^{e}edition, 1969.

PERTURBATIONS IN EXCESSIVE STRUCTURES

by

N. Boboc and Gh. Bucur

Introduction

The study of so called polysuperharmonic functions was one of the topics of classical potential theory. The beginning of the researches in this field may be found in the papers of M.Nicolescu [6].

In his thesis E.Smyrnelis [7] developed an axiomatic approach concerning the concept of the polysuperharmonic function. Other aspects of the theory of polysuperharmonic functions were considered by N.Boboc and P.Mustaţă [1] and by M.Ito [5] specially in connection with the concept of complete superharmonic function.

More recently N.Bouleau [3] showed how the bisuperharmonic functions considered by E.Smyrnelis may be regarded as excessive functions with respect to some resolvents or semigroups of kernels. Thus, the theory of bisuperharmonic functions becomes a special case of the theory of excessive functions with respect to an adequate semigroup or resolvent of kernels.

The aim of this paper is to develop a theory of perturbation of a resolvent or a semigroup of kernels which extends the well known perturbation of a resolvent or a semigroup of operators in a Banach space [4]. If $\mathcal{V}=(V_\lambda)_{\lambda>0}$ is a general resolvent of kernels on a measurable space E and B is an arbitrary kernel on E we define the B-perturbation of $\mathcal{V}$ as being the resolvent $\mathcal{W}=(W_\lambda)_{\lambda>0}$ defined by

$$W_\lambda = V_\lambda \left(\sum_{n=0}^{\infty} (BV_\lambda)^n \right)$$

We characterise the excessive functions with respect to $\mathcal{W}$. For instance we show that a finite positive measurable function s on E is $\mathcal{W}$-excessive iff it is $\mathcal{V}$-excessive and we have, for any $\alpha > 0$

$$\alpha V_\alpha s + V_\alpha B s \leq s$$

A similar treatment is given for the perturbation of a semigroup of kernels on a measurable space.

In the last part of the paper a special case of perturbation is considered. Namely the space E is a direct sum of a finite system $(E_i)_{1 \leq i \leq n}$ of measurable spaces, the resolvent is a direct sum of resolvents on E_i , $i \in \{1,2,\ldots,n\}$ and the kernel B, considered as a matrix of kernels, is of the form $B=(B_{ij})$ where

$$B_{ij} = 0 \quad \text{if} \quad i \leq j$$

This case extends the triangular resolvents considered by N.Bouleau [3]. We give thus some supplementary results concerning the characterisation of excessive functions with respect to a triangular resolvents.

1. Perturbations of a resolvent of kernels

We consider on the measurable space $(E, \mathcal{B})$ the set $\mathcal{F} = \mathcal{F}(\mathcal{B})$ of all positive numerical $\mathcal{B}$-measurable functions on E. Obviously $\mathcal{F}$ is a convex cone with respect to the pointwise algebraic operations.

A kernel V on $(E,\mathcal{B})$ is a map from $\mathcal{F}$ into $\mathcal{F}$ which is countable additive and $V(0)=0$. We say that a property on E holds V almost everywhere (V-a.e.) if the set A of all points $x \in E$ for which the property does not hold is contained in a subset $B \in \mathcal{B}$ such that $V(1_B)=0$. The kernel V is proper if there exists a strictly positive element $f \in \mathcal{F}$ such that V(f) is bounded. The kernel V is bounded if the function V(1) is bounded.

A family $\mathcal{V}=(V_\alpha)_{\alpha>0}$ of kernels on $(E,\mathcal{B})$ such that for any $\alpha, \beta > 0$, $\alpha < \beta$ we have

$$V_\alpha V_\beta = V_\beta V_\alpha$$

$$V_\alpha = V_\beta + (\beta-\alpha) V_\alpha V_\beta$$

is called a resolvent of kernels on $(E,\mathcal{B})$.

An element $s \in \mathcal{F}$ is called $\mathcal{V}$-supermedian if for any $\alpha > 0$ we have

$$\alpha V_\alpha s \leq s$$

The set of all supermedian functions with respect to the resolvent $\mathcal{V}$ is denoted by $\mathcal{S}_{\mathcal{V}}$.

An element $s \in \mathcal{F}$ is called $\mathcal{V}$-excessive is s is finite V_{α_0}-a.e, $s \in \mathcal{S}_{\mathcal{V}}$ and

$$\lim_{\alpha \to \infty} \alpha V_\alpha s = s$$

The set of all $\mathcal{V}$-excessive functions with respect to the resolvent $\mathcal{V}$ is denoted by $\mathcal{E}_{\mathcal{V}}$.

It is known that if $\mathcal{V} = (V_\alpha)_{\alpha>0}$ is a resolvent family of kernels on $(E,\mathcal{B})$ then the map $V_0 : \mathcal{F} \to \mathcal{F}$ defined by

$$V_0(f) = \sup_\alpha V_\alpha(f) = \lim_{\alpha \to 0} V_\alpha(f)$$

is also a kernel on $(E, \mathcal{B})$ called the <u>initial kernel</u> of the rezolvent $\mathcal{V}$.

If $\mathcal{V} = (V_\alpha)_{\alpha>0}$ is a resolvent family of kerneles on $(E,\mathcal{B})$ then for any $\alpha_0 > 0$ the family of kernels

$$\mathcal{V}_{\alpha_0} := (V_{\alpha_0+\alpha})_{\alpha>0}$$

is also a resolvent. If V_{α_0} is a proper kernel then the initial kernel of the resolvent $\mathcal{V}_{\alpha_0}$ is just the kernel V_{α_0}. An element of $\mathcal{S}_{\mathcal{V}_{\alpha_0}}$ (resp $\mathcal{E}_{\mathcal{V}_{\alpha_0}}$) is called α_0-supermedian (resp. α_0-excessive) with respect to the resolvent $\mathcal{V} = (V_\alpha)_{\alpha>0}$ It is easy to see that

$$\mathcal{S}_{\mathcal{V}} = \bigcap_{\alpha_0>0} \mathcal{S}_{\mathcal{V}_{\alpha_0}}, \quad \mathcal{E}_{\mathcal{V}} = \bigcap_{\alpha_0>0} \mathcal{E}_{\mathcal{V}_{\alpha_0}}$$

If $s,t \in \mathcal{F}$ we shall use the notation

$$s \preceq_{\mathcal{S}_{\mathcal{V}}} t \quad (\text{resp. } s \preceq_{\mathcal{E}_{\mathcal{V}}} t)$$

to express that there exists $u \in \mathcal{S}_{\mathcal{V}}$ (resp. $u \in \mathcal{E}_{\mathcal{V}}$) such that s+u=t.

<u>Theorem 1.1</u>. Let $\mathcal{V} = (V_\alpha)_{\alpha>0}$ be a resolvent of kernels on $(E,\mathcal{B})$ and let B be a kernel on $(E,\mathcal{B})$. Then the family of kernels on $(E,\mathcal{B})$, $\mathcal{W} = (W_\alpha)_{\alpha>0}$ where for any α

$$W_\alpha := V_\alpha \left(\sum_{n=1}^{\infty} (BV_\alpha)^n \right)$$

is a resolvent on $(E,\mathcal{B})$. If V(resp. W) denotes the initial kernel of the resolvent $\mathcal{V}$ (resp. $\mathcal{W}$) then we have

$$\mathcal{W}_\alpha = V_\alpha (I + B\mathcal{W}_\alpha) \qquad (\forall)\ \alpha > 0$$
$$\mathcal{W} = V(I + B\mathcal{W})$$

Proof. Let $\alpha, \beta > 0$ be such that $\beta = \alpha + \gamma$, $\gamma > 0$
Obviously we have

$$V_\alpha = V_{\alpha+\gamma} + \gamma V_{\alpha+\gamma} V_\alpha$$

and we want to show that

$$\mathcal{W}_\alpha = \mathcal{W}_{\alpha+\gamma} + \gamma \mathcal{W}_{\alpha+\gamma} \mathcal{W}_\alpha$$

From the definition of $\mathcal{W}_\alpha$ it will be sufficient to check that for any $n \in N$ we have

$$V_\alpha (BV_\alpha)^n = V_{\alpha+\gamma} (BV_{\alpha+\gamma})^n + \gamma V_{\alpha+\gamma} \Big(\sum_{i+j=n} (BV_{\alpha+\gamma})^i V_\alpha (BV_\alpha)^j \Big)$$

We shall verify inductively this formula. For n=0 the preceding equality follows from the fact that $\mathcal{V} = (V_\alpha)_{\alpha>0}$ is a resolvent of kernel on $(E, \mathcal{B})$. Suppose now that the relation holds for n. Hence we have

$$V_\alpha (BV_\alpha)^{n+1} = V_{\alpha+\gamma} (BV_{\alpha+\gamma})^n (BV_\alpha) + \gamma V_{\alpha+\gamma} \sum_{i+j=n} (BV_{\alpha+\gamma})^i V_\alpha (BV_\alpha)^{j+1}$$

Using again the resolvent formula for $(V_\alpha)_{\alpha>0}$ we get

$$V_\alpha (BV_\alpha)^{n+1} = V_{\alpha+\gamma} (BV_{\alpha+\gamma})^n BV_{\alpha+\gamma} + \gamma V_{\alpha+\gamma} (BV_{\alpha+\gamma})^{n+1} V_\alpha +$$
$$+ \gamma V_{\alpha+\gamma} \sum_{i+j=n} (BV_{\alpha+\gamma})^i V_\alpha (BV_\alpha)^{j+1} = V_{\alpha+\gamma} (BV_{\alpha+\gamma})^{n+1} +$$
$$+ \gamma V_{\alpha+\gamma} \sum_{i+j=n+1} (BV_{\alpha+\gamma})^i V_\alpha (BV_\alpha)^j$$

Now we want to prove that

$$W_\alpha W_{\alpha+\gamma} = W_{\alpha+\gamma} W_\alpha .$$

For this it will be sufficient to show inductively that for any $n \in N$ we have

$$V_\alpha \sum_{p+q=n}' (BV_\alpha)^p V_{\alpha+\gamma} (BV_{\alpha+\gamma})^q =$$

$$= V_{\alpha+\gamma} \sum_{p+q=n} (BV_{\alpha+\gamma})^p V_\alpha (BV_\alpha)^q$$

This equality is true for n=0 since

$$V_\alpha V_{\alpha+\gamma} = V_{\alpha+\gamma} V_\alpha$$

Suppose that it is true for n. Multiplying this equality on the left by $V_\alpha B$ and then adding $V_\alpha V_{\alpha+\gamma} (BV_{\alpha+\gamma})^{n+1}$ we get

$$V_\alpha \sum_{p+q=n+1} (BV_\alpha)^p V_{\alpha+\gamma} (BV_{\alpha+\gamma})^q =$$

$$= V_\alpha \sum_{p+q=n} (BV_{\alpha+\gamma})^{p+1} V_\alpha (BV_\alpha)^q + V_\alpha V_{\alpha+\gamma} (BV_{\alpha+\gamma})^{n+1} .$$

Using resolvent formula for $\mathcal{V} = (V_\alpha)_{\alpha>0}$ and the relation

$$V_{\alpha+\gamma} (BV_{\alpha+\gamma})^n + \gamma V_{\alpha+\gamma} \sum_{p+q=n} (BV_{\alpha+\gamma})^p V_\alpha (BV_\alpha)^q = V_\alpha (BV_\alpha)^n$$

proved in the first part of the proof then the second part of the previous equality becomes

$$V_{\alpha+\gamma} \sum_{p+q=n} (BV_{\alpha+\gamma})^{p+1} V_\alpha (BV_\alpha)^q + V_\alpha V_{\alpha+\gamma} (BV_{\alpha+\gamma})^{n+1} +$$

$$+ \gamma V_\alpha V_{\alpha+\gamma} \sum_{p+q=n} (BV_{\alpha+\gamma})^{p+1} V_\alpha (BV_\alpha)^q =$$

$$= V_{\alpha+\gamma} \sum_{p+q=n}' (BV_{\alpha+\gamma})^{p+1} V_\alpha (BV_\alpha)^q +$$

$$+V_\alpha V_{\alpha+\gamma} B\left[V_{\alpha+\gamma}(BV_{\alpha+\gamma})^n + \gamma V_{\alpha+\gamma} \sum_{p+q=n} (BV_{\alpha+\gamma})^p V_\alpha (BV_\alpha)^q\right] =$$

$$= V_{\alpha+\gamma} \sum_{p+q=n} (BV_{\alpha+\gamma})^{p+1} V_\alpha (BV_\alpha)^q + V_\alpha V_{\alpha+\gamma} BV_\alpha (BV_\alpha)^n =$$

$$= V_{\alpha+\gamma} \sum_{p+q=n+1}' (BV_{\alpha+\gamma})^p V_\alpha (BV_\alpha)^q$$

and thus the proof is complete.

<u>Definition</u>. The resolvent $\mathcal{W} = (W_\alpha)_{\alpha>0}$ obtained from the resolvent $\mathcal{V} = (V_\alpha)_{\alpha>0}$ by the relations

$$W_\alpha = V_\alpha \sum_{n=0}^{\infty} (BV_\alpha)^n = \sum_{n=0}^{\infty} (V_\alpha B)^n V_\alpha$$

where B is a kernel on $(E, \mathcal{B})$ is called the B-perturbation of the resolvent $\mathcal{V}$ (or the perturbation of $\mathcal{V}$ with respect to B) and it will be denoted by $\mathcal{V}(B)$.

<u>Remarks</u>. a) From the relations

$$W_\alpha = V_\alpha \left(\sum_{n=0}^{\infty} (BV_\alpha)^n\right) \quad (\forall)\ \alpha > 0$$

we get

$$V_\alpha \le W_\alpha \qquad (\forall)\ \alpha > 0$$

Also if $A \in \mathcal{B}$ then we have

$$V_\alpha(1_A) = 0 \iff W_\alpha(1_A) = 0$$

b) Suppose that $\mathcal{V} = (V_\alpha)_{\alpha>0}$ is a resolvent of kernels such that V_α is proper for any $\alpha > 0$. Then for any $\alpha_0 > 0$ we have

$$V_\alpha = V_{\alpha_0+\alpha}\left(\sum_{n=0}^{\infty}(\alpha_0 V_\alpha)^n\right) \quad (\forall)\ \alpha>0$$

i.e. the resolvent $\mathcal{V}$ is the perturbation of the resolvent $\mathcal{V}_{\alpha_0} = (V_{\alpha_0+\alpha})_{\alpha>0}$ with respect to the kernel $B = \alpha_0 I$.

Theorem 1.2. Let $\mathcal{V}$ be a resolvent of kernels on $(E,\mathcal{B})$ and let B_1, B_2 be two kernels on $(E,\mathcal{B})$. Then we have

$$[\mathcal{V}(B_1)](B_2) = \mathcal{V}(B_1+B_2).$$

Proof. Obviously it is sufficient to show that if U is a kernel on $(E,\mathcal{B})$ then we have

$$U\sum_{n=0}^{\infty}[(B_1+B_2)U]^n = UT\sum_{n=0}^{\infty}(B_2UT)^n$$

where

$$T = \sum_{n=0}^{\infty}(B_1U)^n$$

Firstly we remark that we have, inductively,

$$[(B_1+B_2)U]^n = \sum_{j_0+j_1+\cdots j_k+k=n}(B_1U)^{j_0}B_2U(B_1U)^{j_1}B_2U\cdots B_2U(B_1U)^{j_k}$$

for any $n \in N$.

The assertion follows now from the relation

$$T\sum(B_2UT)^n = \sum_{k=0}^{\infty}\sum_{j_0,j_1,\ldots j_k=0}^{\infty}(B_1U)^{j_0}B_2U(B_1U)^{j_2}B_2U\cdots B_2U(B_1U)^{j_k}$$

<u>Proposition 1.3</u>. Let $\mathcal{W}=(W_\alpha)_{\alpha>0}$ be a resolvent of kernels on $(E,\mathcal{B})$ which is the B-perturbation of the resolvent $\mathcal{V}=(V_\alpha)_{\alpha>0}$ and let s be an element of $\mathcal{F}$. We consider the following assertions:

a) $$V_0 Bs \precsim_{\mathcal{S}_\mathcal{V}} s$$

b) $$\alpha V_\alpha s + V_\alpha Bs \leq s \quad (\forall)\ \alpha>0$$

c) $$s \in \mathcal{S}_\mathcal{W}$$

Then we have a) $\Rightarrow$ b) $\Rightarrow$ c)

<u>Proof</u>. a) $\Rightarrow$ b). Let $s' \in \mathcal{S}_\mathcal{V}$ be such that

$$V_0 Bs + s' = s$$

We have, for any $\alpha>0$.

$$\alpha V_\alpha s' \leq s'$$

Adding in the both parts of this inequality the positive function

$$\alpha V_\alpha (V_0 Bs) + V_\alpha Bs$$

we obtain the required relation i.e.

$$\alpha V_\alpha s + V_\alpha Bs \leq s \quad (\forall)\ \alpha>0$$

b) $\Rightarrow$ c). It is sufficient to show inductively that

$$\alpha V_\alpha \sum_{i=0}^{n} (BV_\alpha)^i s \leq s$$

For n=0 this relation follows from

$$\alpha V_\alpha s \leq \alpha V_\alpha s + V_\alpha Bs \leq s$$

Suppose that the required relation holds for n. We have

$$\alpha V_\alpha \sum_{i=0}^{n} (BV_\alpha)^i s \le s ,$$

$$\alpha V_\alpha \sum_{i=1}^{n+1} (BV_\alpha)^i s = V_\alpha B\left(\alpha V_\alpha \sum_{i=0}^{n} (BV_\alpha)^i s\right) \le V_\alpha B s$$

and therefore

$$\alpha V_\alpha \sum_{i=0}^{n+1} (BV_\alpha)^i s \le \alpha V_\alpha s + V_\alpha B s \le s .$$

Theorem 1.4. Let $\mathcal{W} = (W_\alpha)_{\alpha>0}$ be a resolvent of kernels on $(E, \mathcal{B})$ which is the B-perturbation of the resolvent $\mathcal{V} = (V_\alpha)_{\alpha>0}$ of kernels on $(E,\mathcal{B})$. If $s \in \mathcal{F}$ is finite then the following assertions are equivalent:

a) $s \in \mathcal{E}_{\mathcal{W}}$

b) $s \in \mathcal{E}_{\mathcal{V}}$ and $VBs \precsim_{\mathcal{E}_{\mathcal{V}}} s$

c) $s \in \mathcal{E}_{\mathcal{V}}$ and $\alpha V_\alpha s + V_\alpha B s \le s \quad (\forall)\ \alpha > 0.$

Proof. The relations b) $\Rightarrow$ c), c) $\Rightarrow$ a) follows immediately from the preceding proposition.

a) $\Rightarrow$ b). Let $s \in \mathcal{E}_{\mathcal{W}}$. Obviously, for any $\alpha > 0$, we have $s \in \mathcal{E}_{\mathcal{W}_\alpha}$. Since $\mathcal{E}_{\mathcal{V}} = \bigcap_{\alpha>0} \mathcal{E}_{\mathcal{V}_\alpha}$ it is sufficient to show that

$$s \in \mathcal{E}_{\mathcal{V}_\alpha} \quad \text{and} \quad V_\alpha B s \precsim_{\mathcal{E}_{\mathcal{V}_\alpha}} s .$$

Indeed, for any $n \in N$, we have

$$W_\alpha(n(s - nW_{\alpha+n} s)) = nW_{\alpha+n} s,$$

$$(nW_{\alpha+n} s)_n \uparrow s .$$

If we put

$$f_n = n(s - nW_{\alpha+n}s)$$

then from the inequalities

$$W_{\alpha+\beta} f_n \leq W_{\alpha+\beta} ns \leq \frac{n}{\alpha+\beta} s \qquad (\forall)\ \beta > 0$$

$$V_{\alpha+\beta} W_\alpha f_n \leq W_{\alpha+\beta} W_\alpha f_n \leq \frac{1}{\beta} W_\alpha f_n < +\infty$$

we deduce that $W_\alpha f_n$ is $\mathcal{W}_\alpha$ -excessive and that $W_\alpha f_n$ is also $\mathcal{V}_\alpha$ -excessive.

Since

$$u_{n,\alpha} := W_\alpha f_n - V_\alpha B W_\alpha f_n = V_\alpha f_n$$

it follows that

$$V_\alpha B W_\alpha f_n + u_{n,\alpha} = W_\alpha f_n$$

and therefore, $u_{n,\alpha}$ being $\mathcal{V}_\alpha$ -excessive,

$$V_\alpha B W_\alpha f_n \precsim_{\mathcal{E}_{\mathcal{V}_\alpha}} W_\alpha f_n$$

Since the sequences of $\mathcal{V}_\alpha$ -excessive functions $(V_\alpha B W_\alpha f_n)_n$, $(W_\alpha f_n)_n$ are increasing to $V_\alpha B s$ respectively s we get $s \in \mathcal{E}_{\mathcal{V}_\alpha}$, $V_\alpha B s \in \mathcal{E}_{\mathcal{V}_\alpha}$ and

$$V_\alpha B s \precsim_{\mathcal{E}_{\mathcal{V}_\alpha}} s$$

Corollary 1.5. Let $\mathcal{W} = (W_\alpha)_{\alpha>0}$ be a resolvent of kernels on $(E,\mathcal{B})$ which is a B-perturbation of the resolvent $\mathcal{V} = (V_\alpha)_{\alpha>0}$ of kernels on $(E,\mathcal{B})$. We suppose that for any $f \in \mathcal{F}$ we have

$$f=0 \ \mathcal{V}\text{-a.e} \Rightarrow Bf=0 \ \mathcal{V}\text{-a.e.}$$

Then for any $s \in \mathcal{F}$ which is finite the following assertions are equivalent:

a) $s \in \mathcal{S}_{\mathcal{W}}$

b) $VBs \precsim_{\mathcal{S}_{\mathcal{V}}} s$

c) $\alpha V_\alpha s + V_\alpha Bs \leq s \quad (\forall)\ \alpha > 0$

Proof. The relations b)$\Rightarrow$c)$\Rightarrow$a) follow from Proposition 1.3.

a)$\Rightarrow$b) Suppose $s \in \mathcal{S}_{\mathcal{W}}$ and let $\tilde{s}$ be the function

$$\tilde{s} := \sup_{\alpha>0} \alpha W_\alpha s$$

Obviously $\tilde{s} \in \mathcal{E}_{\mathcal{W}}$, $\tilde{s} = s$ $\mathcal{V}$-a.e. and

$$\tilde{s} \precsim_{\mathcal{S}_{\mathcal{W}}} s, \quad \tilde{s} \precsim_{\mathcal{S}_{\mathcal{V}}} s$$

Since $\tilde{s}$ is finite then, from theorem 1.4, we deduce

$$VBs = VB\tilde{s} \precsim_{\mathcal{E}_{\mathcal{V}}} \tilde{s} \precsim_{\mathcal{S}_{\mathcal{V}}} s$$

Corollary 1.6. Let $\mathcal{W} = (W_\alpha)_{\alpha>0}$ be a resolvent of kernels on $(E,\mathcal{B})$ which is a B-perturbation of the resolvent $\mathcal{V} = (V_\alpha)_{\alpha>0}$ of kernels on $(E,\mathcal{B})$. We suppose that there exists a finite element $s_0 \in \mathcal{S}_{\mathcal{W}}$ such that $s_0 > 0$ $\mathcal{W}$-a.e. Then for any $s \in \mathcal{F}$ the following assertions are equivalent:

a) $s \in \mathcal{E}_{\mathcal{W}}$

b) $s \in \mathcal{E}_{\mathcal{V}}, \ VBs \precsim_{\mathcal{E}_{\mathcal{V}}} s$

c) $s \in \mathcal{E}_{\mathcal{V}}, \ \alpha V_\alpha s + V_\alpha Bs \leq s \quad (\forall)\ \alpha > 0$

The assertion follows from Proposition 1.3 and from Theorem 1.4 using the fact than any $s \in \mathcal{E}_{\mathcal{U}}$ is the limit of an increasing sequence of finite $\mathcal{U}$ -excessive functions.

2. Perturbations of a semigroup of kernels

We remember that a family $P=(P_t)_{t \geq 0}$ of kernels on a measurable space $(E,\mathcal{B})$ is a semigroup (of kernels) if

a) $$P_t P_s = P_{t+s} \quad (\forall) \; s,t \geq 0$$

b) for any $f \in \mathcal{F}$ the function

$$(t,x) \longrightarrow P_t f(x)$$

is a measurable function on the product measurable space $\mathbb{R}_+ \times E$.

We say that a semigroup $P=(P_t)_{t \geq 0}$ is <u>relatively bounded</u> if for any $s \geq 0$ we have

$$\|P\|_s := \sup_{t \leq s,\; x \in E} P_t 1(x) < +\infty$$

If $(P_t)_{t \geq 0}$ is a relatively bounded semigroup on $(E,\mathcal{B})$ then we have

$$\|P\|_s = \sup_{t \leq s} \|P_t\|$$

where $\|P_t\|$ means the norm of the linear operator generated by P_t on the Banach space of real bounded measurable functions on $(E,\mathcal{B})$.

A semigroup $(P_t)_{t \geq 0}$ on $(E,\mathcal{B})$ is called <u>bounded</u> if there exists a positive number M such that

$$\|P_t\| \leq M \quad (\forall) \; t \geq 0$$

We denote by $\|P\|$ the number

$$\|P\| = \sup_{t \geq 0} \|P_t\|$$

Theorem 2.1. Let $P=(P_t)_{t \geq 0}$ be a relatively bounded semigroup of kernels on $(E, \mathcal{B})$ and let B be a bounded kernel on $(E, \mathcal{B})$. Then there exists a unique relatively bounded semigroup $Q=(Q_t)_{t \geq 0}$ of kernels on $(E, \mathcal{B})$ such that

$$Q_t = P_t + \int_0^t P_{t-u} B Q_u \, du \quad (\forall)\, t \geq 0$$

This semigroup satisfies also the relation

$$\int_0^t P_{t-u} B Q_u \, du = \int_0^t Q_{t-u} B P_u \, du$$

More precisely, for any $t \geq 0$, we have

$$Q_t = \sum_{n=0}^{\infty} P_t^{[n]}$$

where

$$P_t^{[0]} := P_t, \quad P_t^{[n+1]} := \int_0^t P_{t-u} B P_u^{[n]} du = \int_0^t P_{t-u}^{[n]} B P_u \, du$$

and

$$\|Q\|_t \leq \|P\|_t \, \mathrm{Exp}\,(t \, \|P\|_t \cdot \|B\|)$$

Moreover, for any bounded function $f \in \mathcal{F}$ for which the function

$$t \to P_t f(x)$$

is continuous on $[0, \infty)$ for any $x \in E$, the function

$$t \to Q_t f(x)$$

is continuous on $[0,\infty)$ for any $x \in E$.

Proof. To show that the family $(Q_t)_{t\geq 0}$ where

$$Q_t = \sum_{n\geq 0} P_t^{[n]}$$

is a semigroup of kernels it is sufficient to prove that, for any $n \geq 0$ we have

$$\sum_{k+l=n} P_t^{[k]} P_s^{[l]} = P_{t+s}^{[n]} \quad (\forall)\ t,s>0$$

This relation is true if n=0. Suppose that the above relation holds for n. We have

$$\sum_{k+l=n+1}' P_t^{[k]} P_s^{[l]} = P_t P_s^{n+1} + \sum_{\substack{(k-1)+l=n\\ k\geq 1}} P_t^{[k]} P_s^{[l]},$$

$$\sum_{\substack{(k-1)+l=n\\ k\geq 1}}' P_t^{[k]} P_s^{[l]} = \int_0^t \sum_{\substack{(k-1)+l=n\\ k\geq 1}}' P_u B P_{t-u}^{[k-1]} P_s^{[l]} du =$$

$$= \int_0^t P_u B \sum_{k+l=n}' P_{t-u}^{[k]} P_s^{[l]} du = \int_0^t P_u B P_{t+s-u}^{[n]} du$$

Since

$$P_t P_s^{[n+1]} = \int_0^s P_{t+u} B P_{s-u}^{[n]} du =$$

$$= \int_t^{t+s} P_u B P_{t+s-u}^{[n]} du$$

we get

$$\sum_{k+l=n+1} P_t^{[k]} P_s^{[l]} = \int_t^{t+s} P_u B P_{t+s-u}^{[n]} du + \int_0^t P_u B P_{t+s-u}^{[n]} du =$$

$$= \int_0^{t+s} P_u B P_{t+s-u}^{[n]} du = P_{t+s}^{[n+1]} .$$

Using the above definition of Q_t we deduce that we have

$$Q_t = P_t + \int_0^t P_u B Q_{t-u} du$$

Further we shall prove inductively that for any $n \geq 0$ we have

$$\int_0^t P_u B P_{t-u}^{[n]} du = \int_0^t P_{t-u}^{[n]} B P_u du$$

For n=0 this relation is trivial. Suppose that it holds for n.

We have

$$\int_0^t P_u B P_{t-u}^{[n+1]} du = \int_0^t \int_0^{t-u} P_u B P_v B P_{t-u-v}^{[n]} dv du,$$

$$\int_0^t P_{t-u}^{[n+1]} B P_u du = \int_0^t \int_0^{t-u} P_v B P_{t-u-v}^{[n]} B P_u dv du$$

Using the induction hypothesis we get

$$\int_0^t P_u B P_{t-u}^{[n+1]} du = \int_0^t \int_0^{t-u} P_u B P_{t-u-v}^{[n]} B P_v dv du$$

Further, using Fubini theorem we deduce, from the last equality, that

$$\int_0^t P_u B P_{t-u}^{[n+1]} du = \int_0^t \int_0^{t-v} P_u B P_{t-u-v}^{[n]} B P_v du dv =$$

$$= \int_0^t P_{t-u}^{[n+1]} B P_u du .$$

The fact that the semigroup $(Q_t)_{t\geq 0}$ is relatively bounded follows from the inequality

$$\sup_{0\leq s\leq t} \| P_s^{[n]} \| \leq (\|P\|_s)^{n+1} (\|B\|)^n \frac{t^n}{n!}$$

We show now the uniqueness of the relatively bounded semigroups $(Q_t)_{t\geq 0}$ for which

$$Q_t = P_t + \int_0^t P_u B Q_{t-u} du \quad (\forall)\ t>0$$

Let $(Q'_t)_{t\geq 0}$ be an other semigroup which satisfies the same relation and let $f\in\mathcal{F}$ be a bounded function. If we denote

$$\varphi(t) = \sup_{x\in E} | Q_t f(x) - Q'_t f(x) |$$

then φ is a bounded function on any compact subset of $[0,\infty)$ and we have

$$\varphi(u) \leq \|P\|_t \|B\| \int_0^u \varphi(s) ds \quad (\forall)\ u\leq t.$$

Grom this relation we deduce, using a standard procedure,

$$\varphi(u) \leq (\|P\|_t)^n (\|B\|)^n \frac{u^n}{n!} \alpha_t^n \quad (\forall)\ u\leq t$$

where

$$\alpha_t := \sup\{ \varphi(u) \mid u\leq t \}$$

and therefore $\varphi = 0,\ Q_t f = Q'_t f.$

Suppose now that f is a bounded function from $\mathcal{F}$ such that for any $x \in E$ the function

$$t \to P_t f(x)$$

is continuous on $[0,\infty)$.

Indeed for n=0 the fact follows from hypothesis. Suppose that the assertion holds for n. From the equality

$$P_t^{[n+1]} f(x) = \left(\int_0^t P_{t-u} B P_u^{[n]} f\, du\right)(x)$$

and using Lebesgue theorem we deduce that the function

$$t \to P_t^{[n+1]} f(x)$$

is continuous on $[0,\infty)$ for any $x \in E$.

The continuity of the function

$$t \to Q_t f(x)$$

on $[0,\infty)$ for any $x \in E$ follows now from the relations

$$Q_t f(x) = \sum_{n=0}^{\infty} P_t^{[n]} f(x)$$

$$P_s^{[n]} f \leq (\|P\|_t)^{n+1} \frac{(\|B\|)^n \|f\| t^n}{n!}, \ (\forall)\, s \in [0,t]$$

<u>Definition</u>. Let $P=(P_t)_{t\geq 0}$ be a relatively bounded semigroup on $(E,\mathcal{B})$ and let B be a bounded kernel on $(E,\mathcal{B})$. The relatively bounded semigroup of kernels $Q=(Q_t)_{t\geq 0}$ such that

$$Q_t = P_t + \int_0^t P_{t-u} B Q_u\, du \qquad (\forall)\ t \geq 0$$

is called the B-<u>perturbation of</u> $P=(P_t)_{t\geq 0}$ and it is denoted by P(B).

Remark. a) Using Theorem 2.1, it is easy to see that if $(Q_t)_{t\geq 0}$ is a B-perturbation of $(P_t)_{t\geq 0}$ then for any $\alpha>0$ the semigroup $Q^{(\alpha)}=(Q_t^{(\alpha)})_{t\geq 0}$ where

$$Q_t^{(\alpha)} = e^{-\alpha t} Q_t \qquad t\geq 0$$

is the B-perturbation of the semigroup $\mathcal{P}^{(\alpha)}=(P_t^{(\alpha)})_{t\geq 0}$ where

$$P_t^{(\alpha)} = e^{-\alpha t} P_t \qquad (\forall)\ t\geq 0$$

b) For any $\alpha>0$ a relatively bounded semigroup $(P_t)_{t\geq 0}$ is αI-perturbation of the semigroup $(P_t^{(\alpha)})_{t\geq 0}$ where $P_t^{(\alpha)}=e^{-\alpha t}P_t$ for any $t\geq 0$.

Theorem 2.2. Let $\mathcal{P}=(P_t)_{t\geq 0}$ be a relatively bounded semigroup of kernels and let B_1 , B_2 be two bounded kernels on $(E,\mathcal{B})$. Then we have

$$\mathcal{P}(B_1+B_2) = \mathcal{P}(B_1)\,(B_2)$$

Proof. We put for any $t\geq 0$

$$Q'_t = (\mathcal{P}(B_1))_t \quad , \quad Q_t = (\mathcal{P}(B_1)(B_2))_t$$

We have, for any $t\geq 0$,

$$Q'_t = P_t + \int_0^t P_u B_1 Q'_{t-u}\, du ,$$

$$Q_t = Q'_t + \int_0^t Q'_u B_2 Q_{t-u}\, du$$

Hence

$$Q_t = Q'_t + \int_0^t [P_u B_2 Q_{t-u} + \int_0^u P_v B_1 Q'_{u-v} B_2 Q_{t-u}\, dv]\, du =$$

$$= Q'_t + \int_0^t P_u B_2 Q_{t-u}\, du + \int_0^t\int_v^t P_v B_1 Q'_{u-v} B_2 Q_{t-u}\, du\, dv =$$

$$= Q'_t + \int_0^t P_u B_2 Q_{t-u} du + \int_0^t \int_0^{t-v} P_v B_1 Q'_w B_2 Q_{t-v-w} dw\, dv =$$

$$= P_t + \int_0^t P_u B_2 Q_{t-u} du + \int_0^t P_v B_1 \left(Q'_{t-v} + \int_0^{t-v} Q'_w B_2 Q_{t-v-w} dw\right) dv =$$

$$= P_t + \int_0^t P_u B_2 Q_{t-u} du + \int_0^t P_v B_1 Q_{t-v} dv = P_t + \int_0^t P_u (B_1 + B_2) Q_{t-u} du .$$

Using now Theorem 2.1 we get for any $t \geq 0$

$$Q_t = (P(B_1 + B_2))_t .$$

Let $P = (P_t)_{t \geq 0}$ be a relatively bounded semigroup of kerneles on $(E, \mathcal{B})$ and for any $\alpha > 0$ let V_α be the kernel on $(E, \mathcal{B})$ defined by

$$V_\alpha f(x) = \int_0^\infty e^{-\alpha t} P_t f(x)\, dt .$$

It is easy to verify that the family $\mathcal{V} = (V_\alpha)_{\alpha > 0}$ of kernels on $(E, \mathcal{B})$ is a resolvent. It is called the resolvent associated with P. It is well known that a function $s \in \mathcal{F}$ is an excessive function with respect to P iff it is excessive with respect to the associated resolvent $\mathcal{V}$.

Proposition 2.3. Let $P = (P_t)_{t \geq 0}$ be a relatively bounded semigroup of kernels on $(E, \mathcal{B})$ and let $\mathcal{V} = (V_\alpha)_{\alpha > 0}$ be its associated resolvent.

If B is a bounded kernel on $(E, \mathcal{B})$ and $Q = (Q_t)_{t \geq 0}$ (resp. $\mathcal{W} = (W_t)_{t > 0}$) is a semigroup (resp. a resolvent) of kernels on $(E, \mathcal{B})$ which is the B-perturbation of P (resp. $\mathcal{V}$) then $\mathcal{W}$ is the resolvent associated with Q.

Proof. Using Theorem 1.1 and Theorem 2.1 it is sufficient to show inductively that for any $n \in N$, for any $f \in \mathcal{F}$ and any $\alpha > 0$ we have

$$V_\alpha (BV_\alpha)^n f = \int_0^\infty e^{-\alpha t} P_t^{[n]} f \, dt$$

This relation is trivial for n=0. Suppose that it hods for n. Then we have

$$\int_0^\infty e^{-\alpha t} P_t^{[n+1]} f \, dt = \int_0^\infty e^{-\alpha t} \int_0^t P_u B P_{t-u}^{[n]} f \, du \, dt = \int_0^\infty \int_u^\infty e^{-\alpha t} P_u B P_{t-u}^{[n]} f \, dt \, du =$$

$$= \int_0^\infty \left(e^{-\alpha u} P_u B \int_u^\infty e^{-\alpha (t-u)} P_{t-u}^{[n]} f \, dt\right) du = \int_0^\infty e^{-\alpha u} P_u B \int_0^\infty e^{-\alpha t} P_t^{[n]} f \, dt \, du =$$

$$= \int_0^\infty e^{-\alpha u} P_u B V_\alpha (BV_\alpha)^n du = V_\alpha (BV_\alpha)^{n+1}$$

Proposition 2.4. Let $Q=(Q_t)_{t \geq 0}$ be the B-perturbation of a relatively bounded semigroup $P=(P_t)_{t \geq 0}$ of kernels on $(E, \mathcal{B})$ and let $v \in \mathcal{F}$ be such that

$$P_t v + \int_0^t P_\lambda B v \, d\lambda \leq v \quad (\forall)\, t \geq 0$$

Then v is Q-supermedian function. If moreover v is a P-excessive function then v is also a Q-excessive function.

Proof. We want to show that

$$Q_t v \leq v \quad (\forall)\, t \geq 0$$

From the definition of $Q = (Q_t)_{t \geq 0}$ (see Theorem 2.1) it is sufficient to prove that for any $n \in N$ we have

$$\sum_{k=0}^{n} P_t^{[k]} v \le v \quad (\forall)\ t \ge 0$$

For n=0 the last inequality follows immediately from hypothesis. Suppose that it is true for n, i.e.

$$\sum_{k=0}^{n} P_\lambda^{[k]} v \le v \quad (\forall)\ \lambda \ge 0$$

We get

$$\sum_{k=0}^{n} P_\lambda B P_{t-\lambda}^{[k]} v \le P_\lambda B v \quad (\forall)\ t \ge 0,\ \lambda \in [0;t]$$

and therefore, by integration, we deduce

$$\sum_{k=0}^{n} \int_0^t P_\lambda B P_{t-\lambda}^{[k]} d\lambda \le \int_0^t P_\lambda B v\, d\lambda ,$$

$$\sum_{k=1}^{n+1} P_t^{[k+1]} v \le \int_0^t P_\lambda B v\, d\lambda ,$$

$$\sum_{k=0}^{n+1} P_t^{[k]} v \le P_t v + \int_0^t P_\lambda B v\, d\lambda \le v$$

The last part of the assertion follows from the inequality

$$P_t v \le Q_t v \le v \quad (\forall)\ t \ge 0$$

<u>Theorem 2.5</u>. Let $Q=(Q_t)_{t\ge 0}$ be the B-perturbation of a relatively bounded semigroup $P=(P_t)_{t\ge 0}$. If $s \in \mathcal{F}$ is finite then the following assertion are equivalent:

a) s is excessive with respect to Q.

b) s is excessive with respect to P and

$$P_t s + \int_0^t P_u B s\, du \le s \quad (\forall)\, t \ge 0$$

Proof. The relation b) $\Rightarrow$ a) follows from Proposition 2.4.

a) $\Rightarrow$ b). We denote by P^{α} the semigroup of kernels on $(E, \mathcal{B})$ defined by

$$P_t^{\alpha} = e^{-\alpha t} P_t$$

and by Q^{α} the semigroup of kernels on $(E, \mathcal{B})$ defined by

$$Q_t^{\alpha} = e^{-\alpha t} Q_t$$

which is the B-perturbation of P^{α}.

It is sufficient to show that

$$P_t^{\alpha} s + \int_0^t P_u^{\alpha} B s\, du \le s \quad (\forall)\, t \ge 0$$

Since s is the limit of the increasing sequence $(\mathbb{W}_{\alpha} f_n)_n$ where

$$f_n = n(s - n \mathbb{W}_{\alpha + n} s)$$

and

$$\mathbb{W}_{\beta} = \int_0^{\infty} e^{-\beta t} Q_t\, dt = \int_0^{\infty} Q_t^{\beta}\, dt \quad (\forall)\, \beta > 0$$

the preceding inequality must be check for s of the form $\mathbb{W}_{\alpha} g$ where $g \in \mathcal{F}$ and $W_{\alpha} g < +\infty$

We have

$$\mathbb{W}_{\alpha} g = \int_0^t Q_u^{\alpha} g\, du + Q_t^{\alpha}(\mathbb{W}_{\alpha} g)$$

Using the formula

$$Q_t^\alpha = P_t^\alpha + \int_0^t P_t^\alpha B\, Q_{t-u}^\alpha \, du .$$

we get

$$Q_t^\alpha (W_\alpha g) = P_t^\alpha (W_\alpha g) + \int_0^t \Big(\int_0^\infty P_u^\alpha B\, Q_{t+v-u}^\alpha g \, dv \Big) du =$$

$$= P_t^\alpha (W_\alpha g) + \int_0^t \Big(\int_{t-u}^\infty P_u^\alpha B\, Q_v^\alpha \, dv \Big) du$$

From

$$\int_0^t P_u^\alpha B\, W_\alpha g \, du = \int_0^t \Big(\int_0^\infty P_u^\alpha B\, Q_v^\alpha g \, dv \Big) du$$

we deduce

$$Q_t^\alpha (W_\alpha g) + \int_0^t \Big(\int_0^{t-u} P_u^\alpha B\, Q_v^\alpha g \, dv \Big) du =$$

$$= P_t^\alpha (W_\alpha g) + \int_0^t P_u^\alpha B\, W_\alpha g \, du$$

The inequality

$$P_t^\alpha (W_\alpha g) + \int_0^t P_u^\alpha B\, W_\alpha g \, du \le W_\alpha g$$

follows now from the preceding considerations using the inequality

$$\int_0^t \Big(\int_0^{t-u} P_u^\alpha B\, Q_v^\alpha g \, dv \Big) du = \int_0^t \Big(\int_0^r P_u^\alpha B\, Q_{t-r}^\alpha g \, du \Big) dr \le$$

$$\le \int_0^t Q_r^\alpha g \, dr$$

3. Perturbations of a direct sum of resolvents or semigroups of kernels

We consider now a finite system $(E_i, \mathcal{B}_i)_{1 \le i \le n}$ of measurable spaces and for any $i \in \{1,2,\dots,n\}$ a resolvent $\mathcal{V}_i = (V_{\alpha,i})_{\alpha>0}$ of kernels on $(E_i, \mathcal{B}_i)$. We denote by $(E, \mathcal{B})$ the direct sum of the measurable spaces $(E_i, \mathcal{B}_i)_{1 \le i \le n}$. A function f on E is completely determined by its restriction to the components $E_1, E_2, \dots, E_n$. Thus a function f on E will be identified with a system $(f_1, f_2, \dots, f_n)$ of functions on $E_1, E_2, \dots, E_n$ respectively. In this way a kernel V on $(E, \mathcal{B})$ will be identified with a matrix of kernels of the form

$$V = (V_{ij})_{1 \le i,j \le n}$$

where V_{ij} is a kernel[1] from $(E_j, \mathcal{B}_j)$ into $(E_i, \mathcal{B}_i)$ and we have

$$Vf = \left(\sum_{i=1}^{n} V_{1i} f_i, \sum_{i=1}^{n} V_{2i} f_i, \dots, \sum_{i=1}^{n} V_{ni} f_i\right)$$

We denote by $\mathcal{V} = (V_\alpha)_{\alpha>0}$ the resolvent of kernels on $(E, \mathcal{B})$ where V_α is the following diagonal matrix.

$$V_\alpha = \begin{pmatrix} V_{\alpha,1}, & 0, & 0, & 0, \cdots & 0 \\ 0, & V_{\alpha,2}, & 0, & 0, \cdots & 0 \\ 0, & 0, & V_{\alpha,3}, & 0, \cdots & 0 \\ \vdots & & & & \\ 0, & 0, & 0, & \cdots, 0, & V_{\alpha,n} \end{pmatrix}$$

[1] A kernel from $(E_j, \mathcal{B}_j)$ into $(E_i, \mathcal{B}_i)$ is a map V_{ij} from $\mathcal{F}(E_i)$ into $\mathcal{F}(E_j)$ which is countable additive and $V_{ij}(0) = 0$.

The resolvent $\mathcal{V}$ is called the <u>direct sum</u> of $(\mathcal{V}_i)_{1\leq i\leq n}$

It is easy to see that a positive measurable function $s=(s_1,s_2,\ldots,s_n)$ on E is $\mathcal{V}$-supermedian (resp. $\mathcal{V}$-excessive) iff for any i=1,2,...,n the function s_i is $\mathcal{V}_i$- supermedian (resp. $\mathcal{V}_i$ -excessive) function.

In this section we try to describe the excessive functions on $(E,\mathcal{B})$ with respect to some special perturbations of $\mathcal{V}$.

<u>Definition</u>. A resolvent $\mathcal{W}=(W_\alpha)_{\alpha>0}$ on $(E,\mathcal{B})$ is called a <u>triangular resolvent associated with</u> $\mathcal{V}$ if for any $\alpha>0$ the kernel W_α has the following triangular form

$$W_\alpha=\begin{pmatrix} V_{\alpha,1}, & 0, & 0, & 0, & \ldots, & 0 \\ W_\alpha^{2,1}, & V_{\alpha,2}, & 0, & 0, & \ldots, & 0 \\ W_\alpha^{3,1}, & W_\alpha^{3,2}, & V_{\alpha,3}, & 0, & \ldots, & 0 \\ \vdots & & & & & \\ W_\alpha^{n,1}, & W_\alpha^{n,2}, & \ldots & & W_\alpha^{n,n-1}, & V_{\alpha,n} \end{pmatrix}$$

<u>Remark 3.1</u>. a) For n=2, such type of resolvents were considered by N.Bouleau [3] in connection with a generalization of the framework for the study of the biharmonic functions.

b) If $\mathcal{W}=(W_\alpha)_{\alpha\geq 0}$ is a triangular resolvent associated with $\mathcal{V}$ then we may write

$$W_\alpha=\begin{pmatrix} V_{\alpha,1}, & 0 \\ T_\alpha, & W'_\alpha \end{pmatrix}$$

where T_α is a kernel from E_1 into $E'=\bigoplus_{i=2}^{n} E_i$ of the form

$$T_\alpha = \begin{pmatrix} W_\alpha^{2,1} \\ W_\alpha^{3,1} \\ \vdots \\ W_\alpha^{n,1} \end{pmatrix}$$

and W'_α is a kernel on E′ given by

$$W'_\alpha = \begin{pmatrix} V_{\alpha,2}, & 0, & 0, & \cdots & 0 \\ W_\alpha^{2,2}, & V_{\alpha,3}, & 0, & \cdots & 0 \\ \vdots & & & & \\ W_\alpha^{n,2}, & W_\alpha^{n,3}, & \cdots & W_\alpha^{n,n-1}, & V_{\alpha,n} \end{pmatrix}$$

It is easy to see that $(W'_\alpha)_{\alpha>0}$ is a triangular resolvent on E′ associated with the direct sum of the resolvents $(\mathcal{V}_i)_{2\le i\le n}$.

Proposition 3.2. Let B a kernel on $(E, \mathcal{B})$ of the following form

$$B = \begin{pmatrix} 0, & 0, & 0, & \cdots & , 0 \\ B_{21}, & 0, & 0, & \cdots & , 0 \\ B_{31}, & B_{32}, & 0, & \cdots & , 0 \\ \vdots & & & & \\ B_{n1}, & B_{n2}, & B_{n3}, & \cdots B_{n,n-1}, & 0 \end{pmatrix}$$

Then the B-perturbation $\mathcal{W} = (W_\alpha)_{\alpha>0}$ of the resolvent $\mathcal{V}$ is a triangular resolvent associated with $\mathcal{V}$. Moreover we have

$$W_\alpha = V_\alpha \left(I + (BV_\alpha) + (BV_\alpha)^2 + \cdots + (BV_\alpha)^{n-1}\right)$$

Proof. The assertion follow from the fact that the kernel $(BV_\alpha)^k$ has the same form as kernel B and from the fact that $(BV_\alpha)^k = 0$ if $k \geq n$.

Remark 3.3. If n=2 and B is of form $\begin{pmatrix} 0, & 0 \\ K, & 0 \end{pmatrix}$ then W_α has the following form

$$W_\alpha = V_\alpha + V_\alpha B V_\alpha = \begin{pmatrix} V_{\alpha,1} & , 0 \\ V_{\alpha,1} K V_{\alpha,2} & , V_{\alpha,2} \end{pmatrix}$$

Just these types of triangular resolvents were considered by N.Bouleau [3].

If $\mathcal{W}', \mathcal{W}''$ are two resolvent of kernels on a measurable space we say that $\mathcal{W}'$ is equivalent with $\mathcal{W}''$ (we write $\mathcal{W}' \sim \mathcal{W}''$) if they give the same excessive functions

$$(\mathcal{E}_{\mathcal{W}'} = \mathcal{E}_{\mathcal{W}''})$$

In [3] it is proved for n=2 a theorem which asserts, in some natural conditions, that any triangular resolvent on $(E, \mathcal{B})$ associated with $\mathcal{V}$ coincides, without an equivalence, with a B-perturbation of $\mathcal{V}$ where B is a triangular kernel on $(E, \mathcal{B})$ whose diagonal components are equal zero.

Using Remark 3.1. b) and the above result of N.Bouleau we can obtain a general theorem for the representation of any triangular resolvent associated with $\mathcal{V}$ as a B-perturbation of $\mathcal{V}$ where B is a triangular kernel as above.

Theorem. 3.4. Let B be a triangular kernel on $(E, \mathcal{B})$ such that

$$\mathcal{B} = (B_{i,j})_{i,j} \quad , \; B_{ij} = 0 \text{ if } i \le j$$

and let $\mathcal{V} = (V_\alpha)_{\alpha > 0}$ be the direct sum of the system of resolvents $(\mathcal{V}_i)_{1 \le i \le n}$. We denote by $\mathcal{W} = (W_\alpha)_{\alpha > 0}$ the B-perturbation of $\mathcal{V}$. Then for any finite positive measurable function $s = (s_1, s_2, \dots s_n)$ on $(E, \mathcal{B})$ the following assertions are equivalent:

1) s is $\mathcal{W}$ -excessive.

2) For any $i \in \{1,2,\dots n\}$ s_i is $\mathcal{V}_i$ -excessive and we have

$$\alpha V_{\alpha,i} s_i + \sum_{j=1}^{i-1} V_{\alpha i} B_{ij} s_j \le s_i \quad (\forall) \;\; \alpha > 0$$

3) For any $i \in \{1,2,\dots n\}$ s_i is $\mathcal{V}_i$ -excessive and we have

$$\sum_{j=1}^{n-1} V_{o,i} B_{ij} s_j \precsim_{\mathcal{E}_{\mathcal{V}_i}} s_i$$

Proof. The assertion follows from Theorem 1.4 using a straightforward calculation.

Corollary 3.5. If n=2 and B is of the form

$$\mathcal{B} = \begin{pmatrix} 0 \,, & 0 \\ K \,, & 0 \end{pmatrix}$$

then a finite positive measurable function $s = (s_1, s_2)$ on $(E, \mathcal{B})$ is $\mathcal{W}$-excessive iff one of the following assertions holds:

i) s_1 and s_2 are $\mathcal{V}_1$ (resp. $\mathcal{V}_2$)-excessive and we have

$$\alpha V_{\alpha,2} s_2 + V_{\alpha,2} K s_1 \leq s_2$$

ii) s_1 and s_2 are $\mathcal{V}_1$ (resp. $\mathcal{V}_2$)-excessive and we have

$$V_{0,2} K s_1 \underset{\mathcal{V}_2}{\preccurlyeq} s_2$$

We consider now for any $i \in \{1,2 \ldots n\}$, $P_i = (P_{t,i})_{t \geq 0}$ a relatively bounded semigroup of kernels on $(E_i, \mathcal{B}_i)$. We denote by $P = (P_t)_{t \geq 0}$ the semigroup of kernels on $(E, \mathcal{B})$ where P_t is the following diagonal matrix

$$P_t = \begin{pmatrix} P_{t,1}, & 0, & 0, & 0, & \ldots & ,0 \\ 0, & P_{t,2}, & 0, & 0, & \ldots & ,0 \\ 0, & 0, & P_{t,3}, & 0, & \ldots & ,0 \\ \vdots & & & & & \\ 0, & 0, & 0, & \ldots & 0, & P_{t,n} \end{pmatrix}$$

The semigroup P is called the <u>direct sum</u> of $(P_i)_{1 \leq i \leq n}$. It is easy to see that $(P_t)_{t \geq 0}$ is a semigroup of kernels on $(E, \mathcal{B})$ and a function $s = (s_1, \ldots, s_n)$ on E is P-supermedian (resp. P-excessive) iff for any $i \in \{1,2,\ldots,n\}$ the function s_i is P_i-supermedian (resp. P_i-excessive).

A semigroup of kernels $Q = (Q_t)_{t \geq 0}$ on $(E, \mathcal{B})$ is called a triangular semigroup if for any $t \geq 0$ we have

$$Q_t = \left(Q_t^{i,j} \right)$$

where

$$Q_t^{i,j} = 0 \qquad \text{if } i < j$$

We can verify that if $(Q_t)_{t \geq 0}$ is a triangular semigroup on $(E, \mathcal{B})$ then for any $i \in \{1,2,\ldots,n\}$ the family $(Q_t^{ii})_{t \geq 0}$ is a semigroup of kernels on (E_i, B_i). If we have $Q_t^{ii} = P_{t,i}$ for any $t \geq 0$ and any $i \in \{1,2,\ldots,n\}$ we say that Q is a <u>triangular semigroup associated with P</u>.

It is easy to see that the resolvent associated with a triangular semigroup Q on $(E, \mathcal{B})$ is a triangular resolvent on $(E, \mathcal{B})$.

<u>Proposition 3.6</u>. Let P be the direct sum of the system of semigroups $(P_i)_{1 \leq i \leq n}$, and let B be a bounded kernel on $(E, \mathcal{B})$ of the form

$$B = (B_{ij})$$

where $B_{ij} = 0$ if $i \leq j$. Then the B-perturbation $Q = (Q_t)_{t \geq 0}$ of the semigroup $P = (P_t)_{t \geq 0}$ is a triangular semigroup associated with P. Moreover Q_t is the solution of the following triangular system:

$$Q_t^{i,i} = P_{t,i} \qquad (\forall)\ i \in \{1,2,\ldots n\}$$

$$Q_t^{ij} = \sum_{j \leq k < i} \int_0^t P_{u,i} B_{ij} Q_{t-u}^{kj} \, du \quad \text{if } i > j$$

<u>Proof</u>. The fact that Q is a triangular semigroup associated with P follows from Theorem 2.1. The last part of the proof follows from a straightforward calculation.

<u>Remark 3.7</u>. For n=2 we have

$$Q_t^{11} = P_{t,1} \quad , \quad Q_t^{22} = P_{t,2}$$

$$Q_t^{21} = \int_0^t P_{u,1} K P_{t-u,2} \, du$$

where

$$\mathcal{B} = \begin{pmatrix} 0, & 0 \\ K, & 0 \end{pmatrix}$$

Theorem 3.8. Let P be the direct sum of the system of semigroups $(P_i)_{1 \leq i \leq n}$ and let B be a bounded kernel on $(E, \mathcal{B})$ of the form

$$\mathcal{B} = (\mathcal{B}_{ij}) \quad \text{where} \quad \mathcal{B}_{ij} = 0 \quad \text{if} \quad i \leq j$$

We denote by $Q=(Q_t)_{t \geq 0}$ the B-perturbation of P. Then a finite positive measurable function $s=(s_1, s_2, \dots s_n)$ on $(E, \mathcal{B})$ is Q-excessive iff for any $i \in \{1,2,\dots,n\}$ s_i is P_i-excessive and

$$P_{t,i} s_i + \sum_{j=1}^{i-1} \int_0^t P_{u,i} B_{ij} s_j \, du \leq s_i$$

for any $t > 0$.

Proof. The assertion follows from Theorem 2.5 using a straightforward calculation.

Corollary 3.9. If n=2 and if B is of the form

$$\mathcal{B} = \begin{pmatrix} 0, & 0 \\ K, & 0 \end{pmatrix}$$

then a finite positive measurable function $s=(s_1, s_2)$ on $(E, \mathcal{B})$ is Q-excessive iff s_1 (resp. s_2) is P_1 (resp. P_2)-excessive and for any $t > 0$ we have

$$P_{t,2} s_2 + \int_0^t P_{u,2} K s_1 \, du \leq s_2 .$$

BIBLIOGRAPHY

1. N.Boboc et P.Mustaţă, Considérations assiomatiques sur les fonctions poly-surharmoniques. Rev.Roum.Math.Pures et Appl. 16(8) 1971, p.1167-1184.

2. N.Boboc et M.Nkomba-Tshola, Element complètement excessifs par rapport à une résolvante. Ann.Fac.Sci.de Kinshasa, Zaïre, Section Math.Phys. 2, 1976, p.1-30.

3. N.Bouleau, Espaces biharmoniques et couplage de processus de Markov. J.Math.Pures et Appl. 58, 1979, p.187-240.

4. E.Hille and R.S.Phillips, Functional Annalysis and Semi-groups. Amer.Math.Soc.Colloq.Publ. 31, 1957.

5. M.Itô, Positive eigen elements for an infinitesimal generator of a diffusion semi-group and their integral representation. Lecture Notes in Math.787, Potential Theory Copenhagen 1979, Proceedings, p.163-184.

6. M.Nicolesco, Les fonctions poly-harmonique Hermann. Paris 1936.

7. E.Smyrnelis, Axiomatique des fonctions biharmoniques. Ann. Inst.Fourier 25(1) 1975, p.35-97; 26(3) 1976, p.1-47.

ON THE RIESZ DECOMPOSITION OF BI-EXCESSIVE MEASURES ON LOCALLY COMPACT GROUPS

CORNEA FLORIN OVIDIU

In this work we introduce the notions of bi-excessive measure, bi-invariant measure and bi-potential with respect to two convolution semigroups on a locally compact group. The aim is to obtain a result giving a decomposition of bi-excessive measures into a bi-potential part and a bi-invariant part, similary with the Riesz decomposition of excessive measures.

In final we indicate how the Riesz decomposition can be obtained from our result. The proof follows the model of the abelian case with some modifications due to the fact that on a non-abelian locally compact group convolution of measures is not a commutative operation.

Let $(G\cdot)$ be a locally compact group. We denote by $M^+(G)$ the set of positive Radon measures on G and by $M_b^+(G)=\{\mu\in M^+(G) \mid \mu$ bounded$\}$

We recall that a family of measures $(\mu_t)_{t>0}\subset M_b^+(G)$ such that:

i) $\mu_t(G)\leq 1 \quad \forall\ t>0$

ii) $\mu_{t+s}=\mu_t*\mu_s \quad \forall\ t,s>0$

iii) $\lim\limits_{t\to 0}\mu_t=\varepsilon_0$ in the vague topology, is called convolution semigroup on G. ε_0 means the Dirac measure concentrated at the neutral element 0 of G.

We denote $\mathcal{K}^+(G)=\{f:G\longrightarrow(0,\infty) \mid f$ continuous, supp f compact$\}$ and $K(G)=\{f:G\longrightarrow\mathbb{C} \mid f$ continuous, supp f compact.$\}$

For $f\in K(G)$ and $\mu\in M^+(G)$ we define

$\langle \check{\mu}, f\rangle =: \langle \mu, \check{f}\rangle$, $\check{f}(x) =: f(x^{-1})$

$\langle \tau_a^r \mu, f\rangle =: \int f(x\cdot a^{-1})\, d\mu(x)$

$\langle \tau_a^l \mu, f\rangle =: \int f(a^{-1}\cdot x)\, d\mu(x)$

$$\left.\begin{array}{l}(f*\mu)(x) =: \int f(x\cdot y^{-1})\, d\mu(y)\\ (\mu * f)(x) =: \int f(y^{-1}\cdot x)\, d\mu(y)\end{array}\right\} \text{ for } x\in G.$$

It is easy to check that the following relations hold:

1. $\langle \mu*\nu, f\rangle = \langle \mu, f*\check{\nu}\rangle = \langle \nu, \check{\mu}*f\rangle$ for every $\mu, \nu \in M^+(G)$, $f\in K(G)$
2. $\widecheck{\mu*\nu} = \check{\nu}*\check{\mu}$ for every $\mu, \nu\in M^+(G)$

Let be $(\mu_t)_{t>0}$ and $(\nu_t)_{t>0}$ two convolution semigroups on G. We put $\mu_0 = \varepsilon_0$, $\nu_0 = \varepsilon_0$

<u>Definition 1.</u> $\xi \in M^+(G)$ is called bi-excessive measure with respect to $(\mu_t)_{t>0}$ and $(\nu_t)_{t>0}$ (or briefly bi-excessive measure) if for every t>0 , $\mu_t * \xi * \nu_t \leq \xi$ and the convolutions exist. $\xi\in M^+(G)$ is called bi-invariant measure if for every t>0, $\mu_t*\xi*\nu_t = \xi$ and the convolutions exist.

<u>Lema 2.</u> If $\xi \in M^+(G)$ is a bi-excessive measure then the map $[0,\infty) \longrightarrow M^+(G)$ is decreasing and
$t \longrightarrow \mu_t*\xi*\nu_t$
continuous from the right.

<u>Proof.</u> From $\mu_{t+s}*\xi*\nu_{t+s} = \mu_t*(\mu_s*\xi*\nu_s)*\nu_t \leq$ $\leq \mu_t*\xi*\nu_t$ for every t,s>0 we deduce that this map is decreasing.

Let $f\in K^+(G)$; then $\langle \xi , f\rangle \le \liminf_{t\to 0}\langle \mu_t * \xi * \nu_t , f\rangle \le$

$\le \limsup_{t\to 0}\langle \mu_t * \xi * \nu_t , f\rangle \le \langle \xi , f\rangle$ and we get $\lim_{t\to 0} \mu_t * \xi * \nu_t = \xi$

If $t_0>0$ we obtain $\mu_{t_0} * \xi * \nu_{t_0} = \lim_{t\to 0} \mu_t * (\mu_{t_0} * \xi * \nu_{t_0}) * \nu_t =$

$= \lim_{t\to 0} \mu_{t+t_0} * \xi * \nu_{t+t_0} = \lim_{t\to t_0} \mu_t * \xi * \nu_t$

We denote by $D^+(\mu,\nu) = \{\sigma\in M^+(G) \mid \mu_t * \sigma * \nu_t$ exists for every $t>0\}$

<u>Definition 3.</u> If $\sigma\in D^+(\mu,\nu)$ and for every $f\in K^+(G)$ $\int_0^\infty \langle \mu_t * \sigma * \nu_t , f\rangle\, dt < \infty$ we call $\int_0^\infty \mu_t * \sigma * \nu_t\, dt$ the bi-potential generated by σ. We denote:

$$\mathcal{K}_\sigma =: \int_0^\infty \mu_t * \sigma * \nu_t\, dt$$

<u>Lema 4.</u> If $\mu\in M^+(G)$, $\mu\ne 0$ then for every $K\subseteq G$ compact set there exists $f,g\in K^+(G)$ such that

$$\left.\begin{aligned}(\mu * f)(x) &\geq 1\\ (g*\mu)(x) &\geq 1\end{aligned}\right\}\quad \text{for every } x\in K.$$

<u>Proof</u>. There exists $h\in K^+(G)$ such that $\langle \mu , h\rangle > 0$

$(\mu * \tau_x^h \check{h})(x) = \int (\tau_x^h \check{h})(y^{-1}x)\, d\mu(y) = \int \check{h}(y^{-1}x\cdot x^{-1})\, d\mu(y) =$

$= \int \check{h}(y^{-1})\, d\mu(y) = \int h\, d\mu = \langle \mu , h\rangle > 0$ for every $x\in G$ and similary:

$(\tau_x^\ell \check{h} * \mu)(x) > 0$ for every $x\in G$.

By compactness arguments there exists $x_1,\dots,x_n\in G$ such that $f_1 =: \sum_{i=1}^{n} \tau_{x_i}^h \check{h}$ satisfies $\mu * f_1 > 0$ on K and we can take f as a positive multiple of f_1.

Analogously there exists $y_1,\dots,y_m\in G$ such that $g_1 =: \sum_{j=1}^{m} \tau_{x_j}^\ell \check{h}$ satisfies $g_1 * \mu > 0$ on K and we take g-a positive multiple of g_1.

From now we suppose that: $\int_0^{\infty} \mu_s * f * \nu_s \, ds < \infty \quad \forall f \in K^+(G)$

Theorem 5. If $\xi \in M^+(G)$ is a bi-excessive measure then there exist $\sigma \in D^+(\mu,\nu)$ and η a bi-invariant measure such that

$$\xi = \mathcal{K}\sigma + \eta$$

σ and η are uniquely determined by the formulas:

$$\sigma = \lim_{t\to 0} \frac{1}{t}\left(\xi - \mu_t * \xi * \nu_t\right)$$

$$\eta = \lim_{t\to 0} \mu_t * \xi * \nu_t$$

(limits are taken in the vague topology).

Proof. The unicity part follows by direct computation.

If $\xi = \int_0^{\infty} \mu_s * \sigma * \nu_s \, ds + \eta$, for every $t>0$ we have:

$$\xi - \mu_t * \xi * \nu_t = \int_0^{\infty} \mu_s * \sigma * \nu_s \, ds + \eta - \int_0^{\infty} \mu_{t+s} * \sigma * \nu_{t+s} \, ds - \mu_t * \eta * \nu_t =$$

$$= \int_0^{\infty} \mu_s * \sigma * \nu_s \, ds - \int_t^{\infty} \mu_s * \sigma * \nu_s \, ds = \int_0^{t} \mu_s * \sigma * \nu_s \, ds$$

Hence: $\lim_{t\to 0} \frac{1}{t}\left(\xi - \mu_t * \xi * \nu_t\right) = \lim_{t\to 0} \frac{1}{t}\int_0^t \mu_s * \sigma * \nu_s \, ds = \sigma$

$$\lim_{t\to\infty} \mu_t * \xi * \nu_t = \lim_{t\to\infty}\left(\int_0^{\infty} \mu_{s+t} * \sigma * \nu_{s+t} \, ds + \eta\right) = \lim_{t\to\infty} \int_t^{\infty} \mu_s * \sigma * \nu_s \, ds + \eta = \eta$$

The existence part is more laborious.

From lema 2 the map $t \longrightarrow \mu_t * \xi * \nu_t$ is decreasing hence there exists $\lim_{t\to\infty} \mu_t * \xi * \nu_t =: \eta$ and it is easy to check that η is a bi-invariant measure.

For every $t>0$, $n\in N$ such that $n>t$ we put

$$\sigma_t = \frac{1}{t}\left(\xi - \mu_t * \xi * \nu_t\right)$$

We observe that for t and n fixed there exists $\int_0^n \mu_s * \sigma_t * \nu_s \, ds < \infty$

Indeed $\mu_s * \sigma_t * \nu_s \le \frac{1}{t} \mu_s * \xi * \nu_s$ for every $s>0$ and hence $\int_0^n \langle \mu_s * \xi * \nu_s, f\rangle ds \le \frac{1}{t}\int_0^n \langle \xi, f\rangle ds < \infty$ for every $f\in K^+(G)$.

$$\int_0^n \mu_s * \sigma_t * \nu_s \, ds = \frac{1}{t}\int_0^n \mu_s * \xi * \nu_s \, ds - \frac{1}{t}\int_0^n \mu_{t+s} * \xi * \nu_{t+s} \, ds =$$

$$= \frac{1}{t}\left(\int_0^n \mu_s * \xi * \nu_s \, ds - \int_t^{n+t} \mu_s * \xi * \nu_s \, ds\right) = \frac{1}{t}\left(\int_0^t \mu_s * \xi * \nu_s \, ds - \int_n^{n+t} \mu_s * \xi * \nu_s \, ds\right)$$

Hence: $(*)$: $\displaystyle\int_0^n \mu_s * \sigma_t * \nu_s \, ds = \frac{1}{t}\left(\int_0^t \mu_s * \xi * \nu_s \, ds - \int_n^{n+t} \mu_s * \xi * \nu_s \, ds\right)$

From $\eta \le \mu_s * \xi * \nu_s \le \xi$ for every $s>0$ we obtain:

$\displaystyle\int_0^n \mu_s * \sigma_t * \nu_s \, ds \le \frac{1}{t}\left(\xi \cdot t - \eta \cdot t\right) = \xi - \eta$ hence there exist:

$\displaystyle\int_0^\infty \mu_s * \sigma_t * \nu_s \, ds < \infty$ for every $t>0$.

We want to show that $(\sigma_t)_{t>0}$ is vaguely bounded. For this purpose we need the following

<u>Lemma</u>. For every compact set $K \subseteq G$ there exists $h \in K^+(G)$ such that $\displaystyle 0 < \int_0^\infty \check{\mu}_s * h * \check{\nu}_s \, ds$ on K

<u>Proof of the lemma</u>. First we show that for every $x \in G$ there exists $g_x \in K^+(G)$ such that $\displaystyle\int_0^\infty \check{\mu}_s * g_x * \check{\nu}_s(x) \, ds > 0$

Indeed suppose that there exists $x_0 \in G$ such that for every $g \in K^+(G)$ we have $\displaystyle\int_0^\infty \check{\mu}_s * g * \check{\nu}_s(x_0) \, ds = 0$

Take $h \in K^+(G)$ such that $h(x_0) \neq 0$. Then $\displaystyle\int_0^\infty \check{\mu}_s * h * \check{\nu}_s(x_0) \, ds = 0$ implies $\check{\mu}_s * h * \check{\nu}_s(x_0) = 0$ for every $s \geq 0$ hence $h(x_0)=0$ contradiction.

Again by compactness arguments there exists $x_1, \ldots, x_n \in G$ such that $\displaystyle h = \sum_{i=1}^n g_{x_i} \in K^+(G)$ and $\displaystyle\int_0^\infty \check{\mu}_s * h * \check{\nu}_s \, ds > 0$ on K $\square$

. Returning to our theorem we obtain: for every $f \in K^+(G)$ there exist $g \in K^+(G)$ such that $\displaystyle f < \int_0^\infty \check{\mu}_s * g * \check{\nu}_s \, ds < \infty$

(by the above lemma). Then : $<\sigma_t, f> \leq \int_0^\infty <\sigma_t, \check{\mu}_s * g * \check{\nu}_s> ds =$ $= \int_0^\infty <\mu_s * \sigma_t * \nu_s, g> ds \leq M$ for every $t>0$ ($M=M_f= <\xi-\eta, g>$) hence $(\sigma_t)_{t>0}$ is vaguely bounded. There exists $(t_i)_{i\in I}$, $t_i \geq 0$, $t_i \longrightarrow 0$ such that $\lim_I \sigma_{t_i} = \sigma$.

The net $(\mu_s * \sigma_{t_i} * \nu_s)_{i\in I}$ is vaguely bounded. Indeed let $f \in K^+(G)$. For every $g \in K^+(G)$ such that $g \leq \check{\mu}_s * f * \check{\nu}_s$ we have $<\sigma_{t_i}, g> \leq M$ (σ_{t_i} being bounded) hence:

$$<\mu_s * \sigma_{t_i} * \nu_s ; f> \leq M$$

Now we prove that $\mu_s * \sigma_{t_i} * \nu_s \xrightarrow{i} \mu_s * \sigma * \nu_s$ for every $s>0$. For every $f, g \in K^+(G)$ such that $g \leq \check{\mu}_s * f * \check{\nu}_s$ we have $<\sigma, g> = \lim_i <\sigma_{t_i}; g> \leq \liminf_i <\sigma_{t_i}, \check{\mu}_s * f * \check{\nu}_s>$ $\leq \liminf_i <\mu_s * \sigma_{t_i} * \nu_s, f>$. Taking the supremum over g $<\mu_s * \sigma * \nu_s, f> \leq \liminf_i <\mu_s * \sigma_{t_i} * \nu_s, f>$ hence it is enough to check that : $\limsup_i <\mu_s * \sigma_{t_i} * \nu_s, f> \leq$ $\leq <\mu_s * \sigma * \nu_s, f>$

We observe that for every $K \subseteq G$ compact, if we denote

$$\alpha_s^K = \mu_s|_K , \quad \beta_s^K = \mu_s - \alpha_s^K , \quad \gamma_s^K = \nu_s|_K , \quad \delta_s^K = \nu_s - \gamma_s^K$$

and $f \in K^+(G)$ we have :

$$\limsup_i <\mu_s * \sigma_{t_i} * \nu_s, f> \leq \limsup_i <\alpha_s^K * \sigma_{t_i} * \gamma_s^K, f> +$$
$$+ \limsup_i <\alpha_s^K * \sigma_{t_i} * \delta_s^K ; f> + \limsup_i <\beta_s^K * \sigma_{t_i} * \gamma_s^K, f> +$$
$$+ \limsup_i <\beta_s^K * \sigma_{t_i} * \delta_s^K, f>$$

Since $<\mu_s*(\xi-\eta)*\delta_s^K, f> \searrow 0$ as $K\nearrow G$ - for every $\varepsilon>0$ there exist $K_1\subset G$ compact such that $<\mu_s*(\xi-\eta)*\delta_s^{K_1}, f> < \frac{\varepsilon}{3}$ hence $<\mu_s*\sigma_{t_i}*\delta_s^{K_1}, f> < \frac{\varepsilon}{3}$ for every i (because any i $<\mu_s*\sigma_{t_i}*\delta_s^{K_1}, f> \leq <\mu_s*(\xi-\eta)*\delta_s^{K_1}; f>$)

Similary since $<\beta_s^K*(\xi-\eta)*\nu_s, f> \searrow 0$ as $K\nearrow G$ - there exist $K_2\subset G$ compact such that :

$$<\beta_s^{K_2}*\sigma_{t_i}*\nu_s, f> < \frac{\varepsilon}{3} \qquad \text{for every i}$$

Since $0 \leq <\beta_s^K*\sigma_{t_i}*\delta_s^K, f> \leq <\beta_s^K*(\xi-\eta)*\delta_s^K; f>$ and: if $K\nearrow G$ then $<\beta_s^K*(\xi-\eta)*\delta_s^K, f> \searrow 0$, we obtain there exist $K_3\subset G$ compact such that:

$$<\beta_s^{K_3}*\sigma_{t_i}*\delta_s^{K_3}; f> < \frac{\varepsilon}{3} \quad \text{for every i.}$$

Take $K_0 = K_1 \cup K_2 \cup K_3$, then K_0 is a compact set of G and

$$<\alpha_s^{K_0}*\sigma_{t_i}*\delta_s^{K_0}; f> \leq <\mu_s*\sigma_{t_i}*\delta_s^{K_1}, f> < \frac{\varepsilon}{3} \quad \text{for every i}$$

$$<\beta_s^{K_0}*\sigma_{t_i}*\gamma_s^{K_0}; f> \leq <\beta_s^{K_2}*\sigma_{t_i}*\nu_s, f> < \frac{\varepsilon}{3} \quad \text{for every i}$$

$$<\beta_s^{K_0}*\sigma_{t_i}*\delta_s^{K_0}; f> \leq <\beta_s^{K_3}*\sigma_{t_i}*\delta_s^{K_3}; f> < \frac{\varepsilon}{3} \quad \text{for every i.}$$

But $\alpha_s^{K_0}*\sigma_{t_i}*\gamma_s^{K_0} \xrightarrow{i} \alpha_s^{K_0}*\sigma*\gamma_s^{K_0}$ vaguely.

Hence: $\limsup_i <\mu_s*\sigma_{t_i}*\nu_s; f> \leq <\alpha_s^{K_0}*\sigma*\gamma_s^{K_0}, f> + \varepsilon$ for every $\varepsilon>0$

$$\limsup_i <\mu_s*\sigma_{t_i}*\nu_s; f> \leq <\mu_s*\sigma*\nu_s; f> + \varepsilon$$

for every $\varepsilon > 0$, and we proved that:

$$\limsup_i <\mu_s*\sigma_{t_i}*\nu_s; f> \leq <\mu_s*\sigma*\nu_s, f>$$

From the above results we obtain

$$\int_0^n \mu_s * \sigma_{t_i} * \nu_s \, ds \xrightarrow{i} \int_0^n \mu_s * \sigma * \nu_s \, ds \quad \text{vaguely}$$

Now let $t = t_i$ in (*) and take limit over i and then over n. We obtain the desired result

$$\xi = \mathcal{K}_\sigma + \eta \qquad \square$$

<u>Remark</u>. Let $(\mu_t)_{t>0}$ be a tranzient convolution semi-group on a locally compact abelian group and denote $\mathcal{K} = \int_0^\infty \mu_t \, dt$ his potential kernel. Consider the semi-group $(\nu_t)_{t>0} = (\varepsilon_0)_{t>0}$ Then $D^+(\mu, \nu) = \{\sigma \in M^+(G) \mid \mu_t * \sigma \text{ exists for every } t>0\}$

$\mathcal{K}_\sigma = \mathcal{K} * \sigma$, the bi-excessive measure ξ becomes excessive in the sense of [1], hence we obtain the Riesz decomposition theorem for excessive measures.

Theorem 5 can be succesfully used to prove the following intereseting result.

<u>Definition 6</u>. For $\xi \in M^+(G)$ put

$$T^* \xi = \int_0^\infty \mu_t * \xi * \nu_t \, dt$$

Denote $\mathcal{M}_0 = \{\xi \in M^+(G) \mid \xi \neq 0, \ \text{supp}\,\xi \text{ compact}, \langle T^*\xi, f\rangle < \infty$ for every $f \in K(G)\}$.

Let $\sigma \in \mathcal{M}_0$ and $\omega \subseteq G$ open. A measure $\sigma' \in \mathcal{M}_0$ is called a T^* balayed of σ on ω if i) $T^*\sigma' \leq T^*\sigma$ ii) $T^*\sigma'|_\omega = T^*\sigma|_\omega$ iii) $\text{Supp}\,\sigma' \subseteq \overline{\omega}$

<u>Proposition 7</u>. If $\omega \subseteq G$ is an open relatively compact set and for every $f \in K(G)$ and every $\xi \in M^+(\overline{\omega})$ (positive measures with support in $\overline{\omega}$) such that $\operatorname{Re}\langle \xi, f\rangle > \alpha$ $(\alpha \in \mathbb{R})$ we have $\operatorname{Re}\langle \mu_t * \xi * \nu_t, f\rangle > \alpha$ for every $t>0$, then for every $\sigma \in \mathcal{M}_0$ there exists a T^* balayed of σ on ω.

The proof is laborious and uses Hahn-Banach theorem and will be done in other papers (to appear).

Denote $D^+(\mu) = \{\tau \in M^+(G) \mid \mu_t * \tau \text{ exists for every } t>0\}$

Define

$$P_t : D^+(\mu) \longrightarrow M^+(G)$$
$$P_t \xi = \mu_t * \xi$$

$$R_t : D^+(\mu) \longrightarrow M^+(G)$$
$$R_t \xi = \xi * \nu_t$$

$$Q_t : D^+(\mu, \nu) \longrightarrow M^+(G)$$
$$Q_t \xi = \mu_t * \xi * \nu_t$$

$\xi \in M^+(G)$ is bi-excessive iff $Q_t \xi \le \xi \quad \forall t>0$.

Assume that $\operatorname{supp} \nu_t \subseteq H$ for every $t>0$, where H is a compact subgroup of G. Denote A,B,C the infinitesimal generators of $(P_t)_{t>0}, (R_t)_{t>0}, (Q_t)_{t>0}$, then:

$$C = A + B$$

Indeed, this follows from the relation

$$\frac{\mu_t * \xi * \nu_t - \xi}{t} = \frac{\mu_t * \xi - \xi}{t} * \nu_t + \frac{\xi * \nu_t - \xi}{t}$$

Bibliography

1. C.Berg, G.Forst - „Potential Theory on Locally Compact Abelian Groups" - Springer Verlag, 1975.

2. J.Deny - "Noyaux de convolution de Hunt et noyaux associés à une famille fondamentale" Ann.Inst.Fourier 12, 643 - 667 (1962).

STANDARD H-CONES AND BALAYAGE SPACES

K. Janßen

Institut für Statistik und Dokumentation der
Universität Düsseldorf
Universitätsstraße 1
4000 Düsseldorf 1 West Germany

Abstract

In this paper we study the relations between standard H-cones (c.f.[2],[3] and [4]) and balayage spaces (c.f.[1]). It turns out, that a standard H-cone S is associated with a balayage space if and only if S can be represented on a Bauer-simplex (i.e. a Choquet-simplex for which the set of extreme points is closed).

O. Notations

In the following let E be a locally compact space with countable base. Let B, C, and C_o be the set of all functions on E which are Borel measurable, continuous, and continuous vanishing at infinity, restectively.

Adding + or b to such symbols gives the subset of all positive or bounded functions, respectively.

In the sequel, V and W always denotes a convex cone of positive numerical functions on E.

Definitions:

a) (E,V) is called a normalized balayage space, iff there exists a submarkovian resolvent $(V_\lambda)_{\lambda\geq o}$ on E satisfying

i) $V_\lambda f \in C_o$ for all $f \in B_b$, $\lambda \geq 0$ and $V_\lambda f \to f$ uniformly (or equivalently: pointwise) on E for $\lambda \to \infty$ for $f \in C_o$,

ii) V is the set of (V_λ)-excessive functions on E (i.e. $V = \{v \in B_+ : \sup_{\lambda>o} \lambda V_\lambda v(x) = v(x)$ for $x \in E\}$

$= \{v \in B_+$: for some $(f_n) \subset {}_+B_b$, $(Vf_n) \uparrow v\})$.

b) $(E,\mathcal{W})$ is called a <u>balayage space</u>, iff there exists some strictly positive $p \in \mathcal{W} \cap C$ such that $(E,\mathcal{V})$ is a normalized balayage space for

$\mathcal{V} := \frac{\mathcal{W}}{p} := \{\frac{w}{p} : w \in \mathcal{W}\}$.

<u>Remark</u>: This is not the original definition of [1], but it is an easy consequence of the results of [1] or [5] that the above notion of "balayage space" coincides with the notion of "standard balayage space" of [1] (the omission of the word "standard" is motivated by later papers by Bliedtner and Hansen).

For all notions and results related with a standard H-cone S we refer to [4].

In particular, there exists a dual standard H-cone S^* and for every weak unit $p \in S$ the set $\{\mu \in S^* : \mu(p) \leq 1\}$ is a Choquet-simplex in S^*.

<u>Definition</u>:

Let S be a standard H-cone, and let $(E,\mathcal{W})$ be a balayage space. We say that S and $(E,\mathcal{W})$ are <u>associated</u>, iff there exists an increasingly dense subcone $\mathcal{W}^1$ of $\mathcal{W}$ with the following properties:

i) $\mathcal{W}^1$ is a standard H-cone with respect to pointwise defined algebraic and order relations,

ii) $\mathcal{W}^1$ is isomorphic with S as a standard H-cone.

<u>Remark</u>: It is obvious from theorem 4.4.6 in [4] that every balayage space is associated with a standard H-cone. In the following we find a characterization of those standard H-cones which are associated with a balayage space.

1. <u>Balayage spaces are associated with nice standard H-cones.</u>

In this section let $(E,\mathcal{W})$ be a balayage space.
Let $p \in \mathcal{W} \cap C$ be strictly positive such that $\mathcal{V} := \frac{\mathcal{W}}{p}$ is a normalized balayage space, let $(V_\lambda)_{\lambda \geq 0}$ be a smooth resolvent which has $\mathcal{V}$ as set of excessive functions according to the definition of a normalized balayage space, and denote

$\mathcal{W}^1 := \{w \in \mathcal{W} : w$ is V_o-a.e. finite$\}$.

<u>Proposition 1</u>:

$\mathcal{W}^1$ is a standard H-cone which is associated with $(E,\mathcal{W})$. Moreover, $p \in \mathcal{W}^1$ is a weak unit and $K_p := \{\mu \in (\mathcal{W}^1)^* : \mu(p) \leq 1\}$ is a Bauer-Simplex. More precisely, there is a homeomorphism $z \to \mu_z$ between the Alexandroff-compactification $E_\Delta := E \cup \{\Delta\}$ of E and the set of extreme points of K_p, which can be defined by $\mu_\Delta := 0$ and $\mu_x(w) := \frac{w(x)}{p(x)}$ for $x \in E$, $w \in \mathcal{W}^1$.

Proof: a) From the definition of a normalized balayage space and from theorem 4.4.6 in [4] we conclude that

$\mathcal{S} := \{\frac{w}{p} : w \in \mathcal{W}^1, \frac{w}{p}$ is V_o-a.e. finite$\}$ is a standard H-cone. Obviously, $v \to p\cdot v$ $(v \in \mathcal{S})$ defines a bijection between $\mathcal{S}$ and $\mathcal{W}^1$ such that $\mathcal{W}^1$ is a standard H-cone which is isomorphic as a standard H-cone with $\mathcal{S}$. Moreover, $\mathcal{W}^1$ is increasingly dense in $\mathcal{W}$, and consequently $\mathcal{W}^1$ and $(E,\mathcal{W})$ are associated. Since p is a weak unit for the standard H-cone $\mathcal{W}^1$, we know from proposition 4.2.4 in [4] that K_p is a metrizable Choquet-simplex.

b) Since $(V_\lambda)_{\lambda \geq o}$ is a very smooth resolvent, the usual Stone-Weierstraß arguments show that $\mathcal{S} \cap \mathcal{C}_o$ is total in $\mathcal{C}_o$ with respect to uniform convergence. Then the usual Hahn-Banach arguments imply easily that for $\mu \in K_p$ there exists a unique Radon measure λ on E such that $\lambda(E) \leq 1$ and $\mu(w) = \int \frac{w}{p} d\lambda$ for all $w \in \mathcal{W}^1$. Conversely, this formula defines an element μ in K_p for every Radon measure λ on E which satisfies $\lambda(E) \leq 1$.

c) In particular, for $z \in E_\Delta$ we have $\mu_z \in K_p$, and $z \to \mu_z$ $(z \in E_\Delta)$ is an injective mapping. Obviously, μ_Δ is an extreme point of K_p. Moreover, μ_x is an extreme point of K_p for $x \in E$: if $\mu_x = \frac{1}{2}(\mu_1 + \mu_2)$ for some μ_1, μ_2 in K_p, then the corresponding Radon measures λ_1 and λ_2 on E satisfy $f(x) = \frac{1}{2}(\int f d\lambda_1 + \int f d\lambda_2)$ for all $f \in \mathcal{S}$, hence for all $f \in \mathcal{C}_o$ according to b) and this implies $\lambda_1 = \lambda_2 = \varepsilon_x$ and consequently $\mu_1 = \mu_2 = \mu_x$.

d) We prove that every extreme point μ of K_p equals μ_z for some $z \in E_\Delta$: If $\mu \neq 0$, then $\mu(p) = 1$, and consequently the corresponding measure λ satisfies $\lambda(E) = 1$. This implies $\lambda = \varepsilon_x$ for some x in E, since otherwise λ has a non-trivial decomposition into Radon measures $\lambda = \frac{1}{2}(\lambda_1 + \lambda_2)$, which in turn gives a non-trivial decomposition of μ according to b) in contradiction with the assumption that μ is an extreme point of K_p.

e) It remains to show that $z \to \mu_z$ $(z \in E_\Delta)$ is continuous. By the definition of the natural topology on $(\mathcal{W}^1)^*$ we have to prove: For every universally continuous w in $\mathcal{W}^1$, the function

$$f_w : x \to \frac{w(x)}{p(x)} \quad (x \in E)$$

belongs to $\mathcal{C}_o$. This property holds trivially if $w = p\,V_o h$ for some $h \in {}_+\mathcal{B}_b$. If w is an arbitrary universally continuous element of $\mathcal{W}^1$, then there exists a sequence $(h_n) \subset {}_+\mathcal{B}_b$ such that $(V_o h_n) \uparrow \frac{w}{p}$; in particular we know for $w_n := p\,V_o h_n$ that $(w_n) \subset \mathcal{W}^1$, $(w_n) \uparrow w$, $(f_{w_n}) \uparrow f_w$ and $(f_{w_n}) \subset \mathcal{C}_o$; since w is universally continuous, the convergence of (f_{w_n}) against f_w is uniform on E, hence $f_w \in \mathcal{C}_o$.

If K_p is a Bauer-simplex, then obviously the set $\{\lambda x : \lambda \in [0,1],\ x$ is an extreme point of $K_p\}$ is closed, hence we obtain from the above proposition and from exercise 4.11 in [4] immediately the following

<u>Consequence</u>: If a standard H-cone $\mathcal{S}$ is associated with a balayage space, then the cone of nearly continuous elements of $\mathcal{S}$ is inf-stable.

<u>Remark</u> From 4.3 in [4] and the above proposition we see that a standard H-cone $\mathcal{S}$ can be represented as a standard H-cone of functions on the non-zero extreme points of a Bauer-simplex given by a weak unit $p \in \mathcal{S}$ if $\mathcal{S}$ is associated with a balayage space. We abreviate this by saying: $\mathcal{S}$ can be represented on a Bauer-simplex.

2. <u>Nice standard H-cones are associated with balayage spaces</u>.

In this section let $\mathcal{S}$ be a standard H-cone which can be represented on a Bauer-simplex, i.e. there exists a weak unit $p \in \mathcal{S}$ such that $K_p := \{\mu \in \mathcal{S}^* : \mu(p) \le 1\}$ is a Bauer-simplex. Let $E := \{\mu \in K_p : \mu \neq 0,\ \mu$ is an extreme point$\}$. For $s \in \mathcal{S}$ define the function $\hat{s}$ on E by $\hat{s}(x) := x(s)$ $(x \in E)$ and let $\mathcal{V} := \{v :$ for some $(s_n) \subset \mathcal{S}$ we have $(\hat{s}_n) \uparrow v\}$.

<u>Proposition 2</u>: $(E,\mathcal{V})$ is a normalized balayage space associated with $\mathcal{S}$.

<u>Proof</u>: Obviously, E is locally compact with countable base. According to theorem 4.2.12 and 4.3 in [4], $\hat{\mathcal{S}} := \{\hat{s} : s \in \mathcal{S}\}$

is a standard H-cone of lower-semicontinuous functions on the saturated set E and $\hat{S}$ is isomorphic with S as a standard H-cone. Obviously, $\hat{p}(x) = 1$ for all $x \in E$, i.e. $1 \in \hat{S}$ and $\hat{K}_1 := \{\hat{\mu} \in \hat{S}^* : \hat{\mu}(1) \leq 1\} \cong K_p$.
Consequently, there is some corresponding resolvent and this will prove the proposition: From theorem 4.4.4, lemma 4.3.7 and corollary 4.4.10 of [4] we find a sequence $(s_n) \subset S_o$ and a resolvent $(V_\lambda)_{\lambda \geq o}$ on E such that $\hat{s}_n \leq \frac{1}{2n}$ for all n, $V_o 1 = \Sigma \hat{s}_n$, and $\hat{S}$ is increasingly dense in the set of all V_o-a.e. finite functions which are excessive with respect to (V_λ). Consequently it remains to prove that $V_\lambda f \in C_o$ for $f \in \mathcal{B}_b$ and $\lambda V_\lambda f \to f$ for $\lambda \to \infty$ for $f \in C_o$.
First of all, the map $\mu \to \mu(s_n)$ is continuous on S^* with respect to the natural topology on S^*, and this map vanishes in O; consequently, $\hat{s}_n \in C_o$ for all n, hence $V_o 1 \in C_o$ as a uniform limit of C_o-functions. Since $V_o h$ is lower-semicontinuous for $h \in {}_+\mathcal{B}_b$, we conclude $V_o h \in C_o$ for all $h \in \mathcal{B}_b$ and consequently $V_\lambda h \in C_o$ for all $h \in \mathcal{B}_b$ and $\lambda \geq 0$ by the resolvent equation.
Finally, let $F := \{f \in C_o : \lambda V_\lambda f \to f$ uniformly$\}$. Obviously F is a vector space which is closed with respect to uniform convergence; from Dini's lemma we know $\hat{S} \cap C_o \subset F$, and from the Stone-Weierstraß theorem we conclude $F = C_o$.

Remark: For a strictly positive $p \in C$ let $W := \{pv : v \in V\}$ Then (E, W) is a balayage space associated with S.

Obviously, the two propositions can be summarized by

Theorem
a) Every balayage space is associated with a standard H-cone which can be represented on a Bauer-simplex.
b) A standard H-cone S is associated with a balayage space if and only if S can be represented on a Bauer-simplex.

Looking at proposition 1 one might ask if K_p is a Bauer-simplex for more general p. The following examples might give some indications:

Example 1: Let $E :=]0,\infty[$, let S be the set of positive, finite, decreasing, and lower-semicontinuous functions on E. Then S is a standard H-cone associated with a balayage space. If $p \in S$ is a strictly positive function which is not continuous at $x_o \in E$, then K_p is not a Bauer-simplex: Define for $x \in E$ $\mu_x : s \to \frac{s(x)}{p(x)}$ $(s \in S)$; then μ_x is an extreme point of K_p for all $x \in E$. If $(x_n) \subset E$ increases strictly to x_o, then (μ_{x_n}) converges to $\alpha\mu_{x_o}$ with $\alpha := \frac{p(x_o)}{\lim p(x_n)} \in]0,1[$, hence K_p is not a Bauer simplex.

Example 2: The assumption $p \in \mathcal{W} \cap \mathcal{C}$ does not imply that K_p is a Bauer simplex:
Let $\mathcal{W}$ be the set of classical positive hyperharmonic functions on a domain $E \subset \mathbb{R}^n$. Let p be a continuous potential on E, such that p equals an extreme potential outside some compact subset of E.

If $\mathcal{S}$ is the set of classical positive superharmonic functions on E, then $\mathcal{S}$ is a standard H-cone associated with $(E,\mathcal{W})$. For $x \in E$ the map $\mu_x : s \to \frac{s(x)}{p(x)}$ $(s \in \mathcal{S})$ belongs to K_p. From the definition of the Martin boundary in [6] it is obvious that the map $x \to \mu_x$ $(x \in E)$ extends to a homeomorphism Φ from $\hat{E}$ to the closure of the non-zero extreme points of K_p, and it is easily seen that for $z \in \hat{E} \setminus E$ $\Phi(z)$ is an extreme point of K_p if and only if z is minimal. Consequently, K_p is not a Bauer-simplex if E is choosen in such a way that the set of minimal points is not closed.

For standard H-cones there is a nice theory of duality. This theory is not compatible with the category of balayage spaces in the following sense: If $(E,\mathcal{W})$ is a balayage space, then there exists an associated standard H-cone $\mathcal{S}$ which has a dual standard H-cone $\mathcal{S}^*$. In general it is not true, that $\mathcal{S}^*$ is associated with a balayage space. This can be seen in the followin

Example 3: Let $E \subset \mathbb{R}^2$ be the "lazy crotch"

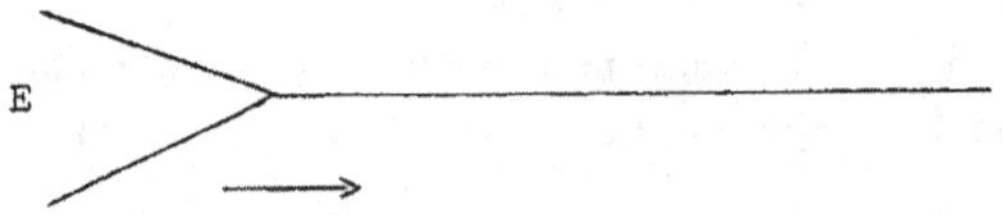

let $\mathcal{S}$ be the set of positive finite lower-semicontinuous functions on E which are "decreasing". Then $\mathcal{S}$ is a standard H-cone associated with a balayage space. Direct computations show that for no weak unit $\mu \in \mathcal{S}^*$ the set $K^*_\mu := \{s \in \mathcal{S}^{**} : s(\mu) \le 1\}$ is a Bauer-simplex. Moreover, in this example the set of nearly continuous elements of $\mathcal{S}^*$ is not inf-stable.

Bibliography

1. Bliedtner,J., Hansen,W.: Markov Processes and Harmonic Spaces. Z.Wahrscheinlichkeitstheorie verw. Gebiete 42 309-325 (1978)

2. Boboc,N., Bucur,Gh., Cornea,A.: H-Cones and Potential Theory. Ann.Inst.Fourier 25, 71-108 (1975)

3. Boboc,N., Cornea,A.: Cônes convexes ordonnés. H-cônes et adjoints de H-cônes. C.R.Acad.Sci.Paris 270, 596-599 (1970)

4. Boboc,N., Bucur,Gh., Cornea,A., Höllein,H.: Order and Convexity in Potential Theory: H-Cones. Lecture Notes in Math.853.Berlin-Heidelberg-New York: Springer (1981)

5. Dembinski,V., Janßen,K.: Standard Balayage Spaces and Standard Markov Processes. Lecture Notes in Math.787, 84-105. Berlin-Heidelberg-New York: Springer (1980)

6. Helms,L.L.: Introduction to Potential Theory. New York-London-Sydney-Toronto: Wiley (1969).

Applications of quasi Dirichlet bounded harmonic functions

at most an enumerable infinite number of analytic curves clustering nowhere in R. Let U(z) be a harmonic function such that

$\varliminf_{M\to\infty} \frac{D(\min(M, |U(z)|)}{M} \leqq \alpha < \infty$. Then we call U(z) a Q.D.B.H.[1] (quasi Dirichlet bounded harmonic function). A positive singular Q.D.B.H. is called a G.G. (generalized Green function). Let U(z) be a Q.D.B.H. Then $U(z) = U_1(z) - U_2(z)$, where $U_i(z)$ is a positive Q. D. B.H. with the same α and $U_i(z) = V_i(z) + S_i(z)$, where $V_i(z)$ is a positive Q.D.B.H. and $S_i(z)$ is a G.G. with the same α. Especially a G.G. has the following properties[1]

1). Let G(z) be a G.G. Then $D(\min(M,G(z)) = M\mathfrak{M}$ for any M and $\int_{C_\delta} \frac{\partial}{\partial n} G(z)ds = \mathfrak{M}$ (is denoted by $\mathfrak{M}(G(z))$ for any level curve C_δ, and $G_\delta = \{z : G(z) > \delta\}$ is an SO_g.

2). Let U(z) be a G.G. Then $U(z) \geqq V(z) \geqq 0$ implies that V(z) is a G.G. and $\mathfrak{M}(U(z)) \geqq \mathfrak{M}(V(z))$.

In this note we discuss some applications.

Application 1. We often use E and I operations to induce grobal properties from local ones. At first we show a G.G. has a property under operations.

Theorem 1. Let $U(z)$ be a G.G. in a domain G with $\mathfrak{M}(U(z)) = 2\pi$. Then $E[U(z)]$ is a G.G. in R with $\mathfrak{M}(E[U(z)]) \leqq 2\pi$. Let $V(z)$ be a G.G. in R with $\mathfrak{M}(V(z)) \leqq 2\pi$. Then $I[V(z)] = 0$ or a G.G. in G with $\mathfrak{M}\, I[V(z)] \leqq 2\pi$ and also $IE[U(z)] = U(z)$.

Let Γ be a disk in G. Let $U'(z) = H_U^{G-\Gamma}$, i.e. $U'(z)$ is the least positive superharmonic function in $G - \Gamma$ with boundary value $U(z)$ on $\partial\Gamma$. Then clearly $D(U'(z)) < \infty$ and $U'(z) \leqq \max_{z \in \partial\Gamma} U(z) = M'$. Evidently $U''(z) = U(z) - U'(z)$ is singular in $G - \Gamma$ and

$$\varlimsup_{M\to\infty} \frac{D(\min(M, U''(z))}{M} \leqq \varlimsup_{M\to\infty} \frac{D(U(z) - U'(z)) \text{ on } G^{M+M'}}{M} \leqq 2\pi,$$

where $G^{M+M'} = \{z : U(z) < M + M'\}$.

Hence $U''(z)$ is a G.G.. in $G - \Gamma$ with $\mathfrak{M}(U''(z)) \leqq 2\pi$. Let $U_n^M(z)$ be a P.H. (positive harmonic function) in $R_n - \Gamma - G''_M$: $G''_M = \{z \in G - \Gamma, U''(z) > M\}$ such that $U_n^M(z) = M$ on $\partial G''_M \cap R_n$, $= U''(z)$ on $G''_\varepsilon \cap \partial R_n - G''^M$, $= \varepsilon$ on $\partial R_n - G''_\varepsilon + \partial\Gamma$. By $U_n^M(z) \geqq U''(z)$ and $U_n^M(z) = U''(z)$ on $\Lambda = \partial G_M \cap R_n + (G''_\varepsilon - G''_M) \cap \partial R_n$, we have $\frac{\partial}{\partial n} U_n^M(z) \geqq \frac{\partial}{\partial n} U''(z)$ on Λ (inner normal), whence

$$0 < \int_{\partial\Gamma} \frac{\partial}{\partial_n} U_n^M(z)ds \leqq \int_{\partial\Gamma+\partial R_n - G''_\varepsilon} \frac{\partial}{\partial_n} U_n^M(z)ds = -\int_{\Lambda} \frac{\partial}{\partial_n} U_n^M(z)ds$$

$$\leqq -\int_{\Lambda} \frac{\partial}{\partial_n} U''(z)ds = \int_{\partial G_\varepsilon \cap R_n} \frac{\partial}{\partial_n} U''(z)ds \leqq \int_{\partial G''_\varepsilon} \frac{\partial}{\partial_n} U''(z)ds$$

$$= \mathfrak{M}(U''(z)) \leqq 2\pi. \qquad (1)$$

Evidently $U_n^M(z) \nearrow U^M(z) : n \to \infty$, $U^M(z) \nearrow U^*(z) : M \to \infty$. By 1) $U^*(z) < \infty$ and $U^*(z) \geqq U''(z)$ in $G - \Gamma$. Since $R \notin O_g$, there exists an H.M. (harmonic measure of Γ) $w(\Gamma, z, R - \Gamma) = w(z)$ such that $w(z)$ is a P.H. in $R - \Gamma$, $w(z) = 1$ on Γ and $0 < w(z) < 1$. Let L be a number such that

$$\max_{z \in \partial\Gamma} \frac{\partial}{\partial_n} U^*(z) < L(-\frac{\partial}{\partial_n} w(z)) \text{ on } \partial\Gamma \text{ and } L \geqq M', \ L > \varepsilon.$$

Then $\infty > U^*(z) + Lw(z)$ is superharmonic in R and $\geqq U''(z)$ in $G - \Gamma$, $\geqq \varepsilon + L \geqq U''(z) + M' \geqq U(z)$ on $\partial\Gamma$. Whence $U^*(z) + Lw(z) \geqq U(z)$ in G and $E[U(z)] \leqq U^*(z) + Lw(z) < \infty$.

Let $U_n^M(z)$ be a P.H. in $R_n - G_M : G_M = \{z \in G : U(z) > M\}$ such that $U_n^M(z) = U(z)$ on $\partial G_M \cap R_n + \partial R_n \cap (G - G_M)$, $= 0$ on $\partial R_n - G$. Since $U_n^M(z) \geqq U(z)$, $U_n^M(z) = U(z)$ on $\Lambda = \partial R_n \cap (G_N - G_M) = \partial R_n \cap \partial(G_N^M)$, where $G_N^M = \{z : U_n^M(z) > N\}$, $G_N = \{z : U(z) > N\}$, $2\pi = \mathfrak{M}(U(z)) = \int_{\partial G_N} \frac{\partial}{\partial_n} U(z)ds \geqq$

$$\int_{\partial G_N \cap R_n} \frac{\partial}{\partial_n} U(z)ds = -\int_{\partial G_M \cap R_n + \Lambda} \frac{\partial}{\partial_n} U(z)ds \geqq \int_{\partial G_M \cap R_n + \Lambda} -\frac{\partial}{\partial_n} U_n^M(z)ds = \int_{\partial G_N^M \cap R_n} \frac{\partial}{\partial_n} U_n^M(z)ds,$$

since $\overline{R}_n \cap \overline{G}_N^M$ is compact, $D(\min(L,U_n^M(z)) = \int_0^L \int_{C_\rho} \frac{\partial}{\partial_n} U_n^M(z)dsd\rho \leqq 2\pi L$,: $C_\rho = \{z: U_n^M(z) = \rho\}: \rho < L < M$.

Now $U_n^M(z) \nearrow U^M(z), U^M(z) \nearrow E[U(z)] = \tilde{U}(z)$. Let Ω be a compact domain such that $\overline{\Omega} \subset \{z: \tilde{U}(z) < L\}$. Then for any L, there exsists a number M_L such that for any $M > M_L$, there exists a number n(M) such that $\Omega \subset \{U_n^M(z) < L\} = G_n^M : n \geqq n(M)$. Whence

$$D_\Omega(\tilde{U}(z)) \leqq \underline{\lim}_{M\to\infty} \underline{\lim}_{n\to\infty} D_{G_n^M}(U_n^M(z)) \leqq 2\pi L.$$

Let $\Omega \to \{z: \tilde{U}(z) < L\}$. Then $D(\min(L, \tilde{U}(z)) \leqq 2\pi L$.

On the other hand, it is well known E[U(z)] is singular or 0. Thus E[U(z)] is a G.G. with $\mathfrak{m}(E[U(z)] \leqq 2\pi$. Evidently $IE[U(z)] = U(z)$. Let V(z) be a G.G. in R. Then it is also a Q.D.B.H. in G and $V(z) = A_1(z) + A_2(z)$ in G. where $A_1(z)$ is the singular part. We see at once $I[V(z)] = A_1(z)$ and $A_1(z)$ is a G.G. with $\mathfrak{m}(A_1(z)) < \infty$.

Application 2. We suppose Kerékjártó's topology is defined on R+Δ, where Δ is the set of all the boundary components. We suppose every neighbourhood of a boundary component is a domain whose relative boundary is compact. Let G be a compact or non compact domain. We denote by $\tilde{H}_U^G$ the harmonic function in G such that $\tilde{H}_U^G = U(z)$ on ∂G and $\tilde{H}_U^G$ has the M.D.I. (minimal Dirichlet integral) among all continuous functions with the same boundary value on ∂G. Evidently $U(z) = \tilde{H}_U^G$ implies $U(z) = \tilde{H}_U^{G'}$ for any domain $G' \subset G$. Suppose U(z) is defined in a neighbourhood $\mathcal{V}(\mathfrak{p})$ of boundary component $\mathfrak{p}$. If $U(z) = \tilde{H}_U^{\mathcal{V}(\mathfrak{p})}$, we say U(z) has M.D.I. property at $\mathfrak{p}$. Let p_0 be a fixed point of $R - R_0$ (an end of a Riemann surface R) and let

$G(z,p_0)$ be a Green function of $R-R_0$. If there exists a sequence of neighbourhood $\{\mathcal{V}_n(p^*)\}$ of p^* such that

$$\overline{\lim_{n=\infty}} \inf_{z\in\partial\mathcal{V}_n(p^*)} G(z,p_0) > 0,$$

then we say p^* is completely irregular.

Class $\mathcal{H}$. In the following we consider P.H., sU(z) in $R-R_0$. If U(z) satisfies, U(z)=0 on ∂R_0, U(z) is finite except p^*, i.e. $U(z)\leqq \max_{z\in\partial\mathcal{V}(p^*)} U(z)$ in $R-R_0-\mathcal{V}(p^*)$, $D_{R-R_0-\mathcal{V}(p^*)}(U(z)) < \infty$ for any $\mathcal{V}(p^*)$ of p^*, U(z) has M.D.I. property at every $p(\neq p^*)$. Then we say $U(z)\in\mathcal{H}$. Let D be a domain such that ∂D is regular and compact. We denote by $U_D(z)$ the positive harmonic function in $R-R_0$ such that $U_D(z)=U(z)$ on D and $U_D(z)=\tilde{H}_U^{R-R_0-D}$ in $R-R_0-D$. If for any domain above mentioned, $U_D(z)\leqq U(z)$, we say U(z) is an H.F.S.P.[3] (harmonic and full superharmonic). Let U(z) be H.F.S.P. If $V(z)=aU(z): 0\leqq a\leqq 1$, for any V(z) such that both V(z) and $U(z)-V(z)\geqq 0$ are H.F.S.P,s, we say U(z) is N-minimal.[3]

Lemma. Let U(z) be a P.H. and p^* be completely irregular. Then U(z) is H.F.S.P, if $U(z)\in\mathcal{H}$.

Proof. Let D be a domain whose ∂D is compact, and D does not tend to p^*. Let $\Delta(D)$ be the set of the boundary components of which D is a nieghbourhood. Since $\Delta-\Delta(D)-\Delta(\mathcal{V}(p^*))=F$ is compact, we can find a finite number of neighbourhood $\mathcal{V}(p_i)$'s such that $\sum^{n_0}\mathcal{V}(p_i)\cap\mathcal{V}(p^*)=0$, $\sum^{n_0}\mathcal{V}(p_i)\supset F$ and U(z) has M.D.I. property in $\mathcal{V}(p_i)$. Put $G=\sum_i^{n_0}\mathcal{V}(p_i)$ and let G' be another open set such that $\partial G'$ is compact, $\overline{G}'\supset F$, $\partial G\cap\partial G'=0$. Let $U_n(z)$ be a P.H. in $R_n-D-R_0-\mathcal{V}(p^*)$ such that $U_n(z)=U(z)$ on $\partial R_0+\partial\mathcal{V}(p^*)+\partial D$, $\frac{\partial}{\partial n}U_n(z)=0$ on $\partial R_n-\mathcal{V}(p^*)$. Then $U_n(z)\to$

$\tilde{H}_U^{R-R_0-\mathcal{V}(p^*)-D}=U^*(z)$. Let $U_n'(z)$ be a P.H. in $G\cap R_n$ ($G\cap R_n$ is note necessarily connected) such that $U_n'(z)=U(z)$ on ∂G and $\frac{\partial}{\partial_n}U_n'(z)=0$ on $\partial R_n\cap G$. Then $U_n'(z)\to\tilde{H}_U^G(=U(z)$, because $U(z)$ has M.D.I. property in G). By the maximum principle

$$\max_{z\in\partial G}|U_n(z)-U_n'(z)| \geqq \max_{z\in\partial G'}|U_n(z)-U_n'(z)|.$$

Let $n\to\infty$. Then

$$\max_{z\in\partial G}|U^*(z)-U(z)| > \max_{z\in\partial G'}|U^*(z)-U(z)|, \text{ if } U^*(z) \neq U(z).$$

Consider $U^*(z)$ and $U(z)$ in $R-R-D-\mathcal{V}(p^*)-G'$, by the maximum

$$\max_{z\in\partial G}|U^*(z)-U(z)| < \max_{z\in\partial G'}|U^*(z)-U(z)|, \text{ if } U^*(z) \neq U(z).$$

This is a contradiction. Hence

$$U(z) = \tilde{H}_U^{R-R_0-D-\mathcal{V}(p^*)} \tag{2}$$

If D tends to p^* and ∂D is compact, D must contain a neighborhood of p^* and we have the same conclusion.

Suppose $D\cap\mathcal{V}(p^*)=0$. We can find $\mathcal{V}(p^*)\supset\mathcal{V}'(p^*)$ such that $\bar{\mathcal{V}}(p^*)-\mathcal{V}'(p^*)$ is compact, and there exists a P.H.S(z) in $\mathcal{V}(p^*)-\mathcal{V}'(p^*)$ such that $S(z)=0$ on $\partial\mathcal{V}(p^*)$, $=1$ on $\partial\mathcal{V}'(p^*)$ and $D(S(z))<\infty$. Put $M=\max_{z\in\partial D}U(z)(<\infty)$.

Let $U'(z)=U(z)$ in $R-R_0-D-\mathcal{V}(p^*)$, $=\max(U(z),MS(z))$ in $\bar{\mathcal{V}}(p^*)-\mathcal{V}(p^*)$, $=M$ in $\bar{\mathcal{V}}'(p^*)$. Then $D(U'(z))<\infty$ and $\tilde{H}_{U'}^{R-R_0-D-\mathcal{V}'(p^*)}$ exists. Analogously $\tilde{H}_V^{R-R_0-D-\mathcal{V}'(p^*)}$ exists, where $V(z)$ is a boundary function such that $V=U(z)$ on $\partial R_0+\partial D$, $V=0$ on $\partial\mathcal{V}'(p^*)$.

Evidently

$$U(z) = \tilde{H}_U^{R-R_0-D-\mathcal{V}'(p^*)} \geqq \tilde{H}_V^{R-R_0-D-\mathcal{V}'(p^*)}. \tag{3}$$

Let $\omega(\mathcal{V}(p^*),z,R-R_0-D)=\omega(\mathcal{V}(p^*),z)$ be a P.H. in $R-R_0-D-\mathcal{V}(p^*)$ such that $\omega(\mathcal{V}(p^*),z)=1$ on $\partial\mathcal{V}(p^*)=0$ on $\partial R_0+\partial D$ and has M.D.I. (i.e. capacity[3)] of $\mathcal{V}(p^*)$ relative to $R-R_0-D$. Put $G_\delta=\{z:G(z,p_0)>\delta\}:\delta>0$. Then $\omega(G_\delta\cap(R-R_n),z,R-R_0)\downarrow 0$ as $n\to\infty$. Now since $\partial\mathcal{V}(p^*)\subset G_\delta$ by the fact that p^* is completely irregular, $\omega(\mathcal{V}(p^*)\cap(R-R_n),z,R-R_0-D)\leqq \omega(G_\delta\cap(R-R_n),z,R-R_0)\downarrow 0$ as $n\to\infty$. Hence

$$
\begin{aligned}
0 &\leqq \tilde{H}_{U'}^{R-R_0-D-\mathcal{V}'(p^*)} - \tilde{H}_{V}^{R-R_0-D-\mathcal{V}'(p^*)} \\
&\leqq M\omega(\mathcal{V}'(p^*),z,R-R_0-D) \\
&\leqq M\omega(\mathcal{V}(p^*),z,R-R_0-D)\downarrow 0 \qquad (4)
\end{aligned}
$$

as $\mathcal{V}(p^*)\to p^*$.

$\tilde{H}_U^{R-R_0-D}=\lim U_n(z)$, where $U_n(z)$ is a P.H. in R_n-R_0-D such that $U_n(z)=U(z)$ on $\partial R_0+\partial D$ and $\frac{\partial}{\partial n}U_n(z)=0$ on ∂R_n-D. Now R_n includes $\partial\mathcal{V}(p):\mathcal{V}(p^*)\to p^*$ and $0<U_n(z)\leqq \max_{z\in\partial R_0+\partial D} U(z)=M$ on $\mathcal{V}(p)$. Hence

$$\tilde{H}_{U'}^{R-R_0-D-\mathcal{V}'(p^*)} \geqq U_n(z) \geqq \tilde{H}_{V}^{R-R_0-D-\mathcal{V}'(p^*)}$$

Whence by (4) and (3)

$$
\begin{aligned}
\tilde{H}_U^{R-R_0-D} &= \lim_{\mathcal{V}(p^*)\to p} \tilde{H}_V^{R-R_0-D-\mathcal{V}'(p^*)} \\
&\leqq \lim_{\mathcal{V}(p^*)\to p} \tilde{H}_U^{R-R_0-D-\mathcal{V}'(p^*)} \\
&= U(z).
\end{aligned}
$$

Thus

$$U_D(z) \leqq U(z).$$

If D tends to $\mathfrak{p}$, we have the result. Thus U(z) is an H.F.S.P.

Martin's topologies. Let $N(z,p):p\in R-R_0$ be an N-Green function of $R-R_0$ and let v(p) be a neighbourhood of p. Then $N(z,p)=\tilde{H}^{R-R_0-V(p)}_{N(z,p)}$, $\sup_{z\notin V(p)} N(z) \leqq \sup_{z\in\partial V(p)} N(z,p)$ and $D(\min(M,N(z,p))=2\pi M$. Let $p_i\to\mathfrak{p}^*$ and $N(z,p_i)$ converges to a P.H. $N(z,\overset{*}{p})$. Let $\mathfrak{p}(\neq\mathfrak{p}^*)$ be a boundary component and let $\mathcal{V}(\mathfrak{p})$ be a neighbourhood with compact $\partial\mathcal{V}(\mathfrak{p})$. Then $p_i\notin\mathcal{V}(\mathfrak{p})$, $i\geqq i_0$, implies $N(z,p_i)\leqq M$ in $\mathcal{V}(\mathfrak{p})$ for $i\geqq i_0$ and $\tilde{H}^{\mathcal{V}(\mathfrak{p})}_{N(z,p_i)}\to\tilde{H}^{\mathcal{V}(\mathfrak{p})}_{N(z,\overset{*}{p})}$. Also $N(z,p_i)\leqq\max_{z\in\partial\mathcal{V}(\mathfrak{p}^*)} N(z,p_i)$, $z\notin\mathcal{V}(\mathfrak{p}^*)$ for $p_i\in\mathcal{V}(\mathfrak{p}^*)$ implies $N(z,\overset{*}{p})\leqq\max_{z\in\partial\mathcal{V}^*(\mathfrak{p})} N(z,\overset{*}{p})$, $z\notin\mathcal{V}(\mathfrak{p}^*)$ and $D(N(z,\overset{*}{p}))<2\pi\max_{z\in\partial\mathcal{V}^*(\mathfrak{p}^*)} N(z,\overset{*}{p})$ on $R-R_0-\mathcal{V}(\mathfrak{p}^*)$.

Thus $N(z,\overset{*}{p})\in\mathfrak{N}$ and $N(z,\overset{*}{p})$ is an H.F.S.P.

We suppose α-Martin's topologies[2] (α=N or K) is defined on $R-R_0+\Delta^\alpha$, where Δ^α is set of boundary points. Let $p_i\overset{\alpha}{\to}p$ and $p_i\to\mathfrak{p}^*$. Then we say p lies on $\mathfrak{p}^*$. We denote by $\Delta^\alpha(\mathfrak{p}^*)$ and $\Delta^\alpha_1(\mathfrak{p}^*)$ the boundary points and α-minimal points on $\mathfrak{p}^*$.

Let $U(z)\in\mathfrak{N}$. Then U(z) is an H.F.S.P, whence $U_{\mathcal{V}(\mathfrak{p}^*)}(z)=U(z)$ and U(z) is represented by a positive mass μ on $\Delta^N_1(\mathfrak{p}^*)$ as follows:

$$U(z) = \int N(z,p)d\mu,$$

$$\int d\mu = \frac{1}{2\pi}\int_{\partial R_0}\frac{\partial}{\partial n}U(z)ds \quad\text{and}\quad D(\min(M,U(z)) \leqq 2\pi M\int d\mu.$$

Hence if $U(z)\in\mathfrak{N}$, U(z) is Q.D.B.H.

Let $N(z,p):p\in\Delta^N_1(\mathfrak{p}^*)$. Then there exists a path γ tending to p. γ intersects $\partial\mathcal{V}_i(\mathfrak{p}^*)$ at p_i. We can find a subsequence $\{p_i\}$ such that $G(z,p_i)$ converges to a function $G(z,p)$. Clearly by $G(p_i,p_0)>\delta>0$

$$N(z,p) \geqq G(z,p) > 0 \text{ and } N(p_0,p) > \delta. \qquad (5)$$

Operations I and E^N. For any P.H. in $R-R_0$. We denote the singular part of $U(z)$ by $I[U(z)]$. By (5) $I[N(z,p)]>0$ for $p\in\Delta_n^1(\mathfrak{p}^*)$. In general, let $U(z)\in\mathfrak{N}$. Then $U(z)$ is a Q.D.B.H. and $I[U(z)]$ is a G.G. Clearly $I[U(z)]=\lim_n U_n(z)$, where $U_n(z)$ is a P.H. in $R_n-R_0-\mathcal{V}_n(\mathfrak{p}^*)$ which is the least positive harmonic function with $U_n(z)=U(z)$ on $\partial\mathcal{V}_n(\mathfrak{p}^*)$. Then $U_n(z)\downarrow I[U(z)]$. Let $U(z)\in\mathfrak{N}$. Then $U(z)=\int N(z,p)d\mu(p)$. Since $N(z,p)$ is continuous with respect to $p\in R-R_0+\Delta^N$ and the integration can be approximated by a linear from in every compact set,

$$I[U(z)] = \int_{\Delta_1^N} I[N(z,p)]d\mu \geqq \delta\int d\mu > 0 \text{ at } z = p_0. \qquad (6)$$

Evidently $I[U(z)]\leqq U$ and $I[U(z)]$ is finite except $\mathfrak{p}^*$.
Next let $V(z)$ be a G.G. which is finite except $\mathfrak{p}^*$. We denote the least positive P.H.$\in\mathfrak{N}$ larger that $V(z)$ by $E^N[V(z)]$. Let $V_n(z)=\tilde{H}_V^{R-R_0-\mathcal{V}_n(\mathfrak{p}^*)}$. Then $V_n(z)\geqq V(z)$ and $V_n(z)=V(z)$ on $\partial\mathcal{V}_n(\mathfrak{p}^*)$ and

$$\frac{\partial}{\partial_n} V_n(z) \geqq \frac{\partial}{\partial_n} V(z) \text{ on } \partial\mathcal{V}_n(\mathfrak{p}^*). \qquad (7)$$

Let $\Omega_\varepsilon=\{z:V(z)>\varepsilon\}$. Then $\Omega_\varepsilon\in SO_g$. By $M=\sup_{z\in\partial\mathcal{V}_n(\mathfrak{p}^*)} V(z)<\infty$, $D_{R-R_0-\mathcal{V}_n(\mathfrak{p}^*)}(V(z))<\infty$.

$$\int_{\partial\Omega_\varepsilon\cap(R-R_0-\mathcal{V}_n(\mathfrak{p}^*))} \frac{\partial}{\partial_n} V(z)ds + \int_{\partial\mathcal{V}_n(\mathfrak{p}^*)\cap\Omega_\varepsilon} V(z)ds = 0 \text{ for } \varepsilon < M.$$

On the other hand, $\int_{\partial\Omega_\varepsilon} \frac{\partial}{\partial_n} V(z)ds = \mathcal{M}(V(z)) < \infty$. $0 \geqq \int_{\partial\mathcal{V}_n(p^*)\cap\Omega_\varepsilon} \frac{\partial}{\partial_{n_*}} V(z)ds =$

$-\int_{\partial\mathcal{V}(p^*)\cap\Omega_\varepsilon} \frac{\partial}{\partial_n} V(z)ds \geqq -\mathcal{M}(V(z))$. Let $\varepsilon\downarrow 0$. Then since $\partial\mathcal{V}_n(p^*)$ is compact,

$\partial\mathcal{V}_n(p^*)\cap\Omega_\varepsilon \nearrow \partial\mathcal{V}_n(p^*)$ and $\int_{\partial\mathcal{V}_n(p^*)} \frac{\partial}{\partial_n} V(z)ds \geqq -\mathcal{M}(V(z))$. Since $V_n(z) =$

$\tilde{H}_V^{R-R_0-\mathcal{V}_n(p^*)}$, $\int_{\partial R_0} \frac{\partial}{\partial_n} V_n(z)ds = -\int_{\partial\mathcal{V}(p^*)} \frac{\partial}{\partial_n} V_n(z)ds$. Whence by (7)

$$\int_{\partial R_0} \frac{\partial}{\partial_n} V_n(z)ds \leqq \mathcal{M}(V(z)) \text{ for any } n. \qquad (8)$$

We see at once $V_n(z)\nearrow$ as $n\to\infty$ and by (8) $\lim_n V_n(z) < \infty$. Hence

$$E[V(z)] = \lim_n V_n(z).$$

We see $V_n(z)$ has M.D.I. property at $p \neq p^*$, whence $E^N[V(z)]$ has M.D.I. property at $p \neq p^*$. Also $\sup_{z\notin\mathcal{V}_n(p^*)} V(z) \leqq \max_{z\in\partial\mathcal{V}_n(p^*)} V(z)$ and since $\partial\mathcal{V}_n(p^*)$ is compact, $\sup_{z\notin\mathcal{V}_n(p^*)} E^N[V(z)] \leqq \max_{z\in\partial\mathcal{V}_n(p^*)} E^N[V(z)]$. Hence

$$E^N[V(z)] \in \mathcal{K} \qquad (9)$$

and $E^N[V(z)]$ is an H.F.S.P.

Let $V(z)$ be a G.G. finite except p^*. Then by (9) $E[V(z)] < \infty$ and $\in \mathcal{K}$, whence $IE^N[V(z)]$ is a G.G. Evidently $E^N[V(z)] \geqq V(z)$ and $IE^N[V(z)] \geqq V(z)$.

On the other hand, any P.H. smaller than a G.G. is also a G.G. Whence $E^N[IE^N[V(z)]-V(z)]$ can be defined. Since E^N and I are linear operators, $E^N[IE^N V(z)-V(z)]=E^N IE^N V(z)-E^N V(z)$. On the other hand,

$$E^N[V(z)] \geqq IE^N[V(z)] \geqq V(z) \text{ and}$$

$$E^N[V(z)] = E^N E^N[V(z)] \geqq E^N IE^N[V(z)] \geqq E^N V(z).$$

Hence $0=E^N[IE^N[V(z)]-V(z)]\geqq IE^N[V(z)]-V(z)$ i.e.

$$IE^N[V(z)] = V(z). \qquad (10)$$

Let $V(z)\in\mathcal{K}$. Then by definition, $II[V(z)]\leqq I[V(z)]\leqq E^N I[V(z)]\leqq V(z)$. Hence

$$I[V(z)] = IE^N I[V(z)]. \qquad (11)$$

$E^N I[V(z)]\in\mathcal{K}$ and $V(z)\in\mathcal{K}$ imply $0\leqq V(z)-E^N I[V(z)]\in\mathcal{K}$. Assume $V(z)>IE^N[V(z)]$. Then by (6) $I[V(z)-E^N I[V(z)]]>0$. This contradicts (11). Hence we have

$$E^N I[V(z)] = V(z) \text{ for } V(z) \in \mathcal{K}. \qquad (12)$$

Let $U(z)\in\mathcal{K}$ and be N-minimal (or U(z) is a multiple of N(z,p): $p\in\Delta_1^N(\mathfrak{p}^*)$). Let $0<V(z)\leqq I[U(z)]$. Since $I[U(z)]$ is a G.G, V(z) is also a G.G. and $E^N[V(z)]<\infty$. $U(z)\in\mathcal{K}$ implies $0\leqq U(z)-E[V(z)]\in\mathcal{K}$. By the N-minimality of U(z), $E[V(z)]=aU(z):0\leqq a\leqq 1$. Now by (10)

$$V(z) = IE[V(z)] = aI[U(z)].$$

Hence we have

If U(z) is N-minimal, I[U(z)] is K-minimal. (13)

Let $V(z)$ be a K-minimal which is finite except $\mathfrak{p}^*$. Then $V(z)$ is a multiple of a $K(z,p):p\in\Delta_1^K(\mathfrak{p})$ and there exists a curve γ tending to p. We can find a sequence $\{p_i\}:p_i\in\partial\mathcal{V}_i(\mathfrak{p}^*)$ such that $K(z,p_i)\to K(z,p)$ and $G(z,p_i)$ converges to a harmonic function $G(z)$ which is clealy >0 and is a G.G. By the K-minimality of $K(z,p)$, $K(z,p)=\acute{a}G(z):\acute{a}>0$ and $V(z)$ is a G.G, whence $E^N[V(z)]<\infty$ and $\in\mathfrak{N}$.

Let $T(z)$ ($\leqq E^N[V(z)]$) be an H.F.S.P. such that $E^N[V(z)]-T(z)$ is also an H.F.S.P. Then we have $T(z)_{\mathcal{V}(\mathfrak{p}^*)}=T(z)$ by $(E^N[V(z)]_{\mathcal{V}(\mathfrak{p}^*)}=E^N[V(z)]$. Whence $T(z)$ is finite except $\mathfrak{p}^*$ and $\in\mathfrak{N}$.

Now by (10)

$$V(z) = IE^N[V(z)] \geqq I[T(z)].$$

By the K-minimality of $V(z)$, $I[T(z)]=a[V(z)]:0<a\leq 1$. By $T(z)\in\mathfrak{N}$, we have by (12) $T(z)=E^N I[T(z)]$ and $T(z)=E^N I[T(z)]=aE^N[V(z)]$. Thus we have

$$\text{If } V(z) \text{ is K-minimal, } E^N[V(z)] \text{ is N-minimal.} \qquad (14)$$

By (13) and (14) we have the following

Theorem 2. Let $R-R_0$ be an end of a Riemann surface $\notin O_g$ and let $\mathfrak{p}^*$ be a completely irregular boundary component. Then $\Delta_1^N(\mathfrak{p}^*)$ and $\Delta_1^K(\mathfrak{p}^*)$ have one-to-one correspondence.

Remark. If $R-R_0$ is an end of a Riemann surface $\in O_g$, every boundary is completely irregular and $G(z,p)=N(z,p)$ and our assertions is trivial. In the previous paper[3] theorem 2 was proved under the condition, every $\mathfrak{p}$ consists of analytic boundary and we can define a symmetric Riemann surface of $R-R_0$ relative to $\sum_{\mathfrak{p}_i\neq\mathfrak{p}^*}\mathfrak{p}_i$.

If $R-R_0$ is a plane domain, $R-R_0$ is mapped into a circle so as $\partial R_0 \to |w|=1$, $p^* \to w=0$ and other boundary conponents is distributed in $|w|<1$ so that $w=0$ may be irregular if and only if p^* is completely irregular. By M. Brelot's theorem there exists only one K-minimal point on $w=0$, which is a multiple of a fixed G.G. Hence there exists only one N-minimal point on $w=0$. This means $N(z,p_i) \to N(z,p)$ as $p_i \to \{w=0\}$.

Hence we have at once the following

Corollary. Let Ω be the above domain. We can map Ω conformally onto a unit circle with radial slits $\tilde{\Omega}$ by an analytic function $\zeta=f(w)$ so that $\{w=0\} \to \{\zeta=0\}$. Such function $f(w)$ is uniquely determined except rotation.

Let $L(z)$ be the conjugate harmonic function $N(z,p)$. Put $f(z)=\exp(-N(z,p)-iL(z))$. Then $f(z)$ is required function. Since $N(z,p)$: $p\in\Delta_1^N$, $\min(M,N(z,p))=M\omega(V_M,z,R-R):V_M(z)=\{z:N(z,p)>M\}$ and $\int_{\partial V_M} \frac{\partial}{\partial n}N(z,p)ds=2\pi$ for almost M. Hence we have: Length of $C_\gamma=2\pi r$ for almost r where $C_\gamma=\{\zeta:|\zeta|=r\}$.

Application 3. On the regularity of a boundary point. Let Γ be a unit circle and let F be a closed set such that $F\ni p:p=\{z=0\}$. Let $\Omega=\Gamma-F$ and let $G(z,p)$ be a Green function of Γ. Then p is regular relative to Ω if and only if

$$G(z,p) = H^{\Omega}_{G(z,p)} \text{ in } \Omega .$$

This fact is prove by $G(z,p)-\log\frac{1}{|z-p|}$ is harmonic near p.

Such theorem is generalized as follows. Let $R-R_0$ be an end of a Riemann surface $\in O_g$ and $\mathfrak{p}$ be a boundary component. Let F be a closed set. Let $G^{\Omega}(z,p)$ be a Green function of $\Omega=R-R_0-F$. We call $\mathfrak{p}$ is regular or irregular according as $\lim_{p\to\mathfrak{p}} G^{\Omega}(z,p_0)=0$ or not: $p_0\in\Omega$. Let $G(z,p_i)$ be a Green function of $R-R_0$.

Let $\{p_i\}$ such that $G(z,p_i)$ converges to a harmonic function $G(z,\{p_i\})$. Then $G(z,\{p_i\})$ is a G.G. and a Q.D.B.H. in Ω. Suppose $\mathfrak{p}$ is regular. We show $G(z,\{p_i\})=H^{\Omega}_{G(z,\{p_i\})}$. Let $S(z)=G(z,\{p_i\})-H^{\Omega}_{G(z,\{p_i\})}$. Then $S(z)$ is singular and a G.G. such that $\mathfrak{M}(S(z))\leqq 2\pi$ and

$$S(z) = \int_{\overline{G}_\lambda\cap\Delta_1^K} K(z,p)d\mu(p);^{1)}$$

where $G_\lambda=\{z:G^{\Omega}(z,p_0)>\lambda\}$ and the closure is taken relative to K-Martin's topology on Ω. By $S(z)\leqq G(z,\{p_i\})$, $S(z)$ is finite in R, $\mu=0$ except $\mathfrak{p}^*$. By the regularity of $\mathfrak{p}$, $\overline{G}_\lambda\cap\Delta_1^K(\mathfrak{p})$ is void and $S(z)=0$. If $\mathfrak{p}$ is not regular, it is easily proved there exists at least one $G(z,\{p_i\})$ such that $G(z,\{p_i\})\neq H^{\Omega}_{G(z,\{p_i\})}$. This consideration is also applicable to N-Martin's topology and N-Green function and $\tilde{H}^{\Omega}_{N(z,p)}$.

Application 4. On Lindelöfian maps. Let $f(z)$ be an analytic function from R into $\underline{R}\notin O_g$. Let $C(r,p):p\in\underline{R}$ be a local disk and let $G(w,q)$ be a Green function of $C(r,p)$. Let R' be a connected piece of R over $C(r,p)$. Then

$$G(w,q) = \sum_i G(z,p_i) + V_q(z) + W_q(z),\ f(p_i) = q,$$

where $G(z,p_i)$ is a Green function of R' and $V_q(z)$ is quasibounded harmonic function and $W_q(z)$ is singular. We suppose K-Martin's topology is defined on R'. Then we denote by $\Delta_1(G)$ the set of minimal points of R' $(=R+\Delta^K)$ of which G is a fine neighbourhood. Suppose $W_q(z)>0$. Then

$$W_q(z) = \int_{\Delta_1} K(z,p)d\mu(p).$$

We denote by $\tilde{f}(p):p\in\Delta_1^K$ the fine limit at p and $f^{-1}(q)$ the set of p such that $\tilde{f}(p)=q$. We state without proof the following

1) μ has no mass on $(\Delta_1^K-\tilde{f}^{-1}(q))$.

2) Let G be a domain such that $\mu(\Delta_1(G))>0$. Then f(G) is a fine neighbourhood of q. Especially if μ has a point mass at p. Then by 1) $\tilde{f}(p)=q$ and let $\mathcal{V}(p)$ be a fine neighbourhood of p, the $f(\mathcal{V}(p))$ is a fine neighbourhood of q.

3) Let $\mathcal{V}(q)$ be a fine neighbourhood of q, then $\mu(\Delta_1-\Delta_1(f^{-1}(\mathcal{V}q))=0$.

4) Let $0<w(z)$ $(\leqq W_q(z))$ be a G.G. function. The μ of w(z) is positive at most countable points.

5) Let n(w) be the number of points of $\underline{R}$ over w. If $n(w)\leqq n_0<\infty$ in C(r,p), then $W_q(z)=\sum_{i=1}^{m} K(z,p_i):p_i\in(\Delta_1^K\cap\tilde{f}^{-1}(q))$ and $m\leqq n_0$, $W_q(z)$ is a G.G. with $\mathfrak{M}(W_q(z))\leqq m$.

By using application 1 we shall obtain some results from the application 4.

References

1) Z. Kuramochi: On quasi Dirichlet bounded harmonic fanctions, Hokkaido Math. Jour., Vol. VIII, 1-22 (1979).

2) Z. Kuramochi: Potentials on Riemann surfaces, Jour. Fac. Sci. Hokkaido Univ. Vol. 16, 80-148 (1962).

C. Constantinuscu and A. Cornea: Ideale Ränder Riemannscher Flächen, Springer. (1963).

3) Z. Kuramochi: On minimal points of Riemann surface II, Hokkaido Math. Jour., Vol. II, 139-175 (1973).

Department of Mathematics
Hokkaido University
Sapporo, Japan

NATURAL LOCALISATION OF A STANDARD H - CONE

Eugen Popa

Seminarul Matematic, Universitatea "Al.I.Cuza"

6600 - Iaşi

Introduction. The purpose of this note is to prove a theorem of localisation for standard H-cones. Results related to localisation was previously established by many authors, in various situations: Mokobodzki-Sibony [7], Feyel-de la Pradelle [5] in the case of an adapted cone of potentials; Bliedtner-Hansen [1] for standard spaces of balayage (which are special H-cones of functions, cf. [9]), Popa [8] in the case of a standard H-cone of functions. All these results were proved for a locally compact topology. Boboc-Bucur-Cornea [2] or [4], prove a theorem of localisation for a standard H-cone, with respect to the fine topology.

The localisation which we are going to prove is with respect to the natural topology, which need not be locally compact. We remark also that, under mild assumptions on the space of representation, the conditions considered in the theorem are equivalent with the natural sheaf property.

All the notions and notations about H-cones are as in [4]. All topological notions, not specified otherwise, reffers to the natural topology.

Theorem Let S be a standard H-cone of functions on the set X, such that:

1. X is a Baire space with respect to the fine topology.

2. The H-measures form a solid subcone in S^* (i.e. any H-integral, dominated by a H-measure, is itself a H-measure)

If the properties of truncation:

(T) $\forall V \subseteq X$ open, $\forall s,t \in S$ such that $\liminf_{\substack{y \to x \\ y \in V}} s(y) \geq t(x), \forall x \in \partial V$

we have $s' \in S$, where $s' = \min(s,t)$ on V, and $s' = t$ on $X \setminus V$

and adaptation:

(A) $\forall V \subseteq X$ open, $\forall h \in \mathcal{P}(V)$, carr $h = \emptyset \Longrightarrow h$ substractible

holds, then S has the natural sheaf property (i.e. $U \longmapsto \mathcal{P}(U)$ is a sheaf on the natural topology of X).

<u>Remark</u>. Condition 1. holds, for exemple, when X is complete metrisable with respect to the natural topology. Let X_1 be the saturated space of representation for S. Condition 2. holds if, for exemple, $X_1 \setminus X$ is polar. The property of truncation is imposed in all the localisation theorems. For some equivalent conditions see prop.2. For the relations of the condition (A) with the usual adaptation condition, see [4]. Prop.4 contains the converse of the theorem.

<u>Proof</u>. We begin with three preliminary results.

<u>Proposition 1</u>. Let S be a standard H-cone on the set X. Let $F \subseteq X$ be closed and $U \subseteq X$ open, such that $F \subseteq U$. Then there exists an open $V \subseteq X$, such that $F \subseteq V \subseteq \overline{V} \subseteq U$ and $B^{X \setminus V}$ is a balayage on S.

<u>Proof</u>. The natural topology is metrisable, hence there exists a continuous $f: X \to [0,1]$ such that $f(x) = 0, \forall x \in F$ and $f(x) = 1$, $\forall x \in X \setminus U$. For each $\alpha \in (0,1)$, let us denote

$$V_\alpha = \{x \in X \mid f(x) < \alpha\}$$

V_α is an open set, $F \subseteq V_\alpha \subseteq \overline{V}_\alpha \subseteq U$, and the proof from [6] (see also [8]) proves our result.

Let β be the set of all open $V \subseteq X$ such that $B^{X \setminus V}$ is a balayage on S, and $(B^{X \setminus V})^* (\varepsilon_x)$ denoted also by $\varepsilon_x^{X \setminus V}$, is a $\mathbb{H}$-measure on X, for each $x \in V$.

<u>Proposition 2</u>. Let S be a standard H-cone of functions on X. Suppose

that condition 2. from theorem holds. The following conditions are equivalent:

a) For any open $U,V \subseteq X$ such that $\overline{V} \subseteq U$, $\forall s \in \mathcal{S}(V)$, $t \in \mathcal{S}(U)$ such that $\liminf_{\substack{y \to x \\ y \in V}} s(y) \geq t(x)$, $\forall x \in \partial V$, we have $s' \in \mathcal{S}(U)$, where $s'=\min(s,t)$ on V; $s=t$ on $U \setminus V$.

b) Condition (T).

c) $\forall V \in \beta$, $\forall x \in V$, $\varepsilon_x^{X \setminus V}$ is carried by ∂V.

d) For any open $U,V \subseteq X$ such that $\overline{V} \subseteq U$ and $V \in \beta$, $\forall x \in V$, $\hat{\varepsilon}_x^{U \setminus V}$ is carried by ∂V.

Proof. a $\Rightarrow$ b Clear.

b $\Rightarrow$ c Let $W \subseteq X$ be open, such that $\overline{V} \subseteq W$ and let $p \in S$ be a generator. Since $p=B^W p$ on ∂V, for any $s \in S$ such that $s \geq B^W p$ on $X \setminus V$ we have $s \geq p$ on ∂V. Condition b) shows then, that $s' \in S$, where $s'=\min(s,p)$ on V and $s'=p$ on $X \setminus V$. But $s' \geq p$ on $X \setminus V$, hence $s' \geq B^{X \setminus V} p$ on X. Since $s' \leq s$ on V, it follows that $s \geq B^{X \setminus V} p$ on V. Since s was arbitrary, we find $B^{X \setminus V} B^W p \geq B^{X \setminus V} p$. The inverse inequality is obvious, hence $\varepsilon_x^{X \setminus V}(p - B^W p)=0$, $\forall x \in V$, which proves that $\varepsilon_x^{X \setminus V}$ doesn't charges $d(W)$. Since we can find a sequence of open sets $W_n \subseteq X$, such that $\overline{V} = \bigcap \overline{W}_n$, and $b(W_n) \subseteq \overline{W}_n$, it follows that $\varepsilon_x^{X \setminus V}$ doesn't charges $X \setminus \overline{V}$. Since $V \in \beta$, it follows that $\varepsilon_x^{X \setminus V}(p - B^{X \setminus V} p)=0$, $\forall x \in V$, and this proves that $\varepsilon_x^{X \setminus V}$ doesn't charges $(X \setminus \overline{V}) \cup d(X \setminus V) \supseteq X \setminus \partial V$.

c $\Rightarrow$ d Since $\varepsilon_x^{X \setminus V}$ is supposed to be a H-measure carried by ∂V, we can view it as a measure on U. It suffices then to show that

$$\hat{\varepsilon}_x^{U \setminus V}(s_B) = \int s_B \, d\varepsilon_x^{X \setminus V}$$

, for any $s \in S^f$ (where $B=B^{\beta(X \quad U)}$ is the greatest balayage on S, dominated by $B^{X \setminus U}$). Let us recall that the natural topology on U, with respect to $\mathcal{S}(U)$, is finer than the trace on U, of the natural topology on X; hence s_B is naturally measurable. We have:

$$\varepsilon_x^{U\setminus V}(s_B) = \hat{B}^{U\setminus V}(s-Bs)(x) = B^{b(X\setminus U)\cup(U\setminus V)}s(x)-Bs(x) =$$

$$= B^{X\setminus V}s(x)-Bs(x)$$

(using [4]) and:

$$\int s_B d\,\varepsilon_x^{X\setminus V} = \varepsilon_x^{X\setminus V}(s) - \varepsilon_x^{X\setminus V}(Bs) = B^{X\setminus V}s(x)-Bs(x)$$

d ⟹ a We follow the proof of [4, th.5.6.10]). Let s, t be as in a). Since there exists an increasing net $t_i \in \mathcal{S}(U)_0$ such that $t = \bigvee t_i$, then $s = \sup s_i$, where: $s_i = \min(s, t_i)$ on V and $s_i = t_i$ on $U\setminus V$. Hence we can suppose that $t \in \mathcal{S}(U)_0$. For each $\alpha > 0$, let us denote $s_\alpha = \min(s+\alpha, t)$ on V, $s_\alpha = t$ on $U\setminus V$. Since $s = \inf_{\alpha>0} s_\alpha$, and s is i.s.c., it suffices to prove that $s_\alpha \in \mathcal{S}(U)$, $\forall\, \alpha > 0$. Let us denote: $A_\alpha = \{x \in V \mid s(x) + \alpha > t(x)\}$. Since

$$\liminf_{\substack{y\to x\\ y\in V}} s(y) \geq t(x), \ \forall\, x \in \partial V,$$ it follows that $A_\alpha \cup (U\setminus V)$ is an open set. Let B_α be the balayage on $A_\alpha \cup (U\setminus V)$, in the H-cone $\mathcal{S}(U)$. We prove now that $s_\alpha - B_\alpha t \in \hat{S}_{B_\alpha}$. Applying prop.1 to the set $U\setminus A_\alpha$, we find a decreasing sequence $W_n \in \beta$, such that $A_\alpha = \bigcup_{n=1}^{\infty}(U\setminus W_n)$. Hence $B_\alpha s = \sup_n B^{U\setminus W_n}s$, $\forall\, s \in \mathcal{S}(U)$. This proves that the measure $B_\alpha^*(\varepsilon_x)$ is carried by $b(B_\alpha)\cap W$, for any open neighbourhood W of ∂A_α, and any $x \in d(B_\alpha)$. Let us denote by $\hat{B}_\alpha$ the corresponding balayage on V, hence $\hat{B}_\alpha = \hat{B}^{A_\alpha \cap d(B)}$, where $B = B^{\beta(U\setminus V)}$ is the greatest balayage on $\mathcal{S}(U)$, dominated by $B^{U\setminus V}$. For each $s \in S^f$ and $x \in d(B)$ we have:

$$\hat{B}^{A_\alpha \cap d(B)}(s-Bs)(x)+Bs(x) = B_\alpha s(x)$$

hence:

$$\hat{B}^*_\alpha(\varepsilon_x)(s-Bs) = B^*_\alpha(\varepsilon_x)(s-Bs)$$

Since $t|_V$ is the supremum of an increasing net $(s_i)_B$ and V is a neighbourhood for ∂A_α, we have:

$$\hat{B}^*_\alpha(\varepsilon_x)(t|_V) = B^*_\alpha(\varepsilon_x)(t)$$

But $s+\alpha > t$ on $A_\alpha \cap d(B)$. Hence we have:

$$\hat{B}^*_\alpha(\varepsilon_x)(t|_V) = \hat{B}^*_\alpha(\varepsilon_x)(\min(s+\alpha,t)|_V)$$

This proves that:

$s_\alpha(x)-B_\alpha t(x) = (s(x)+\alpha) - \hat{B}^*_\alpha(\varepsilon_x)(\min(s+\alpha,t)|_V)$. From [4,th.5.1.4] it follows that: $(s+\alpha)-\hat{B}^*_\alpha(\varepsilon_x)(\min(s+\alpha,t)|_V) \in \mathcal{S}_{\hat{B}_\alpha}$ hence $s_\alpha - B_\alpha t \in \mathcal{S}_{B_\alpha}$.

It follows that there exists $s_i \in S^f$, such that $s_i - Bs_i \uparrow s_\alpha - B_\alpha t$ on $d(B_\alpha)$. Since $b(B_\alpha) \subseteq \overline{A} \cup (U \setminus V)$ (fine closure) we have in fact $s_i - B_\alpha s_i \nearrow s_\alpha - B_\alpha t$ on U. It follows that $s_i - B_\alpha s_i + B_\alpha t \leq \leq s_\alpha \leq t$ on U, and this proves that $s_i - B_\alpha s_i + B_\alpha t \in \mathcal{S}(U)$. Since $s_\alpha = \sup_i (s_i - B_\alpha s_i + B_\alpha t)$, we have $s_\alpha \in \mathcal{S}(U)$.

<u>Proposition 3</u>. Let S be a standard H-cone of functions on X. If $t_n \in S$ is specifically decreasing and t_1 is finite, then: $\curlywedge_n t_n = \inf_n t_n = \bigwedge_n t_n$.

<u>Proof</u>. The first equality is [4,prop.2.1.6.]. From $t_1 - \inf t_n = \sum_{n=1}^{\infty} (t_n - t_{n+1}) \in S$, it follows that $\inf_n t_n$ is finely continuous, hence $\inf_n t_n = \widehat{\inf_n t_n} = \bigwedge_n t_n$.

We pass now to the proof of the theorem. Let $(U_i)_{i\in I}$ be an open covering for $U \subseteq X$. Let $f: U \longrightarrow \overline{R}_+$ be such that $f|_{U_i} \in \mathcal{S}(U_i)$, $\forall i \in I$. We must show that $f \in \mathcal{S}(U)$. Since $\mathcal{S}(U)$ is a standard H-cone of functions on U, it suffices to prove that $\min(f,s) \in \mathcal{S}(U)$, $\forall s \in \mathcal{S}(U)_o$. Indeed, there exists a weak unit $u \in \mathcal{S}(U)$ and a sequence $s_n \in \mathcal{S}(U)_o$, such that $s_n \uparrow u$. If $t_n = \min(f, ns_n)$, we have $t_n \uparrow f$ and $t_n \in \mathcal{S}(U)$. Using [4, th.3.1.5] it remains to prove that f is finite on a finely dense set. Since X is a Baire space, each U_i has the same property, and $f|_{U_i} \in \mathcal{S}(U_i)$ shows that $f|_{U_i}$ is finite on a finely dense set on U_i. This proves that f is finite on a finely dense set.

Let $x \in U$ and $i \in I$ be such that $x \in U_i$. There exists, by prop.1, $W_{i,x} \in \beta$, such that $x \in W_{i,x} \subseteq \overline{W}_{i,x} \subseteq U_i$. Since the natural topology has a countable base, we can suppose that there exists a countable covering W_n of U, such that $W_n \in \beta$ and $\forall n\ \exists i_n \in I$ such that $\overline{W}_n \subseteq U_{i_n}$. Moreover, we can suppose that, for each $n \in \mathbb{N}$, the set $\{m \in \mathbb{N} \mid W_n = W_m\}$ is infinite.

Let $s \in \mathcal{S}(U)_o$ be fixed. We have $\min(s+f, s) = s \in \mathcal{S}(U)$. If $t \in \mathcal{S}(U)$ and $\min(f+t, s) \in \mathcal{S}(U)$, then we prove that $\min(f+B^{U\setminus W_n}t, s) \in \mathcal{S}(U)$, $\forall n \in \mathbb{N}$. Indeed, let $i_n \in I$ be such that $\overline{W}_n \subseteq U_{i_n}$. Then:

$$\min(f+B^{U\setminus W_n}t, s) = \begin{cases} \min(u,v) & \text{on } U_{i_n} \\ v & \text{on } U\setminus U_{i_n} \end{cases}$$

where $u = \min(f+B^{U\setminus W_n}t, s)$ and $v = \min(f+t, s)$. The equality follows from: $U\setminus U_{i_n} \subseteq U\setminus \overline{W}_n \subseteq b(U\setminus W_n)$, hence $B^{U\setminus W_n}t = t$ on

$U \setminus U_{i_n}$; and $B^{U\setminus W_n}t \leq t$ on U.Moreover, since $B^{U\setminus W_n}t=t$ on ∂U_{i_n} we have $\liminf_{\substack{y\to x\\ y\in U_{i_n}}} u(y) \geq v(x), \forall x \in \partial U_{i_n}$.Prop2,c $\Rightarrow$a shows that

$\min(f+B^{U\setminus W_n}t,s) \in \mathcal{S}(U)$.Hence,if $u_n = B^{U\setminus W_n}B^{U\setminus W_{n-1}}\dots B^{U\setminus W_1}s$, we have $u_n \in \mathcal{S}(U)$ and $\min(f+u_n,s) \in \mathcal{S}(U)$.

We prove next that $\inf_n u_n=0$.Let $x \in U$ and $W \subseteq X$ open be such that $x \in W$,and there exists $n \in \mathbb{N}$ such that $W \subseteq W_n$.Hence,for infinitely many $n \in \mathbb{N}$,we have $B^{U\setminus W}u_n=u_n$.Indeed, $u_n \geq B^{U\setminus W}B^{U\setminus W_n}u_n \geq$ $\geq B^{U\setminus W_n}B^{U\setminus W_n}u_n = B^{U\setminus W_n}u_{n-1} = u_n$.Since u_n is decreasing,and $(B^{U\setminus W})^*(\varepsilon_x)$ is a H-measure,it follows that

$$\inf_n u_n = (B^{U\setminus W})^*(\varepsilon_x)(\inf_n u_n)$$

Let us denote $t_n=(B^{U\setminus W})^*(\varepsilon_x)(u_n)|_W$;we have $t_n \in \mathcal{S}(W)$ and $t_n \preccurlyeq t_{n+1}$.Prop.3 proves that $\curlywedge_n t_n = \bigwedge_n t_n = \inf_n t_n$,which implies $\inf u_n|_W=\inf t_n \in \mathcal{S}(W)$.Hence $\inf_n u_n$ is finely continuous on W, hence on U.We have: $\bigwedge_n u_n = \inf_n u_n$ and this proves that $u = \inf_n u_n \in \mathcal{S}(U)$.

Now,the definition shows that carr $u=\emptyset$ (in $\mathcal{S}(U)$,as H-cone on U).By (A), u is substractible.Since $u \leq s$ and $s \in \mathcal{S}(U)_0 \subseteq \mathcal{P}$ (notations from [3,th.2.4.c)] ,we obtain u=0.It remains to remark that $\min(f,s) = \inf_n \min(f+u_n,s)$,hence denoting $v_n = \widehat{\min(f+u_n,s)}$, we have $v_n \in \mathcal{S}(U)$;hence $\bigwedge_n v_n \in \mathcal{S}(U)$.But $\bigwedge_n v_n = \widehat{\inf_n v_n}$ with respect to the fine topology;since f and s are finely continuous, $\min(f,s) = \inf_n v_n$ is also finely continuous,hence $\min(f,s) \in \mathcal{S}(U)$.

<u>Proposition 4</u>.Let S be a standard H-cone of functions on X.If S has the natural sheaf property,then S satisfies (T) and (A).

<u>Proof</u>.(T)Using the remark in the proof of prop.2,d$\Rightarrow$a,if suffices

to prove that $s_\alpha \in S$, with $t \in S_o$. But $s_\alpha|_{U\setminus\overline{V}} \in \mathcal{S}(U\setminus\overline{V})$ and $s_\alpha|_V \in \mathcal{S}(V)$. Moreover, for each $x \in \partial V$ there exists an open neighbourhood W, such that $s_\alpha|_W = t|_W \in \mathcal{S}(W)$, The natural sheaf property implies $s_\alpha \in S$.

(A) From carr $h = \emptyset$ it follows that, for each $x \in X$ there exists an open neighbourhood $W \in \beta$, such that $h = B^{U\setminus W}h$. If $h \leq s$, with $s \in \mathcal{S}(U)$, then $h = B^{U\setminus W}h \preccurlyeq s$ in $\mathcal{S}(W)$, which shows that there exists $u \in \mathcal{S}(W)$ such that $h+u = s$. The natural sheaf property shows now that $h \preccurlyeq s$ in $\mathcal{S}(U)$.

BIBLIOGRAPHY

[1] Bliedtner J., Hansen W. "Harmonic Spaces and Markov Processes" Z.Wahrsheinlick. 42(4), pg.309-326

[2] Boboc N., Bucur Gh., Cornea A. "H-cones and Potentail Theory" Ann.Inst.Fourier, XXV (3-4) 1975, pg.71-108.

[3] Boboc N., Bucur Gh., Cornea A. "Carrier Theory and Negligible Sets on a Standard H-cone Of Functions" Rev.Roumaine Math. Pures et appl. XXV (2) 1980, pg.163-198

[4] Boboc N., Bucur Gh., Cornea A. "Order and Convexity in Potential Theory: H-cones" LNM 853, Springer 1981

[5] Feyel D., de la Pradelle A. "Cônes en dualité. Applications aux fonctions de Green" LNM 518, Springer 1976

[6] Lukes J., Netuka I. "The Wiener Type Solution of the Dirichlet Problem in Potential Theory" Math.Ann.244(1976), pg.173

[7] Mokobodzki G., Sibony D. "Principe du minimum et maximalité en Théorie du potentiel" ann.Inst.Fourier, XVII (1967) pg.401

[8] Popa E. "Localisation and Product of H-cones" preprint INCREST no.10/1979 (to appear in Rev.Roum.Math.Pures at appl.)

[9] Popa E. "Standard H-cones - Standard Spaces of Balayage" (ta appear in Rev.Roum.Math.Pures appl)

ON SUB-MARKOV RESOLVENTS. THE RESTRICTION TO AN OPEN SET AND THE DIRICHLET PROBLEM

by

Lucreţiu STOICA

Abstract This paper deals with sub-Markov resolvents $(V_\lambda)_{\lambda>0}$ on a locally compact space E with countable base. The resolvent has some special properties the main of which are the following:

1° $V_\lambda(C_c(E))\subset C(E)$,

2° There exists a standard process on E associated to the resolvent (V_λ).

For an open set $U\subset E$ we study the resolvent $(V'_\lambda)_{\lambda>0}$ on U associated by killing the process on CU. Namely we give sufficient conditions (expressed by the existence of barrier functions) which imply that the resolvent $(V'_\lambda)_{\lambda>0}$ has properties of the type 1° (see Theorems 3.2 and 4.2). This problem is closely connected to the probabilistic Dirichlet problem (see Proposition 4.1 and Corollary 4.4).

Thanks are do to K.Janssen who made evident an error of the author.

1. Let E be a locally compact space with a countable base and let S be a convex cone of lower semicontinuous nonnegative numerical functions on E such that the constant function 1 belongs to S. Denote by S^* the family of all universally measurable nonnegative numerical functions f such that $\mu(f)\leq f(x)$ for each $x\in E$ and each measure μ which fulfils the inequalities $\mu(s)\leq s(x)$, for all $s\in S$. Obviously S^* is a convex cone stable under infimum. Further let T be another convex cone such that $S\subset T\subset S^*$. If $f:E\longrightarrow R$ is an arbitrary function we shall use the notation

$$^{T}Rf = \inf\{s\in T / f\leq s\},$$

provided the set in brackets is not empty. The function ^{T}Rf is called the reduite function. If $f\leq 0$ we have $^{T}Rf=0$.

We shall denote by D(T) the family of all continous functions f which have the following properties:

1° there exists $s\in T$ such that $|f|\leq s$,

2° $\inf\{^{T}R(|f|\chi_{CK})/K$ compact set$\}=0$

It is easy to see that D(T) is a vector lattice which contains the space $C_c(E)$. We also note that $^{T}Rf<\infty$ for each $f\in D(T)$.

The next result was proved in [6, p.76] in the case $T = S^{*}$. However the proof given there is still valid in the more general case stated here.

<u>Theorem 1.1.</u>

Let $f:E\longrightarrow R$ be an upper semicontinuous function such that there exists a function $g\in D(T)$ with $f\leq g$. Then for each point $x\in E$ there exists a nonnegative measure μ such that

1° $\mu(s)\leq s(x)$ for each $s\in S$,

2° $^{T}Rf(x) = \mu(f)$.

2. Further let $(V_\lambda)_{\lambda>0}$ be a sub-Markov resolvent of kernels on E satisfying the following two conditions:

(2.1) $V_\lambda(C_c(E))\subset C_b(E)$, $\lambda>0$

(2.2) $\lim\limits_{\lambda\to\infty} \lambda V_\lambda f(x) = f(x)$, for each $f\in C_c(E)$ and $x\in E$.

Sometimes we shall discuss consequences of the following supplementary condition:

(2.1') $V_0 = \lim_{\lambda\to\infty} V_\lambda$ is a finite kernel and $V_0(C_c(E)) \subset C_b(E)$.

We shall explicity mention whenever condition (2.1') will be assumed in a statement.

A function $s:E \to [0,\infty]$ will be called α - excessive ($\alpha \geq 0$) provided:

1^0 s is universally measurable,

2^0 $\lambda V_{\lambda+\alpha} s \leq s$, $\lambda > 0$

3^0 $s(x) = \lim_{\lambda\to\infty} \lambda V_{\lambda+\alpha} s(x)$ for each $x \in E$.

Let us note that each lower semicontinuous function $s:E \to [0,\infty]$ satisfying conditions 1^0 and 2^0 is α - excessive. As a consequence let us note that the infimum of two lower semicontinuous excessive functions is also an excessive function.

For $\alpha \geq 0$, the family of all α - excessive functions is a convex cone that will be denoted by $\mathcal{E}_\alpha$. The sub-cone of all real valued continuous excessive functions will be denoted by $\mathcal{E}_{\alpha c}$. Following G.Mokobodzki [4 ch III] we next present some basic properties related to the cones $\mathcal{E}_\alpha$ and $\mathcal{E}_{\alpha c}$. We have $V_\alpha(C_c^+(E)) \subset \mathcal{E}_{\alpha c}$ for $\alpha > 0$ and if (2.1') is valid, this relation still holds for $\alpha = 0$. The monotone class theorem shows that $V_\alpha f \in \mathcal{E}_{\alpha c}^*$ for each Borel nonnegative function f, and hence the same is true for f universally measurable and nonnegative. By standard arguments on excessive functions it follows $\mathcal{E}_\alpha \subset \mathcal{E}_{\alpha c}^*$ for each $\alpha > 0$ and if (2.1') is valid,this relation is still true for $\alpha = 0$.

<u>Theorem 2.1.</u>

The following statements are true for each $\alpha>0$.

1° If f is a lower semicontinuous function, then the reduite ${}^{\mathcal{E}_\alpha}R f$ is also lower semicontinuous.

2° If f is an upper semicontinuous function and there exists a function $g\in D(\mathcal{E}_{\alpha c})$ such that $f\leq g$, then

$$ {}^{\mathcal{E}_{\alpha c}}R f = {}^{\mathcal{E}_\alpha}R f. $$

Particularly ${}^{\mathcal{E}_\alpha}R f$ is upper semicontinuous.

3° If f belongs to $D(\mathcal{E}_{\alpha c})$, then ${}^{\mathcal{E}_\alpha}R f$ is continuous.

If (2.1') holds, then the above three assertions are also valid for $\alpha = 0$.

The proof of the theorem uses Theorem 1.1 and the method of proof of the similar results from [4 ch.III].

In the reminder of this paper we shall assume that, besides properties (2.1) and (2.2), the resolvent has the following property:

(2.3) $V_\lambda f\in D(\mathcal{E}_{\lambda c})$ for each $\lambda>0$ and $f\in C_c(E)$.

Sometimes we shall need the followind supplementary condition:

(2.3') $V_0 f\in D(\mathcal{E}_{0c})$ for each $f\geq 0$, $f\in C_c(E)$.

Remark. K.Jansen produced an example of a resolvent which has not property (2.3) still satisfying conditions (2.1), (2.2) and the following one:

$$V_\lambda f \in D(\mathcal{E}_\lambda) \quad \text{for each } \lambda > 0 \text{ and } f \geq 0,\ f \in C_c(E).$$

In the work[6 ch.VI.2] the author makes a mistake and consider that this condition is equivalent to (2.3). In fact the arguments given there work only under the assumption that (2.3) hold. Next we retake the point.

We shall explicitely mention whenever condition (2.3') will be assumed.

Next we introduce some technical notation. For each $\alpha \geq 0$ we denote by P_α the convex cone of continuous potentials defined by the following equality:

$$P_\alpha = \mathcal{E}_{\alpha c} \cap D(\mathcal{E}_{\alpha c}).$$

It is not hard to see that the cone P_α is stable under infimum. Consequently the space $P_\alpha - P_\alpha$ is a vector lattice. If $s \in P_\alpha$ we shall denote by $F(s)$ the family

$$F(s) = \{t \in P_\alpha / t \leq s\ ,\ s-t \in C_c(E)\}.$$

Since $s \in D(\mathcal{E}_{\alpha c})$ we see that $\inf F(s)=0$.

The space $(P_\alpha - P_\alpha) \cap C_c(E)$ will be denoted by T_α. This space is a vector lattice. We assert that for each $\alpha > 0$ the uniform closure $\overline{T}_\alpha$ coincides with the space $C_0(E)$ of all continuous functions vanishing to infinity. This will follow from the Stone-Weierstrass theorem once we have proved that T linearly separates the points of E. From condition (2.2) and the resolvent equation it follows that for each $\lambda>0$, the space $V_\lambda(C_c(E))$ linearly separates the points of E. Therefore the same is true for $V_\alpha(\mathfrak{C}_c^+(E))$. Condition (2.3) asserts that this cone is contained in P_λ. Therefore for $x,y\in E$, $x\neq y$ we may choose $s,t\in P_\alpha$ such that $s(x)t(y)\neq s(y)t(x)$. Since $\inf F(s)= =\inf F(t)=0$ we further can choose $s'\in F(s)$ and $t'\in F(t)$ such that $(s-s')(x)(t-t')(y)\neq(s-s')(y)(t-t')(x)$. Since $s-s'$, $t-t'\in T_\alpha$ we see that T_α linearly separates the points of E. Therefore we have $\overline{T}_\alpha= C_o(E)$ for each $\alpha>0$. This relation is still true for $\alpha=0$ provided (2.1') and (2.3') hold.

Further the methods of J.C.Taylor [7] together with Lemmas 2.6 and 2.7 of ch.VI in [6] give us the following:

<u>Theorem 2.2.</u>

There exists a Hunt process $(\Omega,M,M_t,X_t,\theta_t,P^*)$ with state space E such that

$$V_\lambda f(x)=E^*[\int_0^\infty \exp(-\lambda t)f(X_t)dt],$$

for each $f\in B_b(E)$, $x\in E$, $\lambda>0$.

3. In this section we consider a fixed open set U and study the behaviour of the process $(\Omega,M,M_t,X_t,\theta_t,P^*)$ given in Theorem 2.2 in connection with this open set. Namely we are interested in the regularity properties of the resolvent $(V'_\lambda)_{\lambda>0}$ associated to the process killed on CU. On Markov

processes we shall use the notation from [1]. The resolvent $(V'_\lambda)_{\lambda>0}$ is expressed by the following equality:

$$V'_\lambda f(x)=E^x[\int_0^{T_{CU}} \exp(-\lambda t)f(X_t)dt]$$

for each $f\in B_b(U)$ and $x\in U$. It is also known the following relation:

(3.1) $V_\lambda f(x) = V'_\lambda f(x)+E^x[\exp(-\lambda T_{CU})V_\lambda f(X_{T_{CU}})]$,

for each $f\in B_b(E)$ and $x\in U$. By Hunt's theorem, if $f\geq 0$ we have:

(3.2) $E^x[\exp(-\lambda T_{CU})V_\lambda f(X_{T_{CU}})] = {}^{\mathcal{V}_\lambda}R(\chi_{CU}V_\lambda f)(x)$.

The above relations are also valid for $\lambda=0$ with f such that $\{f\neq 0\}$ is relatively compact, provided (2.1') holds.

In general $(V'_\lambda)_{\lambda>0}$ has the following regularity property:

Proposition 3.1.

If f is a nonnegative bounded lower semicontinuous function on U, then $V'_\lambda f$ is also lower semicontinuous for each $\lambda\geq 0$.

Proof

Let $\lambda>0$. In order to prove the theorem it suffices to show that $V'_\lambda f$ is lower semicontinuous provided f is continouous with compact support in U and $f\geq 0$, because the general case with follow expressing a nonnegative lower semicontinuous function as the supremum of an increasing sequence of nonnegative continuous functions with compact support. Lef $f \in C_c(U)$, $f \geq 0$. Since $V_\lambda f$ is continuous, to show that $V'_\lambda f$ is lower semicontinuous, on account of relations (3.1) and (3.2) it is enough to show that ${}^{\mathcal{E}_\lambda}R(\chi_{CU} V_\lambda f)$ is upper semicontinuous. Since $V_\lambda f \in P_\lambda$ we see that $\chi_{CU} V f$ is an

upper semicontinuous function which fulfils the requirements of Theorem 2.1, 2^0 and this theorem leads us to the desired conclusion for $\lambda > 0$. The case $\lambda = 0$ follows from the equality $V'_0 f = \sup_\lambda V'_\lambda f$.

One can easily construct examples such that $V'_\lambda f$ is not continuous for $f \in C_c(E)$ (see the example from page 47 of[6]). We next give conditions to ensure that $V'_\lambda(C_c(E)) \subset C_b(E)$.

Theorem 3.2.

Let us suppose that conditions (2.1') and (2.3') hold and there exist two functions $f, g \in C(E)$ such that $g, f \geq 0$ on E, $f > 0$ on U, $V_0 f \in C(E)$, $g = V_0 f$ on CU and $g \leq {}^{\mathcal{E}_0}R(\chi_{CU} g)$.

Let $\Gamma = \partial U \cap \{f = 0\}$. Then for each function $h \in C_c(E)$ such that $\operatorname{supp} h \cap \Gamma = \emptyset$ and each $\lambda \geq 0$ the function $V'_\lambda(h_{|U})$ is continuous bounded and vanishes on the boundary ∂U.

If there exists a constant $c > 0$ such that $f \geq c$ on U, then for each $h \in C_b(U)$ and each $\lambda \geq 0$ the function $V'_\lambda h$ is continuous and vanishes on the boundary ∂U.

Proof

The assumptions of the theorem imply

$$^{\mathscr{E}_o}R_g \leq {}^{\mathscr{E}_o}R({}^{\mathscr{E}_o}R(\chi_{CU}g)) = {}^{\mathscr{E}_o}R(\chi_{CU}g) = {}^{\mathscr{E}_o}R(\chi_{CU}V_o f).$$

On the other hand we have ${}^{\mathscr{E}_o}R(\chi_{CU}g) \leq {}^{\mathscr{E}_o}Rg$. Therefore

$${}^{\mathscr{E}_o}R_g = {}^{\mathscr{E}_o}R(\chi_{CU}V_o f).$$

Next we want to show that ${}^{\mathscr{E}_o}Rg$ is continuous. This will follow from Theorem 2.1,3^o once we have proved that $g \in D(\mathscr{E}_{oc})$. The assumptions of the theorem show $g \leq V_o f$. Therefore if we prove that $V_o f \in D(\mathscr{E}_{oc})$, it follows $g \in D(\mathscr{E}_{oc})$. Let $x \in E$ and $\varepsilon > 0$. We choose $\varphi \in C_c(E)$ such that $0 \leq \varphi \leq 1$ and $|V_o f(x) - V_o(\varphi f)(x)| < \varepsilon/2$. From (2.1') and (2.3') we know that $V_o(\varphi f) \in P_o$. Hence there exist a compact set K and $s \in \mathscr{E}_{oc}$ such that $s(x) < \varepsilon/2$ and $V_o(\varphi f) \leq s$ on CK. The function $t = V_o((1-\varphi)f) + s$ belongs to $\mathscr{E}_{oc}$ and satisfies $t(x) < \varepsilon$ and $V_o f \leq t$ on CK. This shows that $V_o f$ and consequently g belong to $D(\mathscr{E}_{oc})$. From relations (3.1) and (3.2) with $\lambda = 0$ we deduce that $V_o'(f) = V_o'(f_{|U})$ is continuous and vanishes on ∂U. If $\varphi \in C(U)$, $0 \leq \varphi \leq 1$, from Proposition 3.1 one deduces that $V_o'(\varphi f)$ and $V_o'(1-\varphi)f)$ are lower semicontinuous. Since their sum is continuous it follows that both $V_o'(\varphi f)$ and $V_o'((1-\varphi)f)$ are continuous. We deduce that $V_o'h$ is continuous for any $h \in C(U)$ satisfying $0 \leq h \leq f$.

Further for $\lambda > 0$ we put $f_\lambda = f + \lambda V_o f$ and write

$$V_\lambda f_\lambda = V_\lambda f + \lambda V_\lambda V_o f = V_o f \in C(U).$$

Then similarly we deduce that $V_\lambda'h$ is continuous for any

$h\in C(U)$ satisfying $0\leq h\leq f_\lambda$. If $h\in C_c(E)$, $h\geq 0$ from the complete maximum principle we know that V_0h is bounded. Since $V_0'h\leq V_0h$ we have $V_0'h\in C_b(U)$. The rest of proof is obvious now.

Theorem 3.2'

Assume that conditions (2.1') and (2.3') hold and there exist a function $f\in C(E)$ such that $f\geq 0$ on E, $f>0$ on U and $V_0f\in C(E)$ and a function $g\in C(U)$ satisfying the following relations:

1° $g\leq {}^{\mathcal{E}_0}R(\chi_{CU}V_0f)$ on U and

2° $\inf\{{}^{\mathcal{E}_{0c}}R(\chi_{U\setminus M}(V_0f-g))(x)/M$ closed set, $M\subset U\}=0$, for each point $x\in U$.

Let $\Gamma=\partial U\cap\{f=0\}$. Then for each function $h\in C_c(E)$ such that $\mathrm{supp}\,h\cap\Gamma=\phi$ and each $\lambda\geq 0$, $V_\lambda'(h_{|U})$ is continuous. If there exists a constant $c>0$ such that $f\geq c$ on U, then for each $h\in C_b(U)$ and each $\lambda\geq 0$ the function $V_\lambda'h$ is continuous.

Proof

First we are going to show that ${}^{\mathcal{E}_{0c}}R(\chi_{CU}V_0f)$ is continuous. A simple compacteness argument and condition 2° imply that for each compact set $K\subset U$ and each $\varepsilon>0$ there exist a continuous excessive function $s\in\mathcal{E}_{0c}$ and a closed set $M\subset U$ such that $s<\varepsilon$ on K and $V_0f-g\leq s$ on $U\setminus M$. Further let φ be a continuous function on E such that $0\leq\varphi\leq 1$ on E, $\varphi=0$ on M and $\varphi=1$ on CU. From Theorem 2.1, 3° we know that the function ${}^{\mathcal{E}_0}R(\varphi V_0f)$ is continuous.

We extend the function g on E by putting $g(x)=V_0f(x)$ if $x\in E\setminus U$. Then we can write:

(3.3) $\quad {}^{\mathcal{E}_0}R(\chi_{CU}V_0f) \leq {}^{\mathcal{E}_0}R(\varphi V_0f) \leq {}^{\mathcal{E}_0}R(\varphi g) + {}^{\mathcal{E}_0}R(\varphi(V_0f-g)).$

On the other hand we have

$$ {}^{\mathcal{E}_0}R(\chi_{CU}V_0f) \leq {}^{\mathcal{E}_0}R(\varphi g) \leq {}^{\mathcal{E}_0}R(g) $$

and on account of 1^0 we deduce

$$ {}^{\mathcal{E}_0}R(\varphi g) = R(\chi_{CU}V_0f). $$

Since $\varphi = 0$ on M we have ${}^{\mathcal{E}_0}R(\varphi(V_0f-g)) \leq s < \varepsilon$ on K. Therefore relation (3.3) became

$$ {}^{\mathcal{E}_0}R(\chi_{CU}V_0f) \leq {}^{\mathcal{E}_0}R(\varphi V_0f) \leq {}^{\mathcal{E}_0}R(\chi_{CU}V_0f) + \varepsilon \quad \text{on K.} $$

From these we infer that the reduite function ${}^{\mathcal{E}_0}R(\chi_{CU}V_0f)$ can be uniformly approximated on each compact subset of U with continuous functions, and hence it is continuous on U.

Further the proof of this theorem is similar to the preceding one.

Before going to apply the above results to a more particular case we give the following lema.

3.3. Lemma

Let f be a real continuous function on E such that the family $\{f(X_T)/T$ stopping time$\}$ is uniformly integrable with respect to each measure P^x. Assume that $\mathcal{U}$ is a family of open subsets of U such that $U = \bigcup_{W \in \mathcal{U}} W$ and

$$ E^x[f(X_{T_{CW}})] \leq f(x), \text{ for each } W \in \mathcal{U} \text{ and each } x \in W. $$

Then for each $x \in U$,

$$E^x[f(X_{T_{\complement U}})] \leq f(x).$$

Proof.

For a fixed point $x \in U$, let $\mathcal{A}$ be the family of all stopping times T such that

1° $T \leq T_{\complement U}$ and

2° $E^x[f(X_T)] \leq f(x)$.

For any increasing sequence of stopping times in $\mathcal{A}$, their limit also belongs to $\mathcal{A}$. We consider the quotient $\hat{\mathcal{A}}$ of $\mathcal{A}$ under the relation of P^x-a.s. equality. It is not difficult to see that $\hat{\mathcal{A}}$ is inductive ordered and that any maximal element of $\hat{\mathcal{A}}$ should coincide with (the equivalence class of) $T_{\complement U}$. Using Zorn's lemma we get the desired conclusion.

4. From now on we assume that the resolvent $(V_\lambda)_{\lambda > 0}$ satisfies conditions (2.1), (2.2), (2.3), (2.1') and (2.3') and the process associated to the resolvent has continuous paths. It is proved in [6, ch.VI] that the process has continuous paths if and only if for each open set $W \subset E$, the space $C_c(W)$ is contained in the uniform closure of its image by V_o, i.e. $C_c(W) \subset \overline{V_o(C_c(W))}$. Particularly we shall use the following relation:

(4.1) $\quad C_c(E) \subset \overline{V_o(C_c(E))}$.

First we shall be interested in the probabilistic solution of the Dirichlet problem. Since the process is continuous, $X_{T_{CU}} \in \partial U$-$P^x$, a.s. for each point $x \in U$. If f is a continuous function on ∂U, the probabilistic solution to the Dirichlet problem with boundary function f is the function Hf defined by

$$Hf(x) = E^x[f(X_{T_{CU}})], \ x \in U.$$

Proposition 4.1.

Suppose that for each $f \in C_c(E)$ the function $V_0'(f_{|U})$ is continuous on U. Then $Hg \in C_b(U)$ for each function $g \in C_c(\partial U)$. Morever if $\lim_{\substack{y \to x \\ y \in U}} V_0'(f_{|U})(y) = 0$ for each boundary point $x \in \partial U$ and each $f \in C_c(E)$, then

$$\lim_{\substack{y \to x \\ y \in U}} Hg(y) = g(x) \quad \text{for each } x \in \partial U \text{ and each } g \in C_c(\partial U).$$

Proof.

If $f \in C_c(E)$, then $H((V_0 f)_{|\partial U}) = (V_0 f)_{|U} - V_0'(f_{|U}) \in C_b(U)$. From relation (4.1) we see that each function $g \in C_c(\partial U)$ can be uniformly approximated with functions of the form $(V_0 f)_{|\partial U}$, which implies the assertion of the proposition.

If D is an open set in E and f is a continuous function on D we say that f is superharmonic (resp. subharmonic) on D if for each open set W, with $\overline{W} \subset D$ and each point $x \in W$,

$$E^x[f(X_{T_{CW}})] \leq f(x) \quad (\text{resp.} f(x) \leq E^x[f(X_{T_{CW}})]).$$

For example a function of the form $V_0 g$ is superharmonic (resp.subharmonic) on the interior of the set $\{g \geq 0\}$ (resp.

$\{g\leqslant 0\}$), where g belongs to $C_c(E)$. We note that f is superharmonic on D if and only if (-f) is subharmonic on D and that the infimum of two superharmonic functions on D is also a superharmonic function on D.

In the reminder of this section we shall study the resolvent $(V'_\lambda)_{\lambda>0}$ in the case of a regular boundary ∂U. Namely we assume that for each point $x\in\partial U$, there exists an open neighbourhood U_x of x and two continuous functions φ_x, ψ_x on U_x such that:

- $\varphi_x(x) = 0$,
- $\varphi_x(y)$ 0 for each point $y\in\overline{U}\cap U_x$, $y \neq x$,
- φ_x is superharmonic on $U\cap U_x$,
- $\psi_x > 0$
- ψ_x is subharmonic on $U_x\cap U$.

This regularity property of ∂U allows us to deduce the following result.

<u>Theorem 4.2.</u>

a) For each function $h\in C_c(U)$ and each $\lambda\geqslant 0$ the function $V'_\lambda h$ is continuous, bounded and vanishes on the boundary ∂U.

b) If there exists a continuous function f such that $f\geqslant c$ for some constant $c>0$ and $V_0 f\in C(E)$, then for each $h\in C_b(U)$ and each $\lambda\geqslant 0$ the function $V'_\lambda h$ is continuous and vanishes on the boundary ∂U.

<u>Proof.</u>

If $\{\varphi_n\}$ is a sequence of functions in $C_c(E)$ then the sum $f = \sum_n (1/2^n)(1/\|\varphi_n\|)\varphi_n$ is a continuous function such that

$V_0 f \in C_b(E)$ and obviously we can choose this sequence such that $f>0$ on E. We shall treat both cases a) and b) together. Namely we shall apply Theorem 3.2 using the above constructed function f in the case a) or the function f given by the statement in the case b). The main difficulty lies in constructing the function g having the properties required by Theorem 3.2.

It is given by the following lemma the proof of which retakes word by word a construction of J. P. Roth [5 p.61]. For the interested reader a detailed proof is presented in the Appendix.

Lemma 4.3.

Let h be a nonnegative continuous function on E such that the family $\{h(x_T)/T$ stopping time$\}$ is uniformly integrable for each measure P^x. Then there exists a continuous function g on $\bar{U}$ with the following properties:

1° $g = h$ on ∂U,

2° $g \leq h$ on U,

3° g is subharmonic on U.

For $h = V_0 f$ the above lemma produces a function g on $\bar{U}$. Then we extend g on E by putting $g(x)=V_0 f(x)$ for $x \in CU$. Since g is subharmonic on U by Lemma 3.3 we get $g(x) \leq E^x[g(X_{T_{CU}})]$ for each $x \in U$. Therefore on account of Hunt's theorem we deduce that g fulfils the requirements of Theorem 3.2, which implies the desired conclusion.

Corollary 4.4.

If $g \in C_c(\partial U)$, then $Hg \in C(U)$ and $\lim_{\substack{y \to x \\ y \in U}} Hg(y) = g(x)$ for each point $x \in \partial U$.

Remark

This Corollary may be viewed as a generalisation of Roth's result from [5, IV.4.4].

Appendix. Proof of Lemma 4.3.

We construct the function g as the uniform limit of a sequence $(g_n)_{n\in N}$, having the following properties:

a) g_n is a continuous nonnegative function on $\overline{U}$,

b) g_n is subharmonic on U,

c) $g_n \leq h$ on $\overline{U}$ and $h < g_n + 1/n$ on ∂U if $n \geq 1$

d) $g_n \leq g_{n+1}$ on $\overline{U}$ and $g_{n+1} = g_n$ on the set

$$\{g_n + 1/n \leq h\} \text{ if } n \geq 1.$$

The sequence will be constructed inductively. Our construction of g_{n+1} will work either for the case n=0 when we take $g_0=0$ or for the general case when $n \geq 1$ and we suppose that g_n exists and satisfies a),b),c),d).

For $x \in U$ we choose $\alpha \geq 0$ such that the inequality

$$(*) \qquad h - (1/_{n+1}) < \alpha\psi_x \leq h$$

holds on a neighbourhood of x. We denote by U'_x an open neighbourhood of x on which the inequality (*) holds and such that $\overline{U'_x} \subset U_x \cap \{h < (1/_n) + g_n\}$ if $n \geq 1$ or $\overline{U'_x} \subset U_x$ if n=0. The properties of φ_x allow us to choose another open neighbourhood U''_x of x and a constant $\beta \geq 0$ such that $\overline{U''_x} \subset U'_x$ and

$$\alpha\psi_x - \beta\varphi_x \leq 0 \text{ on } (U'_x \setminus U''_x) \cap \overline{U}.$$

Further denote by U'''_x another open neighbourhood of x on which the following inequality holds

$$h - (1/_{n+1}) < \alpha\psi_x - \beta\varphi_x$$. Assume also that $\overline{U_x'''} \subset U_x''$.

Now define $g_x = \sup(\alpha\psi_x - \beta\varphi_x, g_n)$ on $U_x' \cap \overline{U}$. This function has the following properties:

1° $g_x \leq h$ on $U_x' \cap \overline{U}$,

2° $g_x = g_n$ on $(U_x' \setminus U_x'') \cap \overline{U}$,

3° $h - (1/_{n+1}) < g_x$ on $U_x''' \cap \overline{U}$,

4° g_x is subharmonic on $U_x' \cap U$,

Further by standard topological arguments we can choose a sequence $(x_k)_{k\in N}$ of points in the boundary ∂U and the neighbourhood U'_{x_k}, $k\in N$ such that the following conditions hold:

5° $\partial U \subset \bigcup_k U'''_{x_k}$

6° for each compact set $K \subset E$ the family $\{k\in N / K \cap U'_{x_k} \neq \emptyset\}$ is finite.

Then we define the function g_{n+1} on $\overline{U}$ by putting:

$$g_{n+1}(x) = \begin{cases} \sup\{g_{x_k}(x) / x \in U''_{x_k}\} & \text{if } x \in (\bigcup_k U'_{x_k}) \cap \overline{U}, \\ g_n(x) & \text{if } x \in \overline{U} \setminus \bigcup_k U'_{x_k} \end{cases}$$

The properties a), b), c) and d) for g_{n+1} follows using the properties $1^o, 2^o, 3^o, 4^o, 5^o$ and 6°. We check here only a) and b). If $x \in \overline{U} \setminus \bigcup_k U'_{x_k}$, there exists an open neighbourhood D of x such that $D \cap (\bigcup_k \overline{U}''_{x_k}) = \emptyset$, $D \subset U$. From 2° it follows $g_{n+1} = g_n$ on D.

If $x \in (\bigcup_k U'_{x_k}) \cap \overline{U}$ there exists an open neighbourhood D of x and a finite family $\{k_1, k_2, \ldots, k_i\}$ such that $D \subset \bigcap_{j=1}^{i} U'_{x_{kj}}$ and $D \cap U''_{x_k} = \emptyset$ provided $k \notin \{k_1, k_2, \ldots, k_i\}$. Then $g_{n+1} = \sup(g_{x1}, g_{x2}, \ldots, g_{xi})$ on D. Now we see that g_{n+1} is continuous on $\overline{U}$ and locally subharmonic on U. From Lemma 3.3 it follows that g_{n+1} is subharmonic on U.

From the properties a), b), c) and d) one easily deduces that the limit $g = \lim_{n \to \infty} g_n$ is a continuous function on $\overline{U}$, g=h on ∂U and g is subharmonic on U.

[1] Blumenthal, R.M., Getoor, R.K., Markov Processes and Potential Theory, New York-London, Academic Press, 1968.

[2] Constantinescu, C., Cornea, A., Potential Theory on Harmonic Spaces, Springer, Berlin-Heidelberg-New-York, 1972.

[3] Meyer, P.A., Processus de Markov. Notes Math.26, Springer, Berlin-Heidelberg-New York, 1967.

[4] Mokobodzki, G., Cônes de potentiels et noyaux subordonnés, in vol.Potential Theory, Edizione Cremonese, Roma, 1970.

[5] Roth, J.P., Opérateurs dissipatifs et semi-groupes dans les espaces de fonctions continues, Ann.Inst.Fourier, 26,1-98, 1976.

[6] Stoica, L., Local operators and Markov Processes, Lect.Notes Math. 816, Springer, Berlin-Heidelberg-New York, 1980.

[7] Taylor, J.C., Ray processes on locally compact spaces, Math.Ann.208, 233-248, 1974.

DIVERGENCE OF MULTIPOINT PADE APPROXIMATION

Hans Wallin

0. INTRODUCTION

Let n and ν be non-negative integers. If $\beta_{jn\nu}$, $1 \leq j \leq n+\nu+1$, are given complex numbers and f a given function, we want to determine a rational function $R_{n\nu} = P_{n\nu}/Q_{n\nu}$, $Q_{n\nu} \not\equiv 0$, of type (n,ν) such that $fQ_{n\nu} - P_{n\nu} = 0$ at the points $\beta_{jn\nu}$, $1 \leq j \leq n+\nu+1$; if k of the numbers $\beta_{jn\nu}$, $1 \leq j \leq n+\nu+1$, are equal we require that the corresponding zero has multiplicity k. It is easy to prove that $R_{n\nu}$ exists and is unique. The convergence problem, i.e. to decide whether $R_{n\nu} \to f$, as $n+\nu \to \infty$, was studied in [5] and [6] which also contain references to earlier results on this problem by Montessus de Ballore, Walsh, Kalmár, Saff, Karlsson, Warner, and others.

In §2 we recall a convergence result from [6] for the case when ν is fixed and f is meromorphic with ν poles in a certain domain. The aim of this note is to show in §3 by means of a simple method that this result is best possible for a special class of functions f in the sense that we have divergence outside the domain where the convergence theorem guarantees that $R_{n\nu} \to f$, as $n \to \infty$.

1. DEFINITIONS

Let E be a compact subset of the complex plane $\mathbb{C}$, n and ν non-negative integers and $\beta_{jn\nu} \in E$, $1 \leq j \leq n+\nu+1$, given points, the <u>interpolation points</u>. We introduce the polynomial

$$\omega_{n\nu}(z) = \prod_{j=1}^{n+\nu+1} (z-\beta_{jn\nu})$$

and assume that f is a given function wich is analytic at least at all the interpolation points. We determine polynomials $P_{n\nu}$ of degree at most n and $Q_{n\nu}$ of degree at most ν, $Q_{n\nu} \not\equiv 0$, such that

$$(f(z)Q_{n\nu}(z) - P_{n\nu}(z))/\omega_{n\nu}(z) \text{ is analytic at } \beta_{jn\nu},\ 1 \leq j \leq n+\nu+1. \quad (1.1)$$

Then $R_{n\nu} = P_{n\nu}/Q_{n\nu}$ exists and is unique; $R_{n\nu}$ is the <u>multipoint Padé approximant</u> (or <u>rational interpolant) of type</u> (n,ν) <u>to</u> f <u>and the interpolation points</u> $\beta_{jn\nu}$, $1 \leq j \leq n+\nu+1$. When $\beta_{jn\nu} = 0$ for all j, then $R_{n\nu}$ is the Taylor series of f if $\nu = 0$ and the <u>Padé approximants</u> of f if $\nu \geq 0$.

We shall consider the case when ν is fixed and we introduce the <u>associated measure</u> $\mu_n = \mu_{n\nu}$ to $\beta_{jn\nu}$, $1 \leq j \leq n+\nu+1$, as the probability measure on E assigning the point mass $1/(n+\nu+1)$ to $\beta_{jn\nu}$ for all j. We assume that μ is a probability measure on E and introduce its logarithmic potential

$$u(z;\mu) = \int \log \frac{1}{|z-t|}\, d\mu(t),$$

as well as the potential $u(z;\mu_n)$ of μ_n. Then

$$u(z;\mu_n) = \frac{1}{n+\nu+1} \sum_{j=1}^{n+\nu+1} \log \frac{1}{|z-\beta_{jn\nu}|} = \frac{-1}{n+\nu+1} \log|\omega_{n\nu}(z)|.$$

The convergence of $R_{n\nu}$ is determined among other things by the behaviour of $\omega_{n\nu}$, i.e. of $u(z;\mu_n)$, and hence of μ_n, as $n \to \infty$.

To study this convergence we use the following definition, where $\complement E = \mathbb{C} \setminus E$.

DEFINITION. $\{\mu_n\}$ is (μ,E) - <u>regular</u> if

$$\liminf_{n\to\infty} u(z;\mu_n) \geq u(z;\mu), \quad \text{for } z \in \mathbb{C}, \quad (1.2)$$

and

$$\lim_{n\to\infty} u(z;\mu_n) = u(z;\mu), \quad \text{for } z \in \complement E. \quad (1.3)$$

For the general background we list some properties concerning (1.2) and (1.3) and refer to [6] for proofs:

a) (1.2) holds if and only if, for every real α and every compact set $K \subset ¢$, there exists a constant $n(\alpha,K)$ such that $u(z;\mu) > \alpha$ on $K \Rightarrow u(z;\mu_n) > \alpha$ on K for $n > n(\alpha,K)$.

b) If $\mu_n \to \mu$ in the ω^*-topology (i.e. $\int g d\mu_n \to \int g d\mu$ for all continuous g on E), then $\{\mu_n\}$ is (μ,E)-regular.

c) If E has empty interior, then (1.3) implies (1.2) and that $\mu_n \to \mu$.

d) If $\complement E$ is connected, then (1.2) implies (1.3).

e) (1.3) implies that $u(z;\mu_n) \to u(z;\mu)$ uniformly on compact subsets of $\complement E$.

2. A CONVERGENCE RESULT

We recall that $\nu \geq 0$ is fixed, μ_n is the associated measure to $\beta_{jn\nu} \in E$, E compact, f is a given function which is analytic at $\beta_{jn\nu}$ for all j, $R_{n\nu} = P_{n\nu}/Q_{n\nu}$ is the multipoint Padé approximant of type (n,ν) to f and $\beta_{jn\nu}$, and μ is a probability measure on E. We introduce

$$E_\rho = E_{\mu,\rho} = \{z \in ¢ : u(z;\mu) > \log \frac{1}{\rho}\}, \quad \text{for } \rho > 0.$$

Then $E_\rho \subset E_{\rho'}$, if $\rho < \rho'$, and E_ρ is open since potentials are lower semicontinuous.

The following theorem was proved in [6] and [5] where the history of the theorem and related results was also given.

THEOREM 1. <u>Let</u> μ <u>and</u> $\rho > 0$ <u>be given. Suppose that</u> $\{\mu_n\}$ <u>is</u> (μ,E)-<u>regular and that</u> f <u>is meromorphic in an open set containing</u> $\bar{E}_\rho \cup E$ <u>with exactly</u> ν <u>poles</u> $z_1,\ldots,z_\nu$ $(\neq \beta_{jn\nu})$, <u>counted with their multiplicities, so that</u> $z_j \in E_\rho$ <u>for</u> $1 \leq j \leq \nu$.

Then $R_{n\nu} \to f$, as $n \to \infty$, uniformly and even with geometric degree of convergence on compact subsets of $E_\rho \setminus \{z_j\}_1^\nu$, i.e. for each compact set $K \subset E_\rho \setminus \{z_j\}$,

$$\limsup_{n\to\infty} \{\sup_{z\in K} |f(z) - R_{n\nu}(z)|\}^{1/n} < 1.$$

Furthermore, $R_{n\nu}$ has ν poles in $\mathbb{C}$, if n is large, and these converge to the poles of f, as $n \to \infty$.

Example. In the Padé case, i.e. when $\beta_{jn\nu} = 0$ for all j and $E = \{0\}$, then $\mu_n = \mu$ is the unit mass at 0, $E_\rho = \{z: |z| < \rho\}$, and we get a classical theorem on Padé approximants by Montessus de Ballore [2] which, for $\nu = 0$, gives the basic result on the convergence of Taylor series inside the circle of convergence. Since Taylor series diverge outside the circle of convergence, we may expect that in general $R_{n\nu}$ will diverge outside $\bar{E}_{\rho'}$ if ρ' is the largest ρ such that f is meromorphic with ν poles in $E_\rho \cup E$. In §3 we shall prove a result of that kind.

Remark. Since $E_\rho = \{z: u(z;\mu) > \log 1/\rho\}$, the boundary ∂E_ρ of E_ρ is a subset of $\{z: u(z;\mu) \leq \log 1/\rho\}$. The assumption in Theorem 1 that f is meromorphic with ν poles in $\bar{E}_\rho \cup E$ can be changed to the assumption that f is meromorphic with ν poles in $E_\rho \cup E$ if, for some $\varepsilon > 0$,

$$\partial E_{\rho'} \subset \{z: u(z;\mu) = \log 1/\rho'\}, \quad \text{for } \rho > \rho' > \rho - \varepsilon. \tag{2.1}$$

In fact, if $K \subset E_\rho \setminus \{z_j\}$ is compact, then $u(z;\mu)$, being lower semi-continuous, assumes a minimum value on K and this value is larger then $\log 1/\rho$. This means that $K \subset E_{\rho'} \subset E_\rho$ for some $\rho' < \rho$ such that $z_j \in E_{\rho'}$, $1 \leq j \leq \nu$, if we have fixed a function f which is meromorphic in $E_\rho \cup E$ with ν poles $z_1, \ldots, z_\nu$ such that $z_j \in E_\rho$ for all j. But

$\bar{E}_{\rho'} \subset E_\rho$ according to (2.1). Hence, f is meromorphic in $\bar{E}_{\rho'} \cup E$ with ν poles $z_j \in E_{\rho'}$. By using Theorem 1 with ρ changed to ρ' we get the convergence of $R_{n\nu}$ to f on K which proves what we claimed. The condition (2.1) is satisfied in all "normal" situations, for instance when $E_\rho \supset E$ or $u(z;\mu)$ is a continuous function of z. In these cases the convergence in $E_\rho \setminus \{z_j\}$ in Theorem 1 is, consequently, not affected by singularities on ∂E_ρ.

3. A DIVERGENCE RESULT

If ρ, and hence E_ρ, grows under the condition that f is meromorphic with ν poles in $\bar{E}_\rho \cup E$ (or in $E_\rho \cup E$ if (2.1) is satisfied), then Theorem 1 guarantees convergence of $R_{n\nu}$ to f in $E_\rho \setminus \{z_j\}$ all the time. However, if we reach a new singularity of f we can not expect convergence outside $\bar{E}_\rho$ as is shown by the following theorem where, as usual, μ_n is the associated measure to $\beta_{jn\nu} \in E$, E compact, and $R_{n\nu} = P_{n\nu}/Q_{n\nu}$ is the multipoint Padé approximant to f and $\beta_{jn\nu}$. Except for the set $E_\rho = \{z: u(z;\mu) > \log 1/\rho\}$, $\rho > 0$, we also introduce

$$\Gamma_\rho = \{z: u(z;\mu) = \log 1/\rho\}.$$

THEOREM 2. Let μ and $\rho'' > \rho > \rho' > 0$ be given. Suppose that $\{\mu_n\}$ is (μ,E)-regular and that f is meromorphic in an open set containing $\bar{E}_{\rho'} \cup E$ with ν poles $z_1,\dots,z_\nu$ so that $z_j \in E_{\rho'}$, $1 \le j \le \nu$. Suppose that f has a simple pole $\alpha \in \Gamma_\rho$, possibly poles $\alpha_1,\dots,\alpha_k \in E_{\rho''} \setminus (E_\rho \cup \Gamma_\rho \cup E)$, and that f has no further singularities in $\bar{E}_{\rho''}$. Suppose that α is different from and not a limit point of the interpolation points $\beta_{jn\nu}$. Finally, suppose that $z \in E_{\rho''} \setminus (E_\rho \cup \Gamma_\rho \cup E)$ and that $z \neq \alpha_j$, $1 \le j \le k$.

Then $|R_{n\nu}(z)| \to \infty$, as $n \to \infty$.

Remark. We thus have divergence at $z \in E_{\rho''} \setminus E$ if $u(z;\mu) < \log 1/\rho$. At points $z \in E_{\rho''} \cap E$ where $u(z;\mu) < \log 1/\rho$ we may have divergence or convergence (see for instance [7, Ex 1, p.154]).

Historical remarks. When $\beta_{jn\nu} = 0$, for all j, and $E = \{0\}$ (the Padé case), then $R_{n\nu} \to f$ in $\{z: |z| < \rho\} \setminus \{z_j\}_1^\nu$ by Theorem 1 and the remark in §2 (this is the Montessus de Ballore case; see the example in §2). In this case we have the following version of Theorem 2: If f is meromorphic with ν poles $z_1,\ldots,z_\nu$ in $|z| < \rho$ and f has at least one singularity at $|z| = \rho$, then $\{R_{n\nu}\}$, $n=1,2,\ldots$, diverges in $|z| > \rho$; see Montessus de Ballore [2], Wilson [8], or, for the general version and an attractive proof, Stahl [3,Satz 2.1].

A related result in the Padé case has been given by Vavilov [4] who showed that if $R_{n\nu}$ tends to f in capacity in some neighbourhood of z_0, as $n \to \infty$, then f is meromorphic in $|z| \leq |z_0|$ with at most ν poles.

A divergence result in the multipoint Padé approximation case related to Theorem 2 was proved when μ is the so called equilibrium distribution of E by Karlsson [1,Th.2.2] and follows also by general results by Walsh (see the discussion in [1,p.14].

Finally, Warner [7,Th.3,p.150] has proved a divergence theorem related to Theorem 2 in the polynomial case (i.e. $\nu = 0$) when $\beta_{jn\nu}$ is independent of n.

The proof below differs in method from the proofs of the authors mentioned.

Proof of Theorem 2. Step 1. We introduce $h_\nu(z) = \prod_{j=1}^{\nu} (z-z_j)$. Since f is meromorphic in $\bar{E}_{\rho'} \cup E$ with ν poles belonging to $E_{\rho'}$, it follows from Theorem 1 that the zeros of $Q_{n\nu}$ approach the poles z_j, $1 \leq j \leq \nu$, as $n \to \infty$. By multiplying the numerator and the denominator in $R_{n\nu} = P_{n\nu}/Q_{n\nu}$ by a suitable constant, if necessary, we can assume that

$Q_{n\nu} \to h_\nu$, as $n \to \infty$, uniformly on compact sets.

Step 2. Here we consider the case when we have the pole $\alpha \in \Gamma_\rho$ but no extra poles $\alpha_1, \dots, \alpha_k$. We introduce, except for $h_\nu(z)$ in step 1, $g(z) = f(z) - a/(z-\alpha)$ where $a \neq 0$ is the residu of f at α. By the assumptions and (1.1) the function $h_\nu(fQ_{n\nu} - P_{n\nu})/\omega_{n\nu}$ is analytic in $D \setminus \{\alpha\}$ where D is an open set containing $\bar{E}_{\rho''} \cup E$. Hence, Cauchy's integral formula gives, if Γ is a cycle in $D \setminus (\bar{E}_{\rho''} \cup E)$ with index $\mathrm{ind}_\Gamma(\omega) = 0$ for $\omega \notin D$ and $\mathrm{ind}_\Gamma(\omega) = 1$ for $\omega \in \bar{E}_{\rho''} \cup E$, and γ is a small circle around α with $\mathrm{ind}_\gamma(\alpha) = -1$ and $\mathrm{ind}_\gamma(\omega) = 0$ when $\omega = z$, $\omega = \beta_{jn\nu}$ for some j, or $\omega \notin D$,

$$\frac{h_\nu(z)(fQ_{n\nu} - P_{n\nu})(z)}{\omega_{n\nu}(z)} = \frac{1}{2\pi i}\left(\int_\Gamma + \int_\gamma\right) \frac{h_\nu(t)[g(t)Q_{n\nu}(t) - P_{n\nu}(t)]}{\omega_{n\nu}(t)(t-z)} dt +$$

$$\frac{1}{2\pi i}\left(\int_\Gamma + \int_\gamma\right) \frac{h_\nu(t)Q_{n\nu}(t)a/(t-\alpha)}{\omega_{n\nu}(t)(t-z)} dt. \tag{3.1}$$

By deforming Γ into infinity and γ into $\{\alpha\}$ we realize that the contribution to the right hand member from the term containing $P_{n\nu}(t)$ is zero, since the degree of $h_\nu P_{n\nu}$ is at most $n+\nu$ and the degree of $\omega_{n\nu}(t)(t-z)$ is $n+\nu+2$. Also, in the first term on the right hand side the integral over γ is zero and in the second term the integral over Γ is zero if $n \geq \nu - 1$ which is again seen by deforming Γ into infinity.

What remains of the second term in the right hand member equals $-a\, h_\nu(\alpha)Q_{n\nu}(\alpha)/\omega_{n\nu}(\alpha)(\alpha-z)$ and we shall see that this is the dominating term of the right hand member. In fact, since by step 1, $Q_{n\nu} \to h$ uniformly on compact sets, there exist constants $\varsigma_1 > 0$ and ς_2 so that (3.1) gives, for large n,

$$|f(z) - R_{n\nu}(z)| \geq \varsigma_1 \left|\frac{\omega_{n\nu}(z)}{\omega_{n\nu}(\alpha)}\right| - \varsigma_2 \max_{t \in \Gamma} \left|\frac{\omega_{n\nu}(z)}{\omega_{n\nu}(t)}\right|. \tag{3.2}$$

We now use the assumption that $\{\mu_n\}$ is (μ,E)-regular on the expression

$$\frac{\omega_{n\nu}(z)}{\omega_{n\nu}(\alpha)} = \exp\{(n+\nu+1)(u(\alpha;\mu_n) - u(z;\mu_n))\}.$$

Since, as $n\to\infty$, $u(\alpha;\mu_n)\to u(\alpha;\mu) = \log 1/\rho$ and $u(z;\mu_n)\to u(z;\mu) < \log 1/\rho$, we get, for some $\varepsilon > 0$,

$$\left|\frac{\omega_{n\nu}(z)}{\omega_{n\nu}(\alpha)}\right| \geq \exp\{(n+\nu+1)\varepsilon\}\to\infty, \quad \text{as} \quad n\to\infty.$$

Similarly, since by (1.3), $u(t;\mu_n)\to u(t;\mu) \leq \log 1/\rho''$ uniformly on Γ and $u(z;\mu_n)\to u(z;\mu) > \log 1/\rho''$, we obtain, for some $\varepsilon_1 > 0$,

$$\max_{t\in\Gamma} \left|\frac{\omega_{n\nu}(z)}{\omega_{n\nu}(t)}\right| \leq \exp\{(n+\nu+1)(-\varepsilon_1)\}\to 0, \quad \text{as} \quad n\to\infty.$$

By combining the last two estimates with (3.2) we conclude that $|R_{n\nu}(z)|\to\infty$.

Step 3. We now treat the general case when we also have extra poles at $\alpha_1,\dots,\alpha_k$, and we may assume that these are all different. In this case we form the function $g(z)$ by subtracting from $f(z)$ not only the principal part of f at α as in step 2 but also the principal parts of f at α_j, $1\leq j\leq k$. Furthermore, in the integration in (3.1) we use not only Γ and γ but also small circles γ_j around α_j such that γ_j has index -1 at α_j and 0 at z and at the points of E and $\complement D$. The left hand member of (3.1) is unchanged and if the circles γ_j and γ do not overlap, the only new integrals giving non-zero contributions in the right hand member of (3.1) are a fixed number of integrals of the form

$$\frac{\varsigma}{2\pi i}\int_{\gamma_j}\frac{h_\nu(t)Q_{n\nu}(t)}{(t-\alpha_j)^s\omega_{n\nu}(t)(t-z)}\,dt,$$

for $1\le j\le k$, $1\le s\le s_j$, and different constants ς. In the estimate (3.2) this integral gives a contribution which for some constant ς_3, depending on the radius of γ_j, and large n is less than

$$\varsigma_3\max_{t\in\gamma_j}\left|\frac{\omega_{n\nu}(z)}{\omega_{n\nu}(t)}\right| = \varsigma_3\left|\frac{\omega_{n\nu}(z)}{\omega_{n\nu}(\alpha)}\right|\max_{t\in\gamma_j}\left|\frac{\omega_{n\nu}(\alpha)}{\omega_{n\nu}(t)}\right|. \tag{3.3}$$

Now, if the radius of γ_j is small, $u(t;\mu_n)\to u(t;\mu)$ uniformly on γ_j, by (1.3), and $u(t;\mu)$, $t\in\gamma_j$, is close to $u(\alpha_j;\mu) < \log 1/\rho$. Since $u(\alpha;\mu_n)\to u(\alpha;\mu) = \log 1/\rho$, we get, for some $\varepsilon_2 > 0$,

$$\max_{t\in\gamma_j}\left|\frac{\omega_{n\nu}(\alpha)}{\omega_{n\nu}(t)}\right| < \exp\{(n+\nu+1)(-\varepsilon_2)\}\to 0.$$

This shows that (3.3) is small, for large n, in comparison to the first term in the right member of (3.2). Hence, $|R_{n\nu}(z)|\to\infty$ and Theorem 2 is proved.

REFERENCES

[1] J. Karlsson, Rational interpolation with free poles in the complex plane, Department of Math., University of Umeå, No. 6, 1972.

[2] R. de Montessus de Ballore, Sur les fractions continues algébriques, Bull. Soc. Math., France 30 1902, 28-36.

[3] H. Stahl, Beiträge zum Problem der Konvergenz von Padéapproximierenden, Thesis, Technische Universität Berlin, 1976.

[4] V.V. Vavilov, On the convergence of the Padé approximants of meromorphic functions, Math. USSR Sb. 30, 1976, 39-49 [Russian original, Mat. Sb. 101 (143), 1976].

[5] H. Wallin, Potential theory and approximation of analytic functions by rational interpolation, in Proc. Coll. Complex Anal., Joensuu, Finland, 1978, Lecture Notes in Math. 747, Springer-Verlag, 1979, 434-450.

[6] H. Wallin, Rational interpolation to meromorphic functions, in Proc. Conf. Padé and Rational Approx., Theory and Applic., Amsterdam, 1980, To appear in Lecture Notes in Math.

[7] D.D. Warner, Hermite interpolation with rational functions, Thesis, University of California, San Diego, La Jolla, 1974.

[8] R. Wilson, Divergent continued fractions and non-polar singularities, Proc. London Math. Soc. 30, 1930, 38-57.

Department of Mathematics
University of Umeå
S-901 87 Umeå
Sweden

DILATION OF TWO FACTORIZATIONS

Gr. Arsene and Zoia Ceauşescu

1. The contractive intertwining dilations problem is the following. Let H and H′ be two (complex) Hilbert spaces and let $T\in L(H)$ (1), $T'\in L(H')$ be contractions. Denote by $U\in L(K)$, resp. $U'\in L(K')$, the minimal isometric dilation of T, resp. T′. (For the construction and the structure of the minimal isometric dilation we refer to [12], Chapters I and II.) The operators T, T′, U and U′ will be fixed throughout this note. Let now $A\in L(H,H')$ be a contraction which intertwines T and T′, i.e. $T'A=AT$ (notation: $A\in I(T',T)$). Consider the set

$$CID(A)=\{A_\infty \in I(U',U);\ \|A_\infty\|\le 1 \text{ and } P^{K'}_{H'}A_\infty=AP^{K}_{H}\}\ ^{(2)}.$$

The mentioned problem is to describe the set CID(A) for given T, T′ and A. The interest in this matter is due to various connections with topics from Operator Theory, from classical extrapolations problems and even from Geophysics and Electrical Engineering (see for example [10],[8],[9],[2]). The fact that CID(A) is always nonvoid was proved in 1968 by B.Sz.-Nagy and C.Foiaş [11]. General descriptions of this set are given in [6],[7] (see also [5]); detalied study of these descriptions, including a generalized Schur formula, is done in [3], [4].

The aim of this note is to make some remarks on the description of CID(A) while trying to extend the theory to the following setting: Let $T\in L(H)$, $T'\in L(H')$, $A\in L(H,H')$, $B\in L(H,H')$ be contractions such that $T'A=BT$ (notation: $(A,B)\in I(T',T)$). Consider the set:

$$CID(A,B)=\{(A_\infty B_\infty)\in I(U',U);\ \|A_\infty\|\le 1,\ \|B_\infty\|\le 1,\ P^{K'}_{H'}A_\infty=AP^{K}_{H},\ P^{K'}_{H'}B_\infty=BP^{K}_{H}\}.$$

The problem of describing the set CID(A,B) will be refereed to as the *dilation of the two factorizations* (T′A and BT). Note that while $(A,A)\in I(T',T)$ simply means $A\in I(T',T)$, the set CID(A,A) is bigger than CID(A) (see below). The "hidden" reason is that in CID(A) the two dila-

(1) $L(H,H')$ is the set of all (linear bounded) operators from H into H′; $L(H)$ stands for $L(H,H)$. $A\in L(H,H')$ is a contraction if $\|A\|\le 1$.

(2) For a (linear closed) subspace $F\subset H$, P^{H}_{F} stands for the orthogonal projection of H onto F. The same notation will be used for the adjoint of the inclusion of a Hilbert space into a direct sum of it with another Hilbert space.

tions of A are asked to be *equal*; this condition turns out to be quite strong and some nice features of CID(A) heavily depend on it.

2. The description of the set CID(A,B) will be done following closely the methods from [6], [5] and [3]. First of all, let us remind the "construction in steps" of the minimal isometric dilation $U\in\mathcal{L}(K)$ of the contraction $T\in\mathcal{L}(H)$, see [12]. Consider the defect operator of T $D_T=(I-T^*T)^{1/2}$ and the defect space of T, $\mathcal{D}_T=\overline{D_T(H)}$ (1). Denote by $\mathcal{L}=\overline{(U-T)(H)}$; then $\mathcal{L}$ is a wandering subspace for the isometry U and $\mathcal{L}$ is (isometrically) isomorphic with $\mathcal{D}_T$ via $(U-T)h\to D_Th$, $h\in H$. This implies that $K=H+\mathcal{L}+U\mathcal{L}+\ldots$ (direct sum) is canonnically isomorphic with $H\oplus\mathcal{D}_T\oplus\mathcal{D}_T\ldots$; modulo this isomorphism U is the matrix

$$\begin{bmatrix} T & 0 & & \\ D_T & 0 & & \\ 0 & I & & \\ & & I & \\ & & & \ddots \end{bmatrix}.$$

For each $n\geq 1$, define $H_n=H+\mathcal{L}+\ldots+U^{n-1}\mathcal{L}$, and $T_n=P^K_{H_n}U|H_n$; put also $H_o=H$ and $T_o=T$. Then $\{H_n\}_{n=0}^{\infty}$ is an increasing sequence such that $K=\bigvee_{n=0} H_n$, and $U=(s)-\lim_{n\to\infty} T_nP^K_{H_n}$; moreover, for each $n\geq 0$, $T_n=P^{H_{n+1}}_{H_n}T_{n+1}|H_n$. Using the same structure for $U'\in\mathcal{L}(K')$, consider (for $n\geq 0$) the set

(2.1) $$n\text{-PCID}(A,B)=\{(A_n,B_n)\in\mathcal{I}(T'_n,T_n);\ ||A_n||\leq 1,\ ||B_n||\leq 1,\ P^{H'_n}_{H'}A_n=AP^{H_n}_{H},\ P^{H'_n}_{H'}B_n=BP^{H_n}_{H}\}.$$

If $(A_\infty,B_\infty)\in$ CID(A,B), then the sequence $\{(A_n,B_n)\}_{n=0}^{\infty}$, where $A_n=P^{K'}_{H'_n}A_\infty|H_n$, $B_n\in P^{K'}_{H'_n}B_\infty|H_n$ forms a *chain of n-PCID* for (A,B), i.e. $(A_n,B_n)\in$ n-PCID(A,B) and $A_n=P^{H'_{n+1}}_{H'_n}A_{n+1}|H_n$, $B_n=P^{H'_{n+1}}_{H'_n}B_{n+1}|H_n$, for every $n\geq 0$. Conversely, every chain $\{(A_n,B_n)\}_{n=0}^{\infty}$ of n-PCID's for(A,B) defines an element from CID(A,B) by $A_\infty=(s)-\lim_{n\to\infty} A_nP^K_{H_n}$ and $B_n=(s)-\lim_{n\to\infty} B_nP^K_{H_n}$. Note that for a chain $\{(A_n,B_n)\}_{n\geq 0}$ of n-PCID for (A,B), we have $(A_{n+1},B_{n+1})\in$ 1-PCID (A_n,B_n), $n\geq 0$; this implies that the structure of CID(A,B) is determined by "the first step" 1-PCID(A,B) and a good inductive argument. This will be done in Sections 4 and 5.

(1) The bar always means norm closure.

3. We remind now some useful facts in dilation theory. First, for any contraction $X\varepsilon L(F_1,F_2)$, F_1,F_2 Hilbert spaces, the operator

(3.1)
$$J(X):F_1+\mathcal{D}_{X^*}\to F_2+\mathcal{D}_X$$
$$J(X)=\begin{bmatrix} X & D_{X^*} \\ D_X & -X^* \end{bmatrix}$$

is a unitary operator. This can be checked by direct matrix computations. Secondly, we have the following lemma which appear in this form in [5].

LEMMA 3.1. *Let F and G be Hilbert spaces. Suppose that* $F=F_1\oplus F_2$, *and* $X\varepsilon L(F_1,G)$ *is a contraction. Then the formula*

(3.2)
$$Y=(X,D_{X^*}Z)$$

establishes a one-to-one correspondence between the set of all $Y\varepsilon L(F,K)$ *such that* $Y|H_1=X$ *and* $||Y||\leq 1$, *and the set of all contractions* $Z\varepsilon L(F_2,\mathcal{D}_{X^*})$. *Moreover, the operators*

(3.3)
$$\begin{cases} V(X;Y):\mathcal{D}_X\oplus\mathcal{D}_Z\to\mathcal{D}_Y \\ V(X;Y)(D_X\oplus D_Z)=(D_Y|F_1)\oplus(P^{\mathcal{D}_X}_{\mathcal{D}_X\ominus\overline{D_X(F_1)}}D_Y|F_2) \end{cases}$$

and

(3.4)
$$\begin{cases} V_*(X;Y):\mathcal{D}_{Z^*}\to\mathcal{D}_{y^*} \\ V_*(X;Y)(D_{Z^*}D_{X^*})=D_{Y^*} \end{cases}$$

are unitary operators.

The proof of the formula (3.2) and of its properties uses the fact that $R^*R\leq S^*S$ if and only if $R=CS$, where C is a contraction. The formula (3.4) follows by a direct (and easy) computation. The direct proof of (3.3) is harder; this is the reason why we note the following. Consider the unitary operator

(3.5)
$$\begin{cases} W:G\oplus\mathcal{D}_X\oplus\mathcal{D}_Z\to G\oplus\mathcal{D}_Y \\ W=J(Y)(I\oplus V_*(X,Y))(I\oplus J^*(Z))(J^*(X)\oplus I), \end{cases}$$

where the direct sum operators which apper in W act between $\{(G\oplus\mathcal{D}_X)\oplus\mathcal{D}_Z\ ,\ (F_1\oplus\mathcal{D}_{X^*})\oplus\mathcal{D}_Z\}$, $\{F_1\oplus(\mathcal{D}_{X^*}\oplus\mathcal{D}_Z),\ F_1\oplus(F_2\oplus\mathcal{D}_{Z^*})\}$, $\{(F_1\oplus F_2)\oplus\mathcal{D}_{Z^*}\ ,\ (F_1\oplus F_2)\oplus\mathcal{D}_{Y^*}\}$, $\{F\oplus\mathcal{D}_{Y^*}\ ,\ G\oplus\mathcal{D}_Y\}$, respectively. Straightforward computations show that

$$W(g\oplus 0\oplus 0)=g\oplus 0 \qquad \forall g\varepsilon G$$

and that

$$W(0\oplus D_Xf\oplus 0)=0\oplus D_Yf \qquad \forall f\varepsilon F_1.$$

These facts immediately imply the fact that V(X;Y) from formula (3.3) is a unitary operator; namely

(3.6) $I\oplus V(X,Y)=J(Y)(I\oplus V_*(X,Y))(I\oplus J^*(Z))(J^*(X)\oplus I)$,

where the operator in the left hand side acts between $G\oplus(\mathcal{D}_X\oplus\mathcal{D}_Z)$ and $G\oplus\mathcal{D}_Y$. We will refer to Lemma 3.1 as "Row Lemma" and to its variant obtained by taking the adjoints as "Column Lemma".

4. Let us start the analysis of the set 1-PCID(A,B). From the definition given in (2.1), $(A_1,B_1)\in$1-PCID(A,B) means that

$$\begin{cases} A_1:H+\mathcal{L}\to H'+\mathcal{L}' \\ B_1:H+\mathcal{L}\to H'+\mathcal{L}', \end{cases} \tag{4.1}$$

$$\|A_1\|\le 1, \quad \|B_1\|\le 1, \tag{4.2}$$

$$P_{H'}^{H_1'}A_1=AP_H^{H_1}, \qquad P_{H'}^{H_1'}B_1=BP_H^{H_1}, \tag{4.3}$$

and

$$T_1'A_1=B_1T_1. \tag{4.4}$$

The Column Lemma (applied for $A^o:H+\mathcal{L}\to H'$, $A^o=(A,0)$ and $B^o:H+\mathcal{L}\to H'$, $B^o=(B,0)$) implies that conditions (4.1), (4.2) and (4.3) are equivalent to the fact that

$$\begin{cases} A_1(h+\ell)=Ah+X_1(D_Ah+\ell) \\ B_1(h+\ell)=Bh+Y_1(D_Bh+\ell), \quad h\in H,\ \ell\in\mathcal{L}, \end{cases} \tag{4.5}$$

where $X_1:\mathcal{D}_A+\mathcal{L}\to\mathcal{L}'$ and $Y_1:\mathcal{D}_B+\mathcal{L}\to\mathcal{L}'$ are arbitrary contractions. Moreover $\mathcal{D}_{A_1}$ is isomorphic to $\mathcal{D}_{X_1}$ and $\mathcal{D}_{A_1^*}$ is isomorphic to $\mathcal{D}_{A^*}\oplus\mathcal{D}_{X_1^*}$ (and similarly for B).

The asymmetry between "back position" of A and "front position" of B appears only in the study of the key condition (4.4). Using (4.5), the descriptions of T_1 and T_1' given in Section 2, and that $(A,B)\in\mathcal{I}(T',T)$ it follows that (4.4) is equivalent to:

$$Y_1(D_BTh+(U-T)h)=(U'-T')Ah \quad , h\in H. \tag{4.6}$$

Thus, we obtained that no supplementary conditions are required on X_1, while Y_1 has to satisfy (4.6).

Let us proceed now to the analysis of (4.6). Denote

$$\begin{cases} \mathcal{F}_B=\{D_BTh+(U-T)h;\ h\in H\}^- \\ \mathcal{R}_B=(\mathcal{D}_B+\mathcal{L})\ominus\mathcal{F}_B \end{cases} \tag{4.7}$$

and

$$\begin{cases} \mathcal{F}^A=\{D_Ah\oplus(U'-T')Ah;\ h\in H\}^- \\ \mathcal{R}^A=\mathcal{D}_A\oplus\mathcal{L}'\ominus\mathcal{F}^A. \end{cases} \tag{4.8}$$

Then the operator

$$\begin{cases} \sigma_{A,B}:\mathcal{F}_B\to\mathcal{F}^A \\ \sigma_{A,B}(D_BTh+(U-T)h)=D_Ah\oplus(U'-T')Ah, \quad h\in H \end{cases} \tag{4.9}$$

is a unitary operator (direct computations). Consider also the contraction

$$\begin{cases} \alpha : F_B \to L' \\ \alpha = P_{L'}^{\mathcal{D}_A \oplus L'} \sigma_{A,B} . \end{cases} \tag{4.10}$$

The condition (4.6) means that the contraction $Y_1 : F_B \oplus R_B \to L'$ satisfies $Y_1 | F_B = \alpha$. Using the Row Lemma, it follows that

$$Y_1 = (\alpha, D_{\alpha^*} \tilde{\Gamma}_1), \tag{4.11}$$

where $\Gamma_1 : \tilde{R}_B \to \mathcal{D}_{\alpha^*}$ is an arbitrary contraction. Moreover, $\mathcal{D}_{Y_1}$ and $\mathcal{D}_\alpha \oplus \mathcal{D}_{\tilde{\Gamma}_1}$ as well as $\mathcal{D}^*_{Y_1}$ and $\mathcal{D}_{\tilde{\Gamma}_1^*}$ are cannonically isomorphic. It is easy to see that $\mathcal{D}_{\alpha^*} = P_{L'}^{\mathcal{D}_A \oplus L'} R^A$, and that $P_{L'}^{\mathcal{D}_A \oplus L'}$ is one-to-one between R^A and $\mathcal{D}_{\alpha^*}$. Thus, there exists a contraction

$$\begin{cases} \Gamma_1 : R_B \to R^A \\ P_{L'}^{\mathcal{D}_A \oplus L'} \Gamma_1 = D_{\alpha^*} \tilde{\Gamma}_1 \end{cases} \tag{4.12}$$

and the correspondence $\Gamma_1 \leftrightarrow \tilde{\Gamma}_1$ is one-to-one ($\mathcal{D}_{\Gamma_1}$ and $\mathcal{D}_{\Gamma_1^*}$ being isomorphic to $\mathcal{D}_{\tilde{\Gamma}_1}$ and $\mathcal{D}_{\tilde{\Gamma}_1^*}$, respectively). Summing up this analysis, we have the following.

LEMMA 4.1. *The formulas*

$$\begin{aligned} A_1(h+\ell) &= Ah + X_1(D_A h + \ell) \\ B_1(h+\ell) &= Bh + P_{L'}^{\mathcal{D}_A \oplus L'} (\sigma_{A,B} P_{F_B}^{\mathcal{D}_B + L} + \Gamma_1 P_{R_B}^{\mathcal{D}_B + L})(D_B h + \ell), \quad h \in H, \ \ell \in L, \end{aligned} \tag{4.13}$$

establish a one-to-one correspondence between the set 1-PCID(A,B) *and the set of pairs* (X_1, Γ_1) *where* $X_1 : \mathcal{D}_A + L \to L'$ *and* $\Gamma_1 : R_B \to R^A$ *are arbitrary contractions. Moreover, the operators*

$$\begin{cases} \omega_A : \mathcal{D}_{A_1} \to \mathcal{D}_{X_1} \\ \omega_A (D_{A_1}(h+\ell)) = D_{X_1}(D_A h + \ell), \qquad h \in H, \ \ell \in L, \end{cases} \tag{4.14}$$

$$\begin{cases} \omega_{*A} : \mathcal{D}_{A_1^*} \to \mathcal{D}_{A^*} \oplus \mathcal{D}_{X_1^*} \\ \omega^*_{*A}(D_{A^*} \oplus D_{X_1^*}) = (D_{A_1^*} | H') \oplus (P^{\mathcal{D}_{A_1^*}}_{\mathcal{D}_{A_1^*} \ominus \overline{D_{A_1^*}(H')}} D_{A_1^*} | L') , \end{cases} \tag{4.15}$$

$$(4.16)\qquad \begin{cases} \omega_B : \mathcal{D}_{B_1} \to \mathcal{D}_A \oplus \mathcal{D}_{\Gamma_1} \\ \omega_B(D_{B_1}(h+\ell)) = P^{\mathcal{D}_A \oplus L'}_{\mathcal{D}_A}(\sigma_{A,B} P^{\mathcal{D}_B + L}_{F_B} + \Gamma_1 P^{\mathcal{D}_B+L}_{R_B})(D_B h + \ell) + \\ \qquad\qquad + D_{\Gamma_1} P^{\mathcal{D}_B + L}_{R_B}(D_B h + \ell), \end{cases}$$

$$(4.17)\qquad \begin{cases} \omega_{*B} : \mathcal{D}_{B_1^*} \to \mathcal{D}_{B^*} \oplus \mathcal{D}_{\Gamma_1^*} \\ \omega_{*B}^*(D_{B^*} \oplus D_{\Gamma_1^*} P^{\mathcal{D}_A \oplus L'}_{R^A} | L') = D_{B_1^*} | H' \oplus (P^{\mathcal{D}_{B_1^*}}_{\mathcal{D}_{B_1^*} \ominus \overline{D_{B_1^*}(H')}} D_{B_1^*} | L'), \end{cases}$$

are unitary operators.

The formulas (4.13) follow from (4.5), (4.11) and (4.12); the formulas (4.14)-(4.17) follow from the isomorphisms given by Row and Column Lemmas, from the identifications of $\mathcal{D}_{\Gamma_1}$ and $\mathcal{D}_{\Gamma_1}$ with $\mathcal{D}_{\tilde{\Gamma}_1}$ and resp. $\mathcal{D}_{\Gamma_1^*}$, and from the fact that $D_A h \to D_\alpha(D_B \tilde{T} h + D_T h)$ is a unitary operator between $\mathcal{D}_A$ and $\mathcal{D}_\alpha$.

5. We will show now how Lemma 4.1 and an inductive argument provide the structure of CID(A,B). The possibility of induction will be clear after the identification of the spaces $\mathcal{D}_{A_1}$, R_{B_1} and R^{A_1} in terms of A,B, X_1 and Γ_1.

First of all, the role of the space L for T_1 is played by UL (and similarly for T_1').

From (4.14), $\mathcal{D}_{A_1}$ can be identified with $\mathcal{D}_{X_1}$.

Because $F_{B_1} = \{D_{B_1} T_1 h_1 \oplus (U' - T_1')h_1;\ h_1 \varepsilon H_1\}^- = (D_{B_1} UH)^- + U'L'$, it follows that $R_{B_1} = \mathcal{D}_{B_1} \ominus (D_{B_1} UH)^- \subset \mathcal{D}_{B_1}$. From (4.16), we easily infer that $\omega_B(D_{B_1} UH)^- = \mathcal{D}_A$, so $\omega_B | R_{B_1}$ is a unitary operator from R_{B_1} onto $\mathcal{D}_{\Gamma_1}$.

Finally, let us identify R^{A_1}. For this, consider the unitary operator

$$(5.1)\qquad \begin{cases} \Delta : H_1 \oplus \mathcal{D}_{A^*} \oplus \mathcal{D}_{X_1^*} \to H' \oplus F^{A_1} \oplus R^{A_1} \\ \Delta = (I \oplus U' \oplus I) J(A_1)(I \oplus \omega_{*A}^*), \end{cases}$$

where the components of Δ act respectively between $\{H_1 \oplus (\mathcal{D}_{A^*} \oplus \mathcal{D}_{X_1^*})$, $H_1 \oplus \mathcal{D}_{A_1^*}\}$, $\{H_1 \oplus \mathcal{D}_{A_1^*}$, $H_1' \oplus \mathcal{D}_{A_1}\}$, $\{H' \oplus L' \oplus \mathcal{D}_{A_1}$, $H' \oplus U'L' \oplus \mathcal{D}_{A_1} = H' \oplus F^{A_1} \oplus R^{A_1}\}$. Let us show that

(5.2) $$\Delta(H_1 \oplus \mathcal{D}_{A^*}) = H' \oplus F^{A_1} \quad .$$

Indeed, let $h_1 \varepsilon H$ and $h' \varepsilon H'$ and compute

$$\Delta(h_1 \oplus D_{A^*}h') = (I \oplus U' \oplus I)J(A_1)(h_1 \oplus D_{A_1^*}h') = (I \oplus U' \oplus I)[(A_1h_1 + D^2_{A_1^*}h') \oplus$$

$$\oplus (D_{A_1}h_1 - A_1^*D_{A_1^*}h')] = (I \oplus U' \oplus I) \ (h' + P^{H_1'}_{H'}A_1(h_1 - A_1^*h')) \oplus$$

$$\oplus (P^{H_1'}_{L'}A_1(h_1 - A_1^*h')) \oplus (D_{A_1}(h_1 - A_1^*h'))] =$$

$$= (h' + P^{H_1'}_{H'}A_1(h_1 - A_1^*h')) \oplus ((U' - T_1')A_1(h_1 - A_1^*h')) \oplus$$

$$\oplus (D_{A_1}(h_1 - A_1^*h')).$$

This shows that $\Delta(H_1 \oplus \mathcal{D}_{A^*}) \subset H' \oplus F^{A_1}$; it is immediate that we have in fact the equality. From (5.2) we infer that $\Delta^* | R^{A_1}$ is a unitary operator between R^{A_1} and $\mathcal{D}_{X_1^*}$.

The description of CID(A,B) is given by the following.

THEOREM 5.1. *There exists a one-to-one correspondence between the set* CID(A,B) *and the set of sequences of pairs of contractions* $\{(X_n, \Gamma_n)\}_{n=1}^{\infty}$ *, where:*

$(5.3)_1$ $$\begin{cases} X_1 : \mathcal{D}_A + L \to L' \\ \Gamma_1 : R_B \to R^A \quad , \end{cases}$$

and for every $n \geq 2$

$(5.3)_n$ $$\begin{cases} X_n : \mathcal{D}_{X_{n-1}} \oplus L \to L' \\ \Gamma_n : \mathcal{D}_{\Gamma_{n-1}} \to \mathcal{D}_{X_n^*} \quad . \end{cases}$$

This follows from the description of CID(A,B) by chains of PCID's (see the final of Section 2), from Lemma 4.1, and from the identifications made in this section.

6. Let us make some remarks on Theorem 5.1.

(1) First, it is clear that we always have CID(A,B)=∅. The set CID(A,B) consists of exactly one element only in very special cases: either T' is an isometry ($L' = \{0\}$) or T and A are isometries ($\mathcal{D}_A = L = \{0\}$).

(2) When A=B, the request $A_n = B_n$ transports the nice structure of B_n to $A_n (n \geq 1)$. This implies the nice structure of CID(A) from [6]: CID(A) is in one-to-one correspondence with the set of A-choice sequence (i.e. sequences of contractions $\{\Gamma_n\}_{n=1}^{\infty}$, such that $\Gamma_1 : R_A \to R^A$ and

$\Gamma_n : \mathcal{D}_{\Gamma_{n-1}} \to \mathcal{D}_{\Gamma^*_{n-1}}$ for every $n\geq 2$).

(3) It is quite clear from $(5.3)_1$ that it is not reasonable to expect a nice connection between CID(A,B) and CID(B^*,A^*). This is mainly do to the fact the analogue of L for T^* (denoted by L_*) can be very different from L. It is easy to infer that denoting by $\tilde{U}\varepsilon L(\tilde{K})$ (resp. $\tilde{U}'\varepsilon L(\tilde{K}')$) the minimal *unitary* dilation of T (resp.T'), then $P^{\tilde{K}'}_{H'}\tilde{U}^*B_1^*\tilde{U}'=A^*P^{\tilde{K}}_H$ (so B_1^* gives a 1-PCID for A^*); nothing of this sort is true in general for $\tilde{U}^*A_1^*\tilde{U}'$.

(4) If A and B are isometries, then the set CID(A,B) contains a pair (A_∞,B_∞) with isometric components. Indeed, it is enough to note that if A and B are isometries, then A_1 and B_1 can be chosen isometries. We have:

$$||D_Th||^2=||h||^2-||Th||^2=||Ah||^2-||BTh||^2=||Ah||^2-||T'Ah||^2=$$
$$=||D_{T'}Ah||^2,\ h\varepsilon H.$$

This implies that L can be isometrically embedded in L', so X_1 can be chosen to be isometric. The formula (4.14) implies that A_1 is an isometry. Because B is an isometry, $R_B=\{0\}$, and (4.16) implies that B_1 is an isometry. Moreover, if A and B are unitary operators, then A_1 and B_1 (so A_∞ and B_∞) can be chosen to be unitary operators (see formulas 4.15 and 4.16). We obtain then the following corollary,which can be seen as a generalization of a theorem of Ando [1].

COROLLARY 6.1. *If* $T\varepsilon L(H)$, $T'\varepsilon L(H')$, $A\varepsilon L(H,H')$, $B\varepsilon L(H,H')$ *are contractions such that* $T'A=BT$, *then there exist larger Hilbert spaces* $\tilde{H}\supset H$, $\tilde{H}'\supset H'$ *and unitary operators* $\tilde{T}\varepsilon L(\tilde{H})$, $\tilde{T}'\varepsilon L(\tilde{H}')$, $\tilde{A}\varepsilon L(\tilde{H},\tilde{H}')$, $\tilde{B}\varepsilon L(\tilde{H},\tilde{H}')$ *which dilate* T,T',A *and* B *respectively, and* $\tilde{T}'\tilde{A}=\tilde{B}\tilde{T}$.

(5) The intertwining case can be reduced to the commutant case using a matrical trick (see [5]). Taking

$$\hat{T}'=\begin{bmatrix}0 & T'\\ 0 & 0\end{bmatrix}\ \varepsilon\ L(H'\oplus H')$$
$$\hat{T}=\begin{bmatrix}0 & T\\ 0 & 0\end{bmatrix}\ \varepsilon\ L(H\oplus H)$$
$$\hat{A}=\begin{bmatrix}B & 0\\ 0 & A\end{bmatrix}\ \varepsilon\ L(H\oplus H,\ H'\oplus H'),$$

it is clear $\hat{T}'\hat{A}=\hat{A}\hat{T}$ is equivalent to $T'A=BT$. However, this remark is not useful for reducing the study of CID(A,B) to the study of CID($\hat{A}$), because the isometric dilations of $\hat{T}'$ and $\hat{T}$ contain shifts which ruin the wanted correspondence.

REFERENCE

1. ANDO, T.: On a pair of commutative contractions, *Acta Sci.Math. (Szeged)*, 24(1963), 88-90.

2. ARSENE, GR.; CEAUŞESCU, ZOIA: Contractive intertwining dilations and norm approximation techniques, *Circuits, Systems, and Signal Processing*, to appear.

3. ARSENE, GR.; CEAUŞESCU, ZOIA; FOIAŞ, C.: On intertwining dilations. VII, *Proc.Coll.Complex Analysis, Joensuu*, Lecture Notes in Math. (Springer), 747(1979),24-45.

4. ARSENE, GR.; CEAUŞESCU, ZOIA; FOIAŞ, C.: On intertwining dilations. VIII, *J.Operator Theory*,4(1980), 55-91.

5. CEAUŞESCU, ZOIA: *Operatorial Extrapolations* (Romanian), Thesis, Bucharest, 1980.

6. CEAUŞESCU, ZOIA; FOIAŞ, C.: On intertwining dilations.V, *Acta Sci.Math.(Szeged)*, 40(1978), 9-32.

7. CEAUŞESCU, ZOIA; FOIAŞ, C.: On intertwining dilations.VI, *Rev. Roumaine Math.Pures Appl.*, 23(1978), 1471-1482.

8. FOIAŞ, C.: Contractive intertwining dilations and waves in layered media, *Proc. Internat. Congress of Mathematicians, Helsinki* 1978, vol.2, 605-613.

9. HELTON, J.W.: Broadbanding: gain equalization directly from data, to appear.

10. SARASON, D.: Generalized interpolation in H^∞, *Trans.Amer.Math. Soc.*, 127(1967), 179-203.

11. SZ.-NAGY,B.; FOIAŞ, C.: Dilation des commutants d'opérateurs, *C.R.Acad.Sci.Paris, Série A*, 266(1968), 493-495.

12. SZ.-NAGY,B.; FOIAŞ, C.: *Harmonic Analysis of Operators on Hilbert Spaces*, Akadémiai Kiadó-Budapest, North Holland-Amsterdam, 1970.

Department of Mathematics
INCREST
Bdul Păcii 220, 79622 Bucharest
Romania.

Boundary value problems for systems with Cauchy-Riemannian main part

Heinrich Begehr

1. Introduction.

There are two important boundary value problems in complex analysis which originally were posed by Riemann for analytic functions. One of them was solved, among others, by Hilbert and is therefore called the Hilbert or the Riemann-Hilbert boundary value problem. The second one is the so-called Riemann boundary value problem or the problem of linear conjugacy. Besides analytic functions Hilbert already considered solutions of generalized Cauchy-Riemann systems whose solutions later were called generalized - or pseudo-analytic functions. The classical theory of these problems together with their history are contained in the books of Gakhov [1] and Muskhelishvili [1], but since that time the literature has grown and still grows vastly especially in the Russian language, see e.g. Begehr [1].

One reason for this great interest for these problems is a relation to the theory of singular integral equations which was influenced and has influence onto these problems, see Prössdorf [1], Bojarski [2]. This of course is true for the generalizations onto analytic vectors and matrices too. Here the far reaching papers of Simonenko (see Simonenko [1]) have to be mentioned.

Another generalization of analytic functions, which became important in complex analysis are solutions of Beltrami equations (see Ahlfors [1], Bers [1], Vekua [1]).

The Hilbert boundary value problem in connection with Beltrami equations was considered already by Bers-Nirenberg [1] and Vekua [1]. Recently it was studied for quasilinear Beltrami equations (see Gilbert [1], Begehr-Gilbert [1], [2], Begehr-Hsiao [3], Mamourian [1], Tjurikov [1], Wen [1], Wendland [1]-[4]), which first were considered by Bojarski [1], Vinogradov [1]-[4] and for nonlinear elliptic equations of first order by Bojarski-Iwaniec [1], Naas-Tutschke [1]-[2], Tutschke [1]-[4], Wen [2]-[4].

Here the common results attained by the author in collaboration with Hile and Hsiao (see Begehr-Hile [1], [2], Begehr-Hsiao [1]-[5]) will be summarized.

We consider a differential equation of the form

$$(1) \qquad w_{\bar{z}} = H(z,w,w_z)$$

either in a bounded domain D with a smooth boundary $\Gamma = \partial D$ with Hölder continuously varying tangent or in the complex plane $\mathbb{C}$ up to a single closed curve Γ of the same type as the boundary ∂D. Here H is a function of $z,\bar{z},w,\bar{w},v,\bar{v}$ where (z,w,v) is varying in $D \times \mathbb{C} \times \mathbb{C}$ and $\mathbb{C} \times \mathbb{C} \times \mathbb{C}$ respectively. We are interested in solutions of (1) in D and in $\mathbb{C}-\Gamma$ respectively which fulfil the following boundary condition. On Γ there are given two Hölder continuous functions G,g such that G does not vanish on Γ and

(2) $\quad \mathrm{Re}\, \bar{G}w = g \quad \text{on} \quad \Gamma$

or

(3) $\quad w^+ = Gw^- + g \quad \text{on} \quad \Gamma$

where w^+ and w^- are the boundary values of w from the left and the right hand side on Γ respectively. Problem (2) is the Hilbert and (3) the Riemann boundary value problem. As is known from the analytic case the solutions of (2) and (3) are not uniquely given. This can be achieved by additional side conditions which will be posed later. Moreover the solutions to both problems depend on its index, defined by

$$n := \mathrm{ind}\, G := \frac{1}{2\pi} \int_{\Gamma} d\arg G.$$

Evidently n is an integer. In the cases of negative index the problems in general are unsoluble even for analytic functions. Therefore here nonnegative n is considered. How the remaining case similary could be handled can be found in Wendland [4].

The crucial conditions on H in (1) are the following. In the first variable H has to be measurable, in w and v it has to fulfil Lipschitz conditions where the Lipschitz constants with respect to w and v may depend on z but the last one has to be bounded below one in order that (1) is elliptic.

The method of proofs consists in an imbedding procedure combined with a Newton approximation as is used by Wen [3], Wendland [1], [2] and is based on a priori estimates for solutions of related linear equations. There are results for nonlinear boundary conditions of both types too (see Begehr-Hsiao [1]-[3], Begehr-Hile [1], Tutschke [5], Wolska-Bochenek [1], and others) but the results are unsatisfactory, because they hold only under very restrictive conditions. The desired Lipschitz constants not only have to be less than one but also small enough.

2. A priori estimates for linear equations.

A linear elliptic equation related to (1) is

$$(4) \qquad w_{\bar z} + \mu_1 w_z + \mu_2 \overline{w_z} = aw + b\bar{w} + c$$

where μ_1, μ_2 are measurable functions in $\hat{D} := D \cup \Gamma$ and in $\mathbb{C}$ respectively such that

$$(5) \qquad |\mu_1(z)| + |\mu_2(z)| \le q < 1.$$

Although we could restrict ourselves to the case $\mu_2 = o$, $b = o$ different from Begehr - Hile [2], Begehr - Hsiao [5] this more general equation shall be considered here.

Lemma 1.

Let μ_1, μ_2 be two measurable functions in $\hat{D}$ fulfilling (5), let $a,b,c \in L_p(\hat{D})$ where $p > 2$ is close enough to 2, let $\rho, \tau \in C^{1+\alpha}(\Gamma)\ (o < \alpha < 1)$ and

$$n := \frac{1}{2\pi} \int_\Gamma d\tau \ge o,$$

let $\sigma \in C(\Gamma)$ be nonnegative, and

$$\Sigma := \int_\Gamma \sigma(s)ds > o,$$

$\kappa \in \mathbb{R}$, let $z_k \in D$ and $a_k \in \mathbb{C}$ $(1 \le k \le n)$ be given points. Then equation (4) under the boundary and side conditions

$$(6) \qquad \begin{cases} \operatorname{Re} e^{i\tau} w = \rho \quad \text{on} \quad \Gamma, \\ \frac{1}{\Sigma} \int_\Gamma \operatorname{Im} e^{i\tau} w\sigma ds = \kappa, \\ w(z_k) = a_k \end{cases}$$

is uniquely solvable in $W_p^1(\hat{D})$. Moreover there exist constants $\beta, \gamma_1, \gamma_2, \delta$ depending on $\alpha, p, q, \tau, \sigma$, $||a||_p + ||b||_p$, and D but not on $a,b,c,\mu_1,\mu_2,\rho,\kappa$, and w such that

$$(7) \qquad ||w||_o + ||w_z||_p + ||w_{\bar z}||_p \le \beta ||\rho||_{1,\alpha} + \gamma_1 |\kappa| + \gamma_2 \sum_1^n |a_k| + \delta ||c||_p$$

holds.

Proof.

For the case $\mu_2 = o$, $b = o$ the proof based on a representation formula for W_p^1 functions fulfilling (6) (see Haack-Wendland [1], [2], Wendland [3]) is given in Begehr-Hsiao [5]. The general case can be handled similar to the proof of the next lemma.

Remarks.

The assumption on ρ and on τ in lemma 1 is more than is necessary. They only have to be assumed to be Hölder continuous on Γ (with exponent α, $\frac{1}{2} < \alpha < 1$!). Then the term $\|\rho\|_{1,\alpha}$ has to be replaced by the sum of $\|\rho\|_\alpha$ and the L_p-norm of

$$\int_\Gamma \rho(\zeta) \frac{d\zeta}{(\zeta-z)^2}$$

on Γ which exists for $1 < p < (1-\alpha)^{-1}$ as is shown in Muskhelishvili [1],I,20, Vekua [1], p.22.

The general Riemann problem (1), (3) can be reduced to a special one where G is identical to one by using an analytical factorisation of G (see Gakhov [1], p.96). Similar g can be reduced to zero by adding a proper Cauchy integral.

Lemma 2.

Let μ_1, μ_2 be measurable functions in $\mathbb{C}$ fulfilling (5) and

$$(8) \qquad |\mu_1(z)| + |\mu_2(z)| = O(|z|^{-\varepsilon}) \quad (z \to \infty, \quad o < \varepsilon < 1),$$

let $a,b,c \in L_{p'}(\mathbb{C}) \cap L_p(\mathbb{C}) \cap L_{p,2}(\mathbb{C})$ where

$$\frac{2}{1+\varepsilon} < p' < 2 < p < \frac{4}{2-\varepsilon}$$

and p' and p are close enough to 2, let $\rho \in C^{1+\alpha}(\Gamma)\,(o < \alpha < 1)$. Then there exists a unique solution of (4) fulfilling

$$(9) \qquad \begin{cases} w^+ = w^- + \rho & \text{on } \Gamma, \\ w(\infty) = o. \end{cases}$$

Moreover $w_z, w_{\bar z} \in L_{p'}(\mathbb{C}) \cap L_p(\mathbb{C})$, $w_{\bar z} \in L_{p,2}(\mathbb{C})$ and there exist constants β and δ depending on ε, p', p, q, and $\|a\|_{p',p} + \|b\|_{p',p}$ but not on w, μ_1, μ_2, a,b,c, and ρ such that

$$(10) \quad \|w\|_o + \|w_z\|_{p',p} + \|w_{\bar z}\|_{p',p} \le \beta\|\rho\|_{1,\alpha} + \delta\|c\|_{p',p}.$$

Remarks.
As in lemma 1 ρ only has to be in $C^{\alpha}(\Gamma)$ and then the $C^{1+\alpha}$-norm has to be replaced as indicated above. $\|\cdot\|_{p',p}$ denotes the sum of the $L_{p'}$-norm and the L_p-norm.
In both lemmata p' and p respectively have to be close enough to two. More precisely this means the following. The norm Λ_p of the Π operator,

$$\Pi g(z) := -\frac{1}{\pi}\int\limits_{\mathbb{C}} \frac{g(\zeta)}{(\zeta-z)^2}d\xi d\eta, \qquad g \in L_p(\mathbb{C})\,(1 < p),$$

in $L_p(\mathbb{C})$ (see Vekua [1], p.71) is a continuous function of $p(1 < p)$ and $\Lambda_2 = 1$. In lemma 2 p and p' have to be so close to 2 so that for q from (5) one has $q\max\{\Lambda_{p'},\Lambda_p\} < 1$.
In Begehr-Hsiao [5] it is shown that for $D = \{|z| < 1\}$ $(z_k \in D,\ 1 \le k \le n)$

$$\Pi_n g(z) := \sum_1^n \frac{1}{z-z_n} P_n g(z) - \frac{1}{\pi}\int\limits_D \Big\{g(\zeta)\prod_1^n \frac{z-z_k}{\zeta-z_k}\frac{1}{(\zeta-z_k)^2} +$$

$$+ \overline{g(\zeta)}\prod_1^n \frac{z-z_k}{\bar\zeta-\bar z_k}\frac{1}{(1-\bar\zeta z)^2}\Big\}d\xi d\eta,$$

$$P_n g(z) := -\frac{1}{\pi}\int\limits_D \Big\{g(\zeta)\prod_1^n \frac{z-z_k}{\zeta-z_k}\frac{1}{\zeta-z} + \overline{g(\zeta)}\prod_1^n \frac{z-z_k}{\bar\zeta-\bar z_k}\frac{z}{1-\bar\zeta z}\Big\}d\xi d\eta$$

shows a similar behaviour as Π namely if Λ_{np} denotes the L_p-norm of Π_n then

$$q\Lambda_{np} < 1$$

for p close enough to 2.

Proof of lemma 2.
As was pointed out earlier, it is enough to prove the lemma for $\rho = o$. The case $\mu_2 = o$, $b = o$ is proved in Begehr-Hile [2]. The following existence proof for the general case is based on a communication of Hile.
First let as prove uniqueness. A solution of the homogenous problem would solve

$$w_{\bar z} + \mu w_z = Aw \quad \text{in } \mathbb{C}, \quad w(\infty) = o,$$

where

$$\mu := \begin{cases} \mu_1 + \mu_2 \dfrac{\overline{w_z}}{w_z}, & w_z \neq o \\ \mu_1 \quad , & w_z = o \end{cases},$$

$$A := \begin{cases} a + b\dfrac{\overline{w}}{w}, & w \neq o \\ a \quad , & w = o \end{cases}.$$

But this solution vanishes identically (see Begehr-Hile [2]). To prove existence a solution is looked for in the form

$$w = Tg, \quad g \in L_{p'}(\mathbb{C}) \cap L_p(\mathbb{C}),$$

where T is the Hilbert transformation

$$Tg(z) = -\frac{1}{\pi} \int_{\mathbb{C}} g(\zeta) \frac{d\xi d\eta}{\zeta - z},$$

so that

$$w_{\bar{z}} = g, \quad w_z = \Pi g := -\frac{1}{\pi} \int_{\mathbb{C}} g(\zeta) \frac{d\xi d\eta}{(\zeta - z)^2}.$$

Therefore (4) leads to the integral equation

$$(11) \quad g + \mu_1 \Pi g + \mu_2 \overline{\Pi g} = aTg + b\overline{Tg} + c$$

which in shorter form shall be written as

$$(I + P)g = Qg + c, \quad g = (I + P)^{-1} Qg + (I + P)^{-1} c,$$

where Q is a compact and P a contracting linear operator on $L_{p'}(\mathbb{C}) \cap L_p(\mathbb{C})$. Therefore the Fredholm alternative holds for the last equation. To show that the homogeneous equation only has the trivial solution we consider (11) with $c = o$. Because Tg is bounded the right hand side of (11) is in $L_{p,2}(\mathbb{C})$; because Πg is in $L_{p'}(\mathbb{C}) \cap L_p(\mathbb{C})$ by (8) it follows, Pg is in $L_{p,2}(\mathbb{C})$, so g is in $L_{p,2}(\mathbb{C})$. Thus $w = Tg$ is a solution of (4) with $c = o$ in $\mathbb{C}$ vanishing at infinity. By the uniquess proof this solution is the trivial solution. Therefore (11) is solvable in $L_{p'}(\mathbb{C}) \cap L_p(\mathbb{C})$. To show that the solution is in $L_{p,2}(\mathbb{C})$ too, we observe that the right hand side of (11) as well as Pg is in this space. Therefore $w = Tg$ vanishes at infinity.

To prove (10) we observe that a solution of (4) fulfils

$$w_{\bar{z}} + \mu w_z = Aw + c \quad \text{in} \quad \mathbb{C},$$

when μ and A are defined as in the uniquess proof. Therefore (10) follows from the corresponding inequality for the special case $\mu_2 = o$, $b = o$ as is proved in Begehr-Hile [2].
In the following the preceding results are only needed in the special cases $\mu_2 = o$, $b = o$.

3. The nonlinear equation.

The nonlinear function H is assumed to be measurable in the z variable and to fulfil the following Lipschitz conditions

$$(12) \quad |H(z,w_1,v) - H(z,w_2,v)| \le K(z)|w_1-w_2|,$$

$$(13) \quad |H(z,w,v_1) - H(z,w,v_2)| \le q(z)|v_1-v_2|,$$

where K and q are positive functions and

$$q(z) \le q < 1.$$

The last condition guarantees the ellipticity of equation (1).

Theorem 1.

Let α, p and q be real numbers such that

$$1 < 2\alpha < 2 < p < (1-\alpha)^{-1}, \quad o < q < 1, \quad q\Lambda_{np} < 1.$$

Let H be a measurable function from $D \times \mathbb{C} \times \mathbb{C}$ into $\mathbb{C}$, $H(\cdot,o,o) \in L_p(\hat{D})$, and fulfilling (12) and (13). Let $\tau, \rho \in C^{\alpha}(\Gamma)$, $n = -\text{ind}\, e^{i\tau^p} \ge o$, $\sigma \in C(\Gamma)$ nonnegative, $\sum = \int_\Gamma \sigma ds > o$, $\kappa \in \mathbb{R}$, $z_k \in D$, $a_k \in \mathbb{C}$ $(1 \le k \le n)$. Then equation (1) together with conditions (6) is uniquely solvable.

Outline of proof.

Uniqueness follows at once from (7) if one observes that the difference of two solutions of (1), $w = w_1-w_2$ is a solution of

$$w_{\bar{z}} = \hat{\mu}(z,w_1,w_{1z},w_{2z})w_z + \hat{A}(z,w_1,w_2,w_{2z})w$$

where

$$\hat{\mu}(z,w,v_1,v_2) := \begin{cases} \dfrac{H(z,w,v_1)-H(z,w,v_2)}{v_1-v_2}, & v_1 \neq v_2, \\ o & , v_1 = v_2 \end{cases}$$

$$\hat{A}(z,w_1,w_2,v) := \begin{cases} \dfrac{H(z,w_1,v)-H(z,w_2,v)}{w_1-w_2}, & w_1 \neq w_2 \\ o & , w_1 = w_2 \end{cases}.$$

To prove existence (1) is rewritten into

(14) $w_{\bar{z}} = tR(z,w,w_z) + C(z),$

where $t = 1$ and

$$R(z,w,v) := H(z,w,v) - H(z,o,o),$$

$$C(z) := H(z,o,o).$$

Let us assume solvability of (14), (6) for a t_o, $o \leq t_o < 1$, and arbitrary $C \in L_p(\hat{D})$, which for $t_o = o$ is evident. To show solvability for a certain $t := t_o + \Delta$, $t_o < t \leq 1$ we use Newton's method. Let $w_o(z,t) := w(z,t_o)$ be the solution of (14), (6) for t_o and $w_{n+1}(z,t)$ $(n \in \mathbb{N}_o)$ solve

$$w_{n+1\bar{z}} = t_o R(z,w_{n+1},w_{n+1z}) + \Delta R(z,w_n,w_{nz}) + C(z)$$

together with (6). Convergence of (w_n) can be proved by (7) where here the terms coming from the boundary and side data are irrelevant. Moreover convergence holds for small enough Δ, where the smallness is independent of t_o. Therefore by finitely many steps the solution for $t = 1$ can be got.
The same method can be used to prove the next theorem.

Theorem 2.
Let ε,α,p',p, and q be real numbers such that

$$o < \varepsilon < 1 < 2\alpha < 2,\ \frac{2}{1+\varepsilon} < p' < 2 < p < \min(\frac{4}{2-\varepsilon}, \frac{1}{1-\alpha}),$$

$$o < q < 1, \quad q \max(\Lambda_{p'}, \Lambda_p) < 1.$$

Let H be a measurable function from $\mathbb{C}^3$ into $\mathbb{C}$, $H(\cdot,o,o) \in L_{p'}(\mathbb{C}) \cap L_p(\mathbb{C}) \cap L_{p,2}(\mathbb{C})$, and fulfilling (12) and (13) where

$$q(z) = O(|z|^{-\varepsilon}) \quad \text{as} \quad z \to \infty,$$

$$K \in L_{p'}(\mathbb{C}) \cap L_p(\mathbb{C}) \cap L_{p,2}(\mathbb{C}).$$

Let $\rho \in C^\alpha(\Gamma)$. Then equation (1) together with (9) is uniquely solvable.
For the general Riemann boundary value problem we note the following result (see Begehr-Hile [2]).

<u>Theorem 3.</u>
Let H satisfy (12), (13) and

$$H(z,o,o) \in L_{\infty,loc}(\mathbb{C}), \; |H(z,o,o)| = O(|z|^{-1-\varepsilon-n}) \quad \text{as} \quad z \to \infty,$$

$$K \in L_{\infty,loc}(\mathbb{C}), \; K(z) = O(|z|^{-1-\varepsilon-m}) \quad \text{as} \quad z \to \infty,$$

$$q(z) \le q < 1, \quad q(z) = O(|z|^{-\varepsilon-m}) \quad \text{as} \quad z \to \infty,$$

where $m \ge o$ and n are integers. Let $g, G \in C^\alpha(\Gamma)$, $1 < 2\alpha < 2$ and $G(z) \neq o$ on Γ and $\operatorname{ind} G = n$. Let P be a polynomial of degree $\le m$.
Then there exists a unique solution w of (1), (3) satisfying the asymptotic condition

$$h^{-1}(z)w(z) - P(z) \to o \quad \text{as} \quad z \to \infty,$$

where h is the analytic factorization of G.
If the Newton's iterative $w_n(z,t)$ for the solution of (14) together with (6) or with (9) is considered as an approximative solution of (1), (6) and (1), (9) respectively then one would be interested in an estimation of the difference of $w_n(z,t)$ from the solution $w := w(z,1)$.

Theorem 4.

Let for problem (1), (6)

$$\gamma := (q + \|K\|_p)\delta, \; K_o := \beta\|\rho\|_\alpha + \gamma_1|\kappa| + \gamma_2 \sum_1^n |a_k| + \delta\|H(\cdot,o,o)\|_p$$

and for problem (1), (9)

$$\gamma := (q + \|K\|_{p',p})\delta, \; K_o := \beta\|\rho\|_\alpha + \delta\|H(\cdot,o,o)\|_{p',p}$$

and choose Δ so small that $\gamma\Delta < 1$.
Then under the same assumptions as in theorem 1 and 2 respectively one has

$$\|w(\cdot,1) - w_n(\cdot,t)\|_p \leq \gamma K_o\left[\frac{1-\gamma^n\Delta^n}{1-\gamma\Delta}(1-t) + (1-t_o)\gamma^n\Delta^n\right]$$

and

$$\|w(\cdot,1) - w_n(\cdot,t)\|_{p',p} \leq \gamma K_o\left[\frac{1-\gamma^n\Delta^n}{1-\gamma\Delta}(1-t) + (1-t_o)\gamma^n\Delta^n\right]$$

respectively.

References

Ahlfors, L.

[1] On quasi-conformal mappings. J. Analyse Math. 3 (1954), 1-58.

Begehr, H.

[1] Boundary value problems for analytic and generalized analytic functions. To appear in "Complex Analysis - methods, trends, and applications". Ed. E. Lanckau and W. Tutschke, Akademie-Verlag, Berlin.

Begehr, H. - Gilbert, R.P.

[1] Über das Randwert-Normproblem für ein nichtlineares elliptisches System. Lecture Notes in Math. 561, Springer Verlag 1976, 112-121.

[2] Das Randwert-Normproblem für ein fastlineares elliptisches System und eine Anwendung. Ann. Acad. Sci. Fenn. AI, 3 (1977), 179-184.

[3] Randwertaufgaben ganzzahliger Charakteristik für verallgemeinerte hyperanalytische Funktionen. Appl. Anal. 6 (1977), 189-205.

[4] On Riemann boundary value problems for certain linear elliptic systems in the plane. J. Differential Equations 32 (1979), 1-14.

Begehr, H. - Hile, G.N.

[1] Nonlinear Riemann boundary value problems for a nonlinear elliptic system in the plane, to appear in Math. Z.

[2] Riemann boundary value problems for nonlinear elliptic systems, to the published.

Begehr, H. - Hsiao, G.C.

[1] On nonlinear boundary value problems for an elliptic system in the plane. Lecture Notes in Math. 846, Springer Verlag 1981, 55-63.

[2] Nonlinear boundary value problems for a class of elliptic systems. Komplexe Analysis und ihre Anwendung auf partielle Differentialgleichungen. Martin-Luther-Universität, Halle-Wittenberg 1980, 90-102.

[3] Nonlinear boundary value problems of Riemann-Hilbert type, to appear in Proc. AMS special sesson on elliptic systems in the plane. 87 th annual meeting, San Francisco, January 1981.

[4] A priori estimates for elliptic systems, to be published.

[5] The Hilbert boundary value problem for nonlinear elliptic systems, to be published.

Bers, L.

[1] Function theoretic properties of solutions of partial differential equations of elliptic type. Ann. Math. Studies 33(1954), 69-94.

Bers, L. - Nirenberg, L.

[1] One a representation theorem for linear elliptic systems with discontinuous coefficients and its applications. Conv. Eq. Lin. Derivate Partiali. Trieste 1954. Cremonense, Roma 1955, 111-140.

Bojarski, B.

[1] Generalized solutions of a system of differential equation of the first order of elliptic type with discontinuous coefficients. Math. Sbornik 43 (85) (1957), 451-503.

[2] An abstract problem of linear conjugacy and Fredholm pairs of subspaces. Differential and integral equations. Boundary value problems. Tbilis. Gos. Univ., Tbilisi 1979, 45-60.

Bojarski, B. - Iwaniec, T.

[1] Quasiconformal mappings and nonlinear elliptic equations in two variables. I-II. Bull. Acad. Polon. Sci. 22 (1974), 473-478, 479-484.

Džuraev, A.

[1] Systems of equations of composite type. Nauka, Moscow 1972 (Russian).

Gakhov, I.D.

[1] Boundary value problems. Pegamon, Oxford, 1966.

Gilbert, R.P.

[1] Nonlinear boundary value problems for elliptic systems in the plane. Proc. Int. Conf. Nonlinear Systems Appl., ed. V. Lakshmikantham, 1977, Akademic Press, 97-124.

[2] Verallgemeinerte hyperanalytische Funktionentheorie. Komplexe Analysis und ihre Anwendungen auf partielle Differentialgleichungen.

Martin-Luther-Universität, Halle-Wittenberg 1980, 124-145.

Gilbert, R.P.-Buchanan, J.

[1] The Hilbert problem for hyperanalytic functions. Univ. Delaware Techn. Report 66A.

Haack, W.-Wendland, W.

[1] Vorlesungen über partielle und Pfaffsche Differentialgleichungen. Birkhäuser-Verlag, Basel 1969.

[2] Lectures on partial and Pfaffian differential equations. Pergamon Press, Oxford 1972.

Mamourian, A.

[1] General transmission and boundary value problems for first order elliptic equations in multiply-connected plane domains. Demonstratio Math. 12 (1979), 785-802.

Monahov, V.N.

[1] Boundary value problems with free boundaries for elliptic systems. Isdatel'ctvo Nauka Sib. Otdelenie Novocibirsk 1977. (Russian).

Muskhelishvili, N.I.

[1] Singular integral equations. Noordhoff, Groningen, 1953.

Naas, J. - Tutschke, W.

[1] Some probabilistic aspects in partial complex differential equations. Complex analysis and its applications. Akad. Nauk SSSR, Moscow 1978, 409-412.

[2] On the error in the approximate solution of boundary value problems of nonlinear first order differential equations in the plane. Appl. Anal. 7(1978), 239-246.

Prössdorf, S.

[1] Einige Klassen singulärer Gleichungen. Akademieverlag Berlin 1974 und Birkhäuser Verlag 1974.

Simonenko, I.B.

[1] Some general questions in the theory of the Riemann boundary value problem. Isv. Akad. Nauk SSSR, Ser. Mat. 32 (1968), 1138-1146 (Russian).

Tjurikov, E.V.

[1] The nonlinear Riemann-Hilbert boundary value problem for quasi-linear elliptic systems. Soviet Math. Dokl. 20 (1979), 863-866.

Tutschke, W.

[1] Die neuen Methoden der komplexen Analysis und ihre Anwendung auf nichtlineare Differentalgleichungssysteme. S.-ber. Akad. Wiss. DDR, 17 N (1976).

[2] Lösung nichtlinearer partieller Differentialgleichungssysteme erster Ordnung in der Ebene durch Verwendung einer komplexen Normalform. Math. Nachr. 75 (1976), 283-298.

[3] The Riemann-Hilbert problem for nonlinear systems of differential equations in the plane. Complex analysis and its applications, Akad. Nauk SSSR, Moscow 1978, 537-542 (Russian).

[4] Solutions with prescribed periods on the boundary components for non-linear elliptic systems of first order in multiply connected domains in the plane. Martin-Luther-Universität Halle, Preprint Nr. 27 (1979) 3-9.

[5] Reduction of the problem of linear conjugation for first order nonlinear elliptic systems in the plane to an analogous problem for holomorphic functions. Lecture Notes of Math.,Springer-Verlag, 798 (1980), 446-455.

Vekua, I.N.

[1] Generalized analytic functions. Pergamon, London, 1962.

Vinogradov, V.S.

[1] On a boundary value problem for linear elliptic systems of differential equations of the first order on the plane. Dokl. Akad. Nauk SSSR 118 (1958), 1059-1062 (Russian).

[2] Über die Beschränktheit der Lösungen von Randwertproblemen für lineare elliptische Systeme erster Ordnung in der Ebene. Dokl. Akad. Nauk SSSR 121 (1958), 399-402 (Russian).

[3] Über einige Randwertprobleme für quasilineare elliptische Systeme erster Ordnung in der Ebene. Dokl. Akad. Nauk SSSR 121 (1958), 579-581 (Russian).

[4] A certain boundary value problem for an elliptic system of special form. Differencial'nye Uravnenija 7 (1971), 1226-1234, 1341 (Russian).

Wen, Guo-Chun

[1] On Riemann-Hilbert boundary value problems of elliptic systems of linear partial differential equations of the first order. Acta Math. Sinica 15 (1965), 599-613. (Chinese).

[2] On Riemann-Hilbert problems for nonlinear elliptic systems of first order in the plane. Acta Math. Sinica 23 (1980), 244-255 (Chinese).

[3] Modified Dirichlet problem and quasiconformal mappings for nonlinear elliptic systems of first order. Kexue Tongbao 25 (1980), 449-453.

[4] Function-theoretical properties of solutions for nonlinear elliptic complex equations of first order. Hebei Huagong Xueynan Xuebao Shuxue Zhuanji, 1980, 41-61.

[5] The continuously differentiable solutions for nonlinear elliptic complex equations of first order. Hebei Huagong Xueynan Xuebao, Shuxue, Zhuanji, 1980, 62-83 (Chinese).

Wendland, W.

[1] An integral equation method for generalized analytic functions. Lecture Notes in Math., No. 430, Springer-Verlag 1974, 414-452.

[2] On the imbedding method for semilinear first order elliptic systems and related finite element methods. Continuation Methods, ed. H. Wacker, Academic Press, 1977, 277-336.

[3] Elliptic systems in the plane. Pitman Publishing, Inc., London, 1978.

[4] Numerische Methoden bei Randwertproblemen elliptischer Systeme in der Ebene. Komplexe Analysis und ihre Anwendungen auf partielle Differentialgleichungen. Martin-Luther-Universität, Halle-Wittenberg 1980, 310-348.

v. Wolfersdorf, L.

[1] Monotonicity methods for a class of first order semilinear elliptic systems. Komplexe Analysis und ihre Anwendungen auf partielle Differentialgleichungen. Martin-Luther-Universität, Halle-Wittenberg 1980, 369-373.

Wolska-Bochenek, J.

[1] A compound nonlinear boundary value problem in the theory of pseudo-analytic functions. Demonstratio Math. 4 (1972), 105-117.

Freie Univ. Berlin, I. Math. Institut, Hüttenweg 9, 1000 Berlin 33

A NEW TOOL IN THE CALCULUS OF VARIATIONS: GEHRING's THEOREM

CARLO SBORDONE (*)

1. The problem of the L^p-regularity of the weak derivatives of a quasi-conformal mapping has been first solved by B.V. BOYARSKI [2] in 1955 in the case n = 2.

In his paper [4] F.W. GEHRING gave the same type of result for arbitrary dimension by introducing an important theorem (see lemma 2.1) which has revealed a very useful tool in many regularity problems for solutions of general partial differential equations [1], [5], [6], [7], [9] and in the Calculus of Variations [8].

This theorem was generalized by MEYERS-ELCRAT [9] for the proof of the L^p-regularity of the derivatives of a quasi-regular mapping and then modified by GIAQUINTA-MODICA [7] for the proof of several regularity theorems for the local solutions of non linear equations and systems [5], [7].

Recently also the regularity of local minima of multiple integrals of the Calculus of Variations without differentiability assumptions on the integrand has been proved [1], [6]

In this note we give a generalization of these theorems to general functionals of the form $F = F(A,u)$ defined for A bounded open set in R^n and $u = (u^1,\dots,u^N) \in (H^{m,p}(A))^N = V(A)$ verifying the following conditions ([3]):

(j) $\forall\, u \in V(A)$, $A \longrightarrow F(A,u)$ is trace of a measure

(jj) $u = v$ on $A \Longrightarrow F(A,u) = F(A,v)$

(jjj) $\int_A \sum_{|\alpha|=m} |D^\alpha u|^p \le F(A,u) \le s \left(1 + \int_A \sum_{|\alpha|=m} |D^\alpha u|^p\right)$

(*) Istituto Matematico "R.Caccioppoli" Università di Napoli, Via Mezzocannone 8, 80134 NAPOLI.

2. Let us begin with the following lemma which is a local generalization of lemma 2 of GEHRING [6] and of Prop. 3 of MEYERS-ELCRAT [9] which is due to GIAQUINTA-MODICA [7].

For $t > 0$ we set $B_t = B_t(x_o) = \{x : |x-x_o| < t\}$, and for $f \in L^1_{loc}(R^n)$

$$f_t = \fint_{B_t} f(x)dx = \frac{1}{B_t} \int_{B_t} f(x)dx.$$

LEMMA 2.1 - Let Q be a cube in R^n; $g,h: Q \longrightarrow [0,\infty[$ and $g \in L^q(Q)$, $q>1$, $h \in L^r(Q)$, $r>q$. Assume that $\forall\, x_o \in Q$, and for $0 < R < \frac{1}{2}\,\mathrm{dist}(x_o, \partial Q)$:

$$\fint_{B_R} g^q dx \leq b\left\{ \left(\fint_{B_{2R}} g\, dx\right)^q + \fint_{B_{2R}} h^q dx \right\} + \theta \fint_{B_{2R}} g^q dx ,$$

where $B_R = B_R(x_o)$, $B_{2R} = B_{2R}(x_o)$, $b>1$, $\theta \geq 0$.

Then, there exists $\theta_o = \theta_o(q,r,n)$ such that if $\theta < \theta_o$, we have $g \in L^s_{loc}(Q)$ for $s \in [q, q+\varepsilon]$, and $\forall\, x_o \in Q$, $0 < R < \frac{1}{2}\mathrm{dist}(x_o, \partial Q)$:

$$\left(\fint_{B_R} g^s dx\right)^{1/s} \leq c\left\{ \left(\fint_{B_{2R}} g^q dx\right)^{1/q} + \left(\fint_{B_{2R}} h^s dx\right)^{1/s}\right\},$$

where $c>0$ and $\varepsilon>0$ depend only on b, θ, q, r, n.

Let us now fix a bounded open set $\Omega \subset R^n$ and indicate by $Ap(\Omega)$ the family of all bounded open sets A such that $A \subset\subset \Omega$.

We can prove the following

THEOREM - Let $F : Ap(\Omega) \times (H^{m,p}_{loc}(\Omega))^N \longrightarrow [0,\infty[$ verify (j), (jj), (jjj) of section 1. Let $u \in (H^{m,p}_{loc}(\Omega))^N$ satisfy for any $A \in Ap(\Omega)$

$$(2.1) \qquad F(A,u) \leq F(A,v) \qquad \forall\, v \in u + (H^{m,p}_o(A))^N.$$

Then there exists $\varepsilon = \varepsilon(p,n,m,N,s) > 0$ such that for $|\alpha| \leq m$ $D^\alpha u \in (L^{p+\varepsilon}_{loc}(\Omega))$.

PROOF. Let $x_o \in \Omega$, $0 < R < \frac{1}{2}\,\mathrm{dist}(x_o, \partial\Omega)$ and set for $i=1,\dots,\nu \in N$

$$A_o = B_R , \quad A_i = \{x : \mathrm{dist}(x, B_R) < \frac{iR}{\nu}\}.$$

Choose $\varphi_i \in C^m_o(A_i)$ such that

$$(2.2)\qquad \begin{cases} 0 \leq \varphi_i \leq 1 & i=1,\dots,\nu \\ \varphi_i = 1 \text{ on } A_{i-1} \\ |D^\alpha \varphi_i| \leq M \left(\frac{\nu}{R}\right)^{|\alpha|}, & |\alpha| \leq m. \end{cases}$$

Let P be the polynomial of degree $\leq m-1$ such that

$$\int_{B_{2R}} D^\gamma(u-P) = 0 \qquad \forall |\gamma| \leq m-1 \quad .$$

Set

$$w_i = u - \varphi_i(u-P),$$

then, by (2.1), (j),(jj), we have for i =1,...,ν

$$(2.3)\qquad F(A_i,u) \leq F(A_i-A_{i-1},w_i) + F(A_{i-1},P)$$

and also, by (jjj), for i=1,...,ν

$$(2.4)\qquad \int_{B_R} \sum_{|\alpha|=m} |D^\alpha u|^p \leq s|A_i| + s\int_{A_i-A_{i-1}} \sum_{|\alpha|=m} |D^\alpha w_i|^p .$$

Clearly there exists $c>0$ such that

$$(2.5)\qquad \sum_{|\alpha|=m} |D^\alpha w_i|^p \leq c\sum_{|\alpha|=m} |D^\alpha u|^p + c\sum_{|\alpha|=m}\ \sum_{0\neq\beta\leq\alpha} |D^\beta\varphi_i|^p\ |D^{\alpha-\beta}(u-P)|^p$$

Let us now recall ([9]) that, if $\max\left\{1,\frac{np}{n+p}\right\} \leq \mu < p$ and $|\alpha| = m$, $\alpha\geq\beta\neq 0$, then by Sobolev inequality it follows

$$(2.6)\qquad \frac{1}{R^{|\beta|p}} \int_{B_{2R}} |D^{\alpha-\beta}(u-P)|^p \leq c' R^{-n(p/\mu\ -1)} \left(\int_{B_{2R}} |D^\alpha u|^\mu\right)^{p/\mu} .$$

Since (2.2), (2.4),(2.5) imply

$$(2.7)\qquad \int_{B_R} \sum_{|\alpha|=m} |D^\alpha u|^p \leq s|A_i| + sc\int_{A_i-A_{i-1}} \sum_{|\alpha|=m} |D^\alpha u|^p$$

$$+ Mcs\sum_{|\alpha|=m}\ \sum_{0\neq\beta\leq\alpha} \left(\frac{\nu}{R}\right)^{|\beta|p} \int_{B_{2R}} |D^{\alpha-\beta}(u-P)|^p ,$$

and clearly we have

$$\sum_{i=1}^{\nu} \int_{A_i-A_{i-1}} \sum_{|\alpha|=m} |D^\alpha u|^p \leq \int_{B_{2R}} \sum_{|\alpha|=m} |D^\alpha u|^p ,$$

we deduce from (2.7)

$$(2.8)\qquad \int_{B_R} \sum_{|\alpha|=m} |D^\alpha u|^p \le s|B_{2R}| + \frac{sc}{\nu}\int_{B_{2R}} \sum_{|\alpha|=m} |D^\alpha u|^p + \frac{Mcs}{\nu}\sum_{|\alpha|=m}\sum_{0\neq\beta\leq\alpha}\left(\frac{\nu}{R}\right)^{|\beta|p}\int_{B_{2R}} |D^{\alpha-\beta}(u-P)|^p .$$

By (2.6) the last term can be estimated by

$$c'(\nu,m,s)\, R^n\Big(\sum_{|\alpha|=m} \fint_{B_{2R}} |D^\alpha u|^\mu\Big)^{p/\mu}$$

and so (2.8) becomes

$$\fint_{B_R} \sum_{|\alpha|=m} |D^\alpha u|^p \le s + \frac{sc}{\nu}\fint_{B_{2R}} \sum_{|\alpha|=m} |D^\alpha u|^p + c'(\nu,m,s)\Big(\fint_{B_{2R}} \sum_{|\alpha|=m} |D^\alpha u|^\mu\Big)^{p/\mu}$$

and finally, by lemma 2.1 we have the result.

REFERENCES

[1] H.ATTOUCH-C.SBORDONE, Asymptotic limits for perturbed functionals of Calculus of Variations, Ricerche di Matem. XXIX (1) (1980) 85-124.

[2] B.V.BOYARSKI, Homeomorphic solutions of Beltrami systems, Dokl. Akad. Nauk SSSR ,102 (1955) 661-664.

[3] G.BUTTAZZO-M.TOSQUES, Γ-convergenza per alcune classi di funzionali, Ann. Univ. Ferrara, XXIII (1977) 257-267.

[4] F.W.GEHRING, The L^p integrability of the partial derivatives of a quasiconformal mapping, Acta Math. 130 (1973) 265-277.

[5] M.GIAQUINTA-G.GIUSTI, Non linear elliptic systems with quadratic growth, Manuscripta Math. 24 (1978) 323-349.

[6] M.GIAQUINTA-G.GIUSTI, On the regularity of the minima of variational integrals, to appear.

[7] M.GIAQUINTA-G.MODICA, Regularity results for some classes of higher order non linear elliptic systems, J. fur Reine u. Angew. Math. 311/312 (1979) 145-169.

[8] P.MARCELLINI-C.SBORDONE, On the existence of minima of multiple integrals of the Calculus of Variations, to appear.

[9] N.MEYERS-A.ELCRAT, Some results on regularity for solutions of non linear elliptic systems and quasiregular functions, Duke Math. J. 42 (1), 121-136 (1975).

ADMISSIBLE NONLINEAR PERTURBATION OF DIVERGENCE EQUATIONS

Silviu Sburlan

In this lecture we establish the existence of variational solution on general domains for the nonlinear elliptic equation

$$(1) \qquad A(u) + g(x,u) = f(x) ,$$

where $A(u)$ is the generalized divergence operator of second order and $g(x,u)$ is a nonlinear lower-order perturbing term.

The intensive study of such problems in the last period is stimulated by the mathematical models of elastic equilibrium in continuous mechanics. The principal drawback of these models is that the operators involved are not coercive. To eliminate this restriction we use a perturbation procedure by suitable operators of the divergence operator which makes coercive operators.

Let Ω be a domain in $\mathbb{R}^N$, either bounded or unbounded, $x := (x_1,\dots,x_N)$ be a generic point in Ω and dx be the Lebesgue measure. We consider the generalized divergence operator

$$(2) \qquad A(u) := - \sum_{i=1}^{N} \frac{\partial}{\partial x_i} A_i(x,u(x),\nabla u(x)), \qquad \text{a.a. } x \in \Omega$$

with the following hypotheses upon coefficient functions:

(I_1) Each $A_i : \Omega \times \mathbb{R} \times \mathbb{R}^N \longmapsto \mathbb{R}$ satisfies the Caratheodory conditions (i.e., $A_i(x,t,\xi)$ is measurable in $x \in \Omega$ for all fixed $[t,\xi] \in \mathbb{R} \times \mathbb{R}^N$ and continuous in $[t,\xi]$ for a.a. $x \in \Omega$), and there exists constants $p \in (1,+\infty)$, $c_1 \geq 0$ and a function $k_1 \in L^q(\Omega)$, $q := \frac{p}{p-1}$, such that

$$|A_i(x,t,\xi)| \leq k_1(x) + c_1(|t|^{p-1} + |\xi|^{p-1}), \quad 1 \leq i \leq N ,$$

for a.a. $x \in \Omega$, $\forall$ $[t,\xi] \in \mathbb{R} \times \mathbb{R}^N$;

(I_2) $\displaystyle\sum_{i=1}^{N} (A_i(x,t,\xi) - A_i(x,t,\xi'))(\xi_i - \xi'_i) > 0$ for a.a. $x \in \Omega$, $\forall$ $t \in \mathbb{R}$, $\forall$ $\xi,\xi' \in \mathbb{R}^N$ with $\xi \neq \xi'$;

(I_3) There exists a constant $c_2 > 0$ and a function $k_2 \in L^1(\Omega)$ such that

$$\sum_{i=1}^{N} A_i(x,t,\xi)\xi_i \geq c_2|\xi|^p - k_2(x) \quad \text{for a.a. } x \in \Omega ,$$

$\forall \quad [t,\xi] \in \mathbb{R} \times \mathbb{R}^N$.

These assumptions assure that

$$(3) \qquad a(u,v) := \sum_{i=1}^{N} \int_{\Omega} A_i(\cdot,u,\nabla u) \frac{\partial v}{\partial x_i}\, dx$$

is well defined for u,v in any closed subspace X of $W^{1,p}(\Omega)$ - the Sobolev space of real valued functions $u \in L^p(\Omega)$ whose the first distributional derivatives belong to $L^p(\Omega)$. Moreover $v \longmapsto a(u,v)$ is a bounded linear functional that induces a bounded demicontinuous map $T: X \longmapsto X^*$ by the rule

$$(T(u),v) = a(u,v) .$$

Here (f,v) denotes the value of $f \in X^*$ at $v \in X$. Also, by an important result of Browder, [2], for an arbitrary domain Ω, T is pseudo-monotone, that is,

(PM) Whenever $\{u_j\}$ is a sequence in X weakly convergent to an element u in X , (we write $u_j \rightharpoonup u$), $T(u_j) \rightharpoonup z$ in X^* and $\limsup\, (T(u_j),u_j-u) \leq 0$, then $z = T(u)$ and $(T(u_j),u_j-u) \longrightarrow 0$.

On the lower-order perturbing term $g(x,u)$ we impose the following set of hypotheses:

(II_1) For a.a. $x \in \Omega$ and all $t \in \mathbb{R}$

$$g(x,t) = p(x,t) + r(x,t) ,$$

where g,p,r are Caratheodory functions such that

$$p(x,t)t \geq c_3 |t|^p , \quad \forall\ t \in \mathbb{R} ,$$

with $c_3 > 0$ and

$$|r(x,t)| \leq k_3(x), \quad \text{for a.a. } x \in \Omega , \quad \forall\ t \in \mathbb{R},$$

with $k_3 \in L^q(\Omega) \cap L^1(\Omega)$.

(II_2) $\qquad g_s(x) := \sup \{ |g(x,t)| ;\ |t| \leq s \}$ defines an $L^1(\Omega)$ function for $0 \leq s < +\infty$.

Remark that the term $g(x,u)$ does not induce a mapping from X to X^* because its growth in u is not restricted. To eliminate this drowback Webb, [10], has truncated this term as follows:

$$g_n(x,t) := \chi_n(x) p^{(n)}(x,t) + r(x,t)$$

where $\chi_n(x)$ is the characteristic function of $\{x \in \Omega;\ |x| \leq n\}$ and

$$p^{(n)}(x,t) := \begin{cases} p(x,t) & \text{if } |p(x,t)| \leq n, \\ n\,\dfrac{p(x,t)}{|p(x,t)|} & \text{otherwise} . \end{cases}$$

Then

$$b_n(u,v) := \int_\Omega g_n(x,u(x))v(x)dx \tag{4}$$

is well defined for all $u,v \in X$ and $v \mapsto b_n(u,v)$ defines an element of X^*, say $P_n(u)$. Moreover, under assumptions (II), <u>if</u> $u_n \rightharpoonup u$ <u>in</u> X <u>and</u> $\int_\Omega g_n(.,u_n)u_n dx \leq M$ <u>for all</u> $n \in \mathbb{N}$, <u>then</u> $g(.,u)u \in L^1(\Omega)$ <u>and</u> $g_n(.,u_n) \to g(.,u)$ <u>in</u> $L^1(\Omega)$.

Given an element $f \in X^*$, a function $u \in X$ which satisfies

$$a(u,v) + \int_\Omega g(.,u)vdx = \int_\Omega fv\,dx \tag{5}$$

for all $v \in X \cap L^\infty(\Omega)$ and for $v = u$ is said to be a <u>variational solution</u> of equation (1) with the boundary conditions determined by the subspace X of $W^{1,p}(\Omega)$.

<u>Theorem</u>. Let X be a closed subspace of $W^{1,p}(\Omega)$ with the property that for any $u \in X$ there exists a constant $C > 0$ and a sequence $\{w_k\} \subset X \cap L^\infty(\Omega)$ such that $w_k \to u$ in X and

$$|w_k(x)| \leq C|u(x)| \quad \text{a.a. } x \in \Omega .$$

Then under assumptions (I)-(II) there exists a variational solution $u \in X$ of equation (1) for any $f \in X^*$ such that $g(.,u)$ and $g(.,u)u$ belong to $L^1(\Omega)$.

<u>Proof</u>. The map $T+P_n$ of X into X^* is pseudo-monotone since $u_j \rightharpoonup u$ in X implies, at least on a subsequence, that $u_j(x) \to u(x)$ a.a. $x \in \Omega$, (see [9]), and thus $g_n(.,u_j) \to g(.,u)$ in $L^q(\Omega)$ by Lebesgue's dominated convergence theorem. Consequently $(P_n(u_j),u_j-u) \to 0$ and, since T is pseudo-monotone, it follows that $T+P_n$ is also pseudo-monotone.

The map $T+P_n$ is coercive on X since by (I_3) and (II_1) we have

$$\begin{aligned} ((T+P_n)(u),u) &:= a(u,u) + b_n(u,u) = \\ &= c_2\int_\Omega |\nabla u|^p dx - \int_\Omega k_2 dx + c_3\int_\Omega |u|^p dx + \int_\Omega r(.,u)udx \\ &= c\,\|u\|_{1,p}^p - k(1 + \|u\|_p) , \end{aligned}$$

where $c := \min\{c_2, c_3\}$, $k := \max\{\int_\Omega k_2 dx, \int_\Omega k_3 dx\}$, $\|\cdot\|_p$ is the norm in $L^p(\Omega)$ and $\|\cdot\|_{1,p}$ is the norm in $W^{1,p}(\Omega)$.

Thus, by standard existence theorem for pseudo-monotone maps, (e.g. [6]), $T+P_n$ maps X onto X^* ; in particular, there exists $u_n \in X$ such that $(T+P_n)(u_n) = f$. Moreover, due to the coerciveness of $T+P_n$ we have

$$\|u_n\|_{1,p}^p \leq \frac{k}{c}(1 + \|u\|_p) + \frac{1}{c}(f, u_n),$$

which implies the boundedness of $\{u_n\}$ and hence of $\{T(u_n)\}$. Taking (eventually) a subsequence we may assume that $u_n \rightharpoonup u$ in X and $T(u_n) \rightharpoonup z$ in X^* . So there exists a constant $M > 0$ such that

$$\int_\Omega g_n(\cdot, u_n) u_n \, dx \leq M , \quad \forall \quad n \in \mathbb{N}$$

and hence $g(\cdot,u)u \in L^1(\Omega)$ and $g_n(\cdot, u_n) \to g(\cdot, u)$ in $L^1(\Omega)$. Therefore, passing to the limit as $n \to \infty$, we obtain for all $v \in X \cap L^\infty(\Omega)$

$$(z,v) + \int_\Omega g(\cdot,u)v \, dx = \int_\Omega fv \, dx \quad ,$$

which is convinient to put in the form

$$(f-z,v) = \int_\Omega g(\cdot,u)v \, dx \tag{6}$$

The second step is to show that $z = T(u)$ which, by pseudo-monotonicity of T , is equivalent to

$$\limsup \, (T(u_n), u_n - u) \leq 0 .$$

Now, $(T(u_n), u_n - u) = (T(u_n), u_n) - (T(u_n), u)$; so by (6)

$$\limsup \, (T(u_n), u_n - u) = \limsup \, (f - P_n(u), u_n) - (z,u) \leq$$

$$\leq (f-z,u) - \liminf \int_\Omega g_n(\cdot,u_n)u_n dx \leq (f-z,u) - \int_\Omega g(\cdot,u)u \, dx .$$

Thus for any $w \in X \cap L^\infty(\Omega)$, using (6),

$$\limsup \, (T(u_n), u_n - u) \leq (f-z, u-w) + \int_\Omega g(\cdot,u)(w-u) \, dx .$$

Take now instead of w the sequence $\{w_k\} \subset X \cap L^\infty(\Omega)$ assured by hypothesis. Thus $(f-z, u-w_k) \to 0$ and

$$\int_\Omega g(\cdot,u)w_k \, dx \longrightarrow \int_\Omega g(\cdot,u)u \, dx$$

by dominated convergence because $g(\cdot,u)u \in L^1(\Omega)$. It follows that $\limsup \, (T(u_n), u_n - u) \leq 0$ so that $z = T(u)$ and $(T(u_n), u_n - u) \to 0$. Also, for $v = w_k$ in (6) we obtain, when $k \to \infty$, the proof. □

Remarks:

a) The method of the proof in the Theorem has been done by Brèzis and Browder [1] and it was improved by Webb [10]. However the above proof differs from the mentioned ones in the sense that the coercivity of $T+P_n$ is obtained by the contribution of the both operators. This restriction is imposed by physical conditions on the equations of elastic equilibrium where the divergence part does not induce a coercive map, (e.g. [4],[7]).

b) As is pointed out by Webb, [10], the sequence $\{w_k\}$ which appears in the Theorem can be found in the extreme cases $X:= W_0^{1,p}$ or $X:= W^{1,p}$ by a result of L.I.Hedberg.

c) Asking instead of (I_2) the stronger condition

(I_2') $\sum_{i=0}^{N} (A_i(x,s) - A_i(x,s'))(s_i - s_i') > 0$ for a.a. $x\in\Omega$, $\forall$ $s, s' \in \mathbb{R}\times\mathbb{R}^N$ with $s \neq s'$, where $A_0(x,s):= A_0(x,t,\xi)$ has similar properties as $A_i(x,t,\xi)$, $1 \leq i \leq N$, we obtain that T is a monotone demicontinuous operator and, since $D(T):= X$, it is a maximal one. So its graph, G(T), is a demiclosed set in $X\times X^*$, (e.g. [6,p.105]), and we can renounce at the second step in above proof. In this case we have no further use of the sequence $\{w_k\}\subset X\cap L^\infty(\Omega)$ and with slight modifications we can prove the following

Proposition. Under assumptions (I)-(II) there exists a solution of the variational inequality

$$a(u,v-u) + \int_\Omega g(.,u)(v-u)\,dx = \int_\Omega f(v-u)\,dx, \quad \forall\, v\in K\cap L^\infty$$

where K is a closed convex subset of X, (see [8]).

d) Generally, the uniqueness of the variational solution fails to hold. However we can expect the uniqueness when T is a monotone operator and $g(x,u)$ is differentiable in $u\in\mathbb{R}$ for a.a. $x\in\Omega$ with a positive differential.

REFERENCES

1 H. Brèzis, F.E. Browder - Strongly nonlinear elliptic boundary value problems. Ann.Scuola Norm.Sup.Pisa 5, 563-87, (1978).

2 F.E. Browder - Pseudo-monotone operators and nonlinear elliptic boundary value problems on unbounded domains. Proc.Nat.

Acad.Sci. 74, 2659-61, (1977).

[3] F.E. Browder - Strongly nonlinear elliptic and parabolic problems. Raport at the "First Romanian-American Seminar on Operator Theory", Iassy, (1978).

[4] J. Nečas - Theory of locally monotone operators modeled on the finite displacement theory for hyperelasticity. Beiträge zur Anal. 8, 1o3-14, (1976) .

[5] D. Pascali - Variational elliptic problems on unbounded domains Tech.Hochsch.Darmstadt, Preprint nr.537, (198o).

[6] D. Pascali, S. Sburlan - Nonlinear Mappings of Monotone Type . Ed.Acad.RSR - Sijthoff & Noordhoff Intern.Publ., (1979).

[7] S. Sburlan - Some open problems of finite elastic equilibrium . "Applied nonlinear Functionalanalysis" R.Gorenflo - K.H. Hoffmann (ed), 271-88, Peter Lang -Verlag, (1983).

[8] S. Sburlan - Elliptic strongly nonlinear variational inequalities on unbounded domains. Rev.Roum.Math.Pures Appl. 26, 358-64, (1981).

[9] C.G. Simader - Über schwache Lösungen der Dirichlet Problem für streng nichlineare elliptische Differentialgleichungen. Math.Z. 150, 1-26, (1976).

[10] J.R.L. Webb - Boundary value problems for strongly nonlinear elliptic equations. J.London Math.Soc. (2) 21, 123-32, (198o).

Interpolation and domination by positive definite kernels

F.H. Szafraniec

Abstract. Extending the idea of a recent paper [2] of Fitzgerald and Horn we propose a RKHS (reproducing kernel Hilbert space) approach to solution of some interpolation problems.

1. In a recent paper FitzGerald and Horn have presented [2] a new look at the classical Pick-Nevanlinna problem (or rather at one of its versions). They have observed that the condition

$$\sum_{i,j} \frac{1-w_i\bar{w}_j}{1-z_i\bar{z}_j} c_i\bar{c}_j \geq 0,$$

which quarantees the existence of a suitable analytic function interpolating points $\{z_i, w_i\}$, implies another condition

$$\sum_{i,j} \frac{1}{1-z_i\bar{z}_j} c_i\bar{c}_j \geq |\sum_i w_i c_i|^2.$$

This is what we mean by domination of interpolated data by a kernel; more precisely.

$$\sum_{i,j} K(z_i,z_j)c_i\bar{c}_j \geq |\sum_i w_i c_i|^2$$

with the kernel K defined as $K(u,v) = \frac{1}{1-u\bar{v}}$. This kernel is a particular case of an analytic kernel. Let D be a domain in the complex plain $\mathbb{C}$. A function $K: D \times D \to \mathbb{C}$ is said to be an analytic kernel if K is an analytic function in the first variable and antianalytic in the second. The main part of [2] can be stated as follows.

Theorem (FitzGerald-Horn). Let D be a complex domain and let $\{z_i\} \subset D$, $\{w_i\}$ be two sequences of numbers. Suppose $\{z_i\}$ has an accumulation point in D. Suppose K is an analytic kernel in D such that

$$\sum_{i,j} K(z_i,z_j)\, c_i\bar{c}_j \geq \Big| \sum_i w_i c_i \Big|^2$$

for all finite sequences $c_1,\ldots,c_n$ of complex numbers. Then there exists a (unique) analytic function in D such that

$$f(z_i) = w_i$$

and

$$(1) \qquad \sum_{i,j} K(u_i,u_j)c_i\bar{c}_j \geq \Big|\sum_i f(u_i)c_i\Big|^2$$

for all sequences $u_1,\ldots,u_n$ in D and $c_1,\ldots,c_n$.

2. A kernel K is said to be PD (read: positive definite) if for all finite sequences of $u_1,\ldots,u_n$ and $c_1,\ldots,c_n$

$$\sum_{i,j} K(u_i,u_j)c_i\bar{c}_j \geq 0.$$

From their theorem FitzGerald and Horn have obtained, as a by-product, the following result on continuation of PD analytic kernels, cf [1].

Theorem A. Suppose the sequence $\{z_i\} \subset D$ have an accumulation point in D. Suppose K is an analytic kernel in D. If K is PD on the set $\{z_i\}$, then it is PD on D.

This theorem can be proved independently using a somewhat modified RKHS approach. The crucial point here is that such a sequence $\{z_i\}$ is a set of uniqueness for analytic functions in D(this also ensures uniqueness of f in the conclusion of the FitzGerald Horn theorem). Having this question over, we can suppose K is PD on D and, due to this, consider interpolated data $\{z_\alpha, w_\alpha\}$ of arbitrary cardinality. The result is as follows

<u>Theorem B</u>. Let $\{z_\alpha\} \subset D$ and $\{w_\alpha\}$ be two sets of complex numbers indexed in the same way. Suppose K is a PD analytic kernel on D such that

$$(2) \qquad \sum_{i,j} K(z_{\alpha_i}, z_{\alpha_j}) c_i \bar{c}_j \geqq \left|\sum_i w_{\alpha_i} c_i\right|^2$$

for all finite sequences $u_{\alpha_1}, \ldots, u_{\alpha_n}$ and $c_1, \ldots, c_n$. Then there exists an anylytic function f in D such that

$$(3) \qquad f(z_\alpha) = w_\alpha \quad \text{for all } \alpha \, .$$

Moreover, the inequality (1) holds.

Thus the RKHS approach allows us to split conveniently the FitzGerald-Horn theorem into two, of independent interest: Theorem A and Theorem B. This approach, under circumstances exhibited by FitzGerald and Horn, seems to be quite natural and we would like to extend it somewhere else. But we should point out that the use of the RKHS method for problems <u>close</u> to ours has been known in the literature for a long time(cf. [1] for more details and further references;here we mention Korányi's and Sz.-Nagy's papers [3], [4] and [5] where the question of integral representation of operator monotonic functions is considered).

3. Take a kernel K which is PD and analytic on D. For $u_1, \ldots, u_n$ in D and $c_1, \ldots, c_n$ in C, $\sum_i K(\cdot, u_i) c_i$ is the analytic function in the first variable and the set of all such functions forms a pre-Hilbert space with the inner product

$$(4) \qquad \left\langle \sum_i K(\cdot, u_i) c_i, \sum_j K(-, v_j) d_i \right\rangle = \sum_{i,j} K(v_j, u_i) c_i \bar{c}_j$$

It is easy to show (cf. [1]) that the completion of the space of these functions is a Hilbert space composed of analytic functions in D. Denote this space by H. This is the Hilbert space with the reproducing kernel K and the reproducing property reads explicitely as follows

$$(5) \qquad f(z) = \langle f, K(\cdot, z)\rangle, \quad z \in D$$

and arbitrary f in H.

4. Now we are able to prove Theorem B. Take a function $\Sigma K(\cdot, z_{\alpha_i})c_i \in H$. Define

$$(6) \qquad F(\sum_i K(\cdot, z_{\alpha_i})c_i) = \sum_i \overline{w}_i c_i .$$

The inequality (2) and density argument imply that F is a well defined bounded linear functional on the (closed) subspace of H generated by functions $\sum_i K(\cdot, z_{\alpha_i})c_i$. Moreover $\|F\| \leqq 1$. Put F = 0 on the orthogonal complement of this subspace. The Riesz representation theorem provides us with an analytic function f in H such that

$$f(g) = \langle g, f\rangle \quad , \quad g \in H.$$

This, combined with (5) and (6), gives

$$\overline{w}_\alpha = F(K(\cdot, z_\alpha)) = \langle K(\cdot, z_\alpha), f\rangle = \overline{f(z_\alpha)}$$

which is nothing else than the desired interpolating property (3) of f.

The inequality (1) follows from the reproducing property (5) and the fact $\|f\| = \|F\| \leqq 1$.

5. To prove Theorem A we can modify the RKHSpace in the following way: take, instead of all functions $\sum_i K(\cdot,u_i)\,c_i$, only those $\sum_i K(\cdot,z_i)c_i$. Since $\{z_i\}$ is a set of uniqueness for analytic functions in D, one can prove that what is defined by (4) remains still the inner product and that as the competion of the set of functions $\sum_i K(\cdot,z_i)c_i$ we get again a space of analytic functions in D.

6. We conclude with the notice that the same approach can lead to <u>an operator valued version</u> of the FitzGerald-Horn Theorem (not of Theorem B).

References

[1] W.F.Donoghue, Jr., Monotone matrix functions and analytic continuation, Springer, Berlin-New York-Heidelberg, 1974.

[2] C.H. FitzGerald, R.A.Horn, On quadratic and bilinear forms in function theory, The Johns Hopkins University, Department of Mathematical Sciences, Technical Report No.305, June, 1979.

[3] A.Korányi, On a theorem of Löwner and its connection with resolvents of self adjoint transformations, Acta Sci.Math., 17 (1956), 63-70.

[4] -----, B.Sz.-Nagy, Operatortheoretische Behandlung und Verallgemeinerung eines Problemkreis in der Komplexen Functionentheorie, Acta Math., 100 (1958), 171-202.

[5] B.Sz.-Nagy, Remarks to the preceding paper of A.Korányi , Acta Sci, Math., 17 (1956), 71-75.

Uniwersytet Jagielloński

Instytut Matematyki

ul.Reymonta 4, PL-30059 Kraków

ON THE REGULARITY OF THE BOUNDARY MEASURES

by Silviu Teleman

The aim of this Note is to prove some new topological properties of the boundary measures; namely, roughly speaking, that the boundary measures are inner regular (i.e., by closed compact measurable subsests, where compactness, closedness is meant with respect to the Choquet topology). Stronger results are obtained for the pure states space of a C^* - algebra.

1. Let E be any Hausdorff locally convex topological real vector space and $K \subset E$ any non-empty compact convex subset. We shall use the notations introduced in [6] , as well as many of the results we have presented there.

We recall that for any bounded function $f:K \longrightarrow \mathbb{R}$ the function $\bar{f}:K \longrightarrow \mathbb{R}$ is defined by

$$\bar{f} = \inf\{h;\ f \leqslant h,\ \ h \in A(K)\},$$

where the infimum is computed point-wise. Then $\bar{f}$ is the smallest concave upper semicontinuous function majorizing f (see [6] , p.12; [5], §3; [4] , Ch.XI, D18).

PROPOSITION 1. *For any bounded upper semi-continuous function* $f:K \longrightarrow \mathbb{R}$ *we have*

$$\bar{f}(x) = \sup\{\mu(f);\ \mu \sim \varepsilon_x\}, \quad x \in K.$$

PROOF. In this equality μ runs over the compact convex set $\mathcal{M}^1_x(K)$ of all Radon probability measures μ, whose barycenter $b(\mu)=x$.

a) If we define

$$\varphi(x) = \sup\{\mu(f); \mu \sim \varepsilon_x\}, x \in K,$$

then φ obviously is bounded and it is easy to prove that φ is concave; on the other hand, it is easy to prove that $f \leq \varphi$.

b) φ is upper semi-continuous. Indeed, let us first remark that, since the mapping

$$\mathcal{M}^1_+(K) \ni \mu \longmapsto \mu(f_0),$$

on the set $\mathcal{M}^1_+(K)$ of all Radon probability measures on K, is continuous, for any $f_0 \in C(K;\mathbb{R})$, the mapping

(1) $$\mathcal{M}^1_+(K) \ni \mu \longmapsto \mu(f)$$

is upper semi-continuous. Let now $\alpha \in \mathbb{R}$ and define $L_\alpha = \{x \in K; \varphi(x) \geq \alpha\}$. Let $(x_i)_{i \in I}$ be a net in L_α and assume that $x_i \to x$ in K. Let $\varepsilon > 0$ be given. Then we have

$$\varphi(x_i) \geq \alpha > \alpha - \varepsilon,$$

and, therefore, for any $i \in I$, there exists a $\mu_i \in \mathcal{M}^1_{x_i}(K)$, such that

(2) $$\mu_i(f) > \alpha - \varepsilon, \qquad i \in I.$$

Passing to a subnet, if necessary, we can assume that $\lim_{i \in I} \mu_i = \mu$ exists in $\mathcal{M}^1_+(K)$. From (2) and from the fact that the mapping in (1) is upper semi-continuous, we infer that

(3) $$\mu(f) \geq \alpha - \varepsilon;$$

since we have $\lim_{i\in I} b(\mu_i)=b(\mu)$, from (3) we infer that $\varphi(x)\geq\alpha-\varepsilon$, for any $\varepsilon>0$, and, therefore $\varphi(x)\geq\alpha$. It follows that $x\in L_\alpha$, and this shows that L_α is closed ; i.e., φ is upper semi-continuous.

c) If $h\in A(K)$ and $f\leq h$, then

$$\mu(f)\leq\mu(h)=h(b(\mu))=h(x),$$

for any $\mu\in\mathcal{M}^1_x(K)$, $x\in K$. We infer that

$$\varphi(x)\leq h(x),\quad f\leq h\in A(K), x\in K,$$

and, therefore, we have

$$\varphi\leq\bar{f}\,; \tag{4}$$

since $\bar{f}$ is the smallest concave upper semi-continuous function majorizing f, from (4) we immediately infer that $\varphi=\bar{f}$, and the Proposition is proved.

REMARK. Proposition 1 is a slight extension of Proposition 3.1 from [5], where it is stated for a continuous function.

COROLLARY. For any bounded upper semi-continuous function $f:K\to\mathbb{R}$ we have

$$\bar{f}(x)=f(x),\qquad x\in \operatorname{ex} K.$$

PROOF. This is an immediate consequence of the preceding Proposition and of H.Bauer's Theorem (see [5], Proposition 1.4; [6], Proposition 1.3).

PROPOSITION 2. For any bounded upper semi-continuous function $f:K\to\mathbb{R}$ and any decreasing net $(f_\alpha)_{\alpha\in A}$ of bounded upper semi-continuous functions on K, such that $f_\alpha\downarrow f$ point-wise on K, we have $\bar{f}_\alpha\downarrow\bar{f}$ point-wise on K.

<u>PROOF.</u> It is obvious that $(\overline{f}_\alpha)_{\alpha\in A}$ is a decreasing net, such that $\overline{f} \leq \lim_{\alpha\in A} \overline{f}_\alpha$. Let then $\varepsilon>0$ and $x\in K$ be given; there exists a $h\in A(K)$, such that

(1) $h(x) < \overline{f}(x) + \varepsilon$ and $f(y) < h(y)$, $\forall\, y \in K$.

Let $K_\alpha = \{y; f_\alpha(y)-h(y) \geq 0\}$, $\alpha\in A$; since we have

$$\inf\{f_\alpha(y)-h(y); \alpha\in A\} < 0, \qquad \forall\, y\in K,$$

we infer that $K_\alpha \downarrow \emptyset$ and, therefore, we can find an $\alpha_0 \in A$, such that $K_{\alpha_0} = \emptyset$.
(because the sets K are compact).

We infer that $f_{\alpha_0}(y) \leq h(y)$, $y\in K$, and, therefore, we have

$$\mu(f_{\alpha_0}) \leq \mu(h) = h(b(\mu)), \qquad \mu\in \mathcal{M}_+^1(K).$$

From Proposition 1 and from (1) we infer that

$$\overline{f}_{\alpha_0}(x) \leq h(x) < \overline{f}(x)+\varepsilon ,$$

and this implies that

$$\inf\{\overline{f}_\alpha(x); \alpha\in A\} \leq \overline{f}(x), \qquad x\in K.$$

The Proposition is proved.

<u>COROLLARY</u>. <u>For any bounded upper semi-continuous function</u> $f: K \longrightarrow \mathbb{R}$ <u>and any measure</u> $\mu\in\mathcal{M}_+^1(K)$, <u>which is maximal with respect to the Choquet order relation, we have</u>

$$\mu(\overline{f}) = \mu(f).$$

<u>PROOF</u>. There exists a decreasing net $(f_\alpha)_{\alpha\in A}$ of continuous functions $f_\alpha : K \to \mathbb{R}$, such that $f_\alpha \downarrow f$ point-wise on K. If

$\mu \in \mathcal{M}_+^1(K)$ is maximal with respect to the Choquet order relation, then we have

(1) $$\mu(\bar{f}_\alpha) = \mu(f_\alpha), \qquad \alpha \in A,$$

(see [5], Proposition 4.2. ; [6], Lemma 1.2).

From (1) and from Proposition 2, by taking into account the τ-continuity of the measure μ, we infer that $\mu(\bar{f}) = \mu(f)$, and the Corollary is proved.

LEMMA 1. *Let X be any compact space, $F' \subset X$ a G_δ-subset and μ a positive Radon measure on X. Then, for any $\varepsilon > 0$ there exists a compact Baire measurable subset $D \subset F'$, such that*

$$\mu(F') - \varepsilon < \mu(D).$$

PROOF. Let $F' = \bigcap_{n=0}^{\infty} G_n$, where $G_n \subset X$ are open subsets. Since μ is regular, there exists a compact subset $D_0 \subset F'$, such that

$$\mu(F') - \varepsilon < \mu(D_0).$$

For any $n \in \mathbb{N}$ we can find a continuous function $f_n : X \to [0,1]$, such that

$f_n(x) = 1$, for $x \in D_0$, and $f_n(x) = 0$, for $x \in \complement G_n$.

Let $X_n = \{x \in X;\ f_n(x) = 1\}$. Then $D = \bigcap_{n=0}^{\infty} X_n$ is a compact Baire measurable subset of X, such that $D_0 \subset D \subset \bigcap_{n=0}^{\infty} G_n = F'$, and the Lemma is proved.

We shall denote by $\mathcal{B}_o(X)$ the σ-algebra of the Baire measurable subsets of the topological space X, i.e., the σ-algebra of subsets of X, which is generated by the set of all closed G_δ-subsets of X, whereas $\mathcal{B}(X)$ will stand for the σ-algebra of the Borel measurable subsets of X, which is generated by the set of all closed subsets of X.

When several topologies are considered on X, a special mark will indicate the topology to which these σ-algebras correspond.

2. For any function $f:K \to \mathbb{R}$ we shall denote $z(f) = \{x \in K;\ f(x)=0\}$ and $u(f)=\{x \in K;\ f(x) = 1\}$.

Let now $F \subset K$ be a compact subset of K. Then χ_F is an upper semi-continuous function, whereas $\overline{\chi}_F$ is a concave upper semi-continuous function. It is easy to see that $F'=z(\overline{\chi}_F)$ is, therefore, a measure extremal G_δ - subset of K (see [6], p.26 and p.39).

LEMMA 2. $u(\overline{\chi}_F) = \overline{co}(F)$.

PROOF. From $\chi_F \leq \overline{\chi}_F \leq 1$ we infer that

(1) $$F= u(\chi_F) \subset u(\overline{\chi}_F).$$

On the other hand, $u(\overline{\chi}_F)$ is a compact convex subset of K; therefore, from (1) we infer that

$$\overline{co}(F) \subset u(\overline{\chi}_F).$$

Let now $x_0 \in K \setminus \overline{co}(F)$. Since the mapping $\mathcal{M}^1_{x_0}(K) \ni \mu \to \mu(F)$ is upper semi-continuous on the compact space $\mathcal{M}^1_{x_0}(K)$, we infer that there exists a $\mu_0 \in \mathcal{M}^1_{x_0}(K)$, such that

$$\mu_o(F) = \sup \{\mu(F);\ \mu \sim \varepsilon_{x_o}\}.$$

If we had $x_o \in u(\overline{\chi}_F)$, then, with Proposition 1, we would infer that $\mu_o(F)=1$, and, therefore, $x_o=b(\mu_o) \in \overline{co}(F)$, a contradiction. It follows that $x_o \notin u(\overline{\chi}_F)$ and, therefore,

$$u(\overline{\chi}_F) \subset \overline{co}(F).$$

The Lemma is proved.

PROPOSITION 3. *For any compact subset* $F \subset K$ *we have*

a) $F \cap F' = \emptyset$ *and* $F \cup F' \supset \text{ex } K$;

b) $\mu(F) + \mu(F') = 1$, *for any Radon probability measure* $\mu \in \mathcal{M}_+^1(K')$, *which is maximal for the Choquet order relation.*

PROOF. a) If $x \in F$, then $\overline{\chi}_F(x) = 1$, and therefore, $x \notin F'$ (as above, we have $F'=z(\overline{\chi}_F)$); for any $x \in \text{ex } K$, if $x \notin F$, we have $\overline{\chi}_F(x)=\chi_F(x)=0$, by the Corollary to Proposition 1; it follows that $x \in F'$.

b) Let $\mu \in \mathcal{M}_+^1(K)$ be a Choquet maximal Radon probability measure on K. By the Corollary to Proposition 2 we have

$$\mu(F) = \mu(\chi_F) = \mu(\overline{\chi}_F), \tag{1}$$

and, therefore, if we take into account Lemma 2 above and Proposition 1.10, b) from [6], we infer that

$$\begin{aligned}\mu(\overline{\chi}_F) &= \mu(u(\overline{\chi}_F)) + \mu((1-\chi_{u(\overline{\chi}_F)})\overline{\chi}_F) = \\ &= \mu(\overline{co}(F)) + \mu((1-\chi_{u(\overline{\chi}_F)})\overline{\chi}_F) = \\ &= \mu(F) + \mu((1-\chi_{u(\overline{\chi}_F)})\overline{\chi}_F).\end{aligned} \tag{2}$$

From (1) and (2) we infer that

$$\mu((1-\chi_{u(\overline{\chi}_F)})\overline{\chi}_F)=0,$$

and this implies that $\mu(F')=1-\mu(F)$. The Proposition is proved.

3. Let now $D\subset K$ be a Baire measurable subset and $h_n\in A(K)$, $n\in\mathbb{N}$, a sequence of affine continuous real functions on K, such that D be $\{h_n;\, n\in\mathbb{N}\}$ - measurable ; it follows that

$$x_o\in D,\quad x\in K,\quad \text{and}\quad h_n(x)=h_n(x_o),\quad n\in\mathbb{N}\Rightarrow x\in D$$

(such a sequence can always be found; see [6] , §1.5). Let $\mu\in\mathcal{M}_+^1(K)$ be a Choquet maximal Radon probability measure on K. By virtue of Lemma 1.2 from [6] , we have

$$\mu(\overline{h_n^2})=\mu(h_n^2),\qquad n\in\mathbb{N}.$$

We infer that, for any $n\in\mathbb{N}$, there exists a sequence $(h_{nm})_{m\in\mathbb{N}}$, $h_{nm}\in A(K)$, $m\in\mathbb{N}$, such that

$$\overline{h_n^2}\leq h_{nm},\ m\in\mathbb{N},\qquad\text{and}\ \mu(h_n^2)=\inf_{m\in\mathbb{N}}\mu(h_{no}\wedge\ldots\wedge h_{nm}).$$

If we denote $\mathcal{F}_o=\{h_n;n\in\mathbb{N}\}$ and $\mathcal{F}_1'=\{h_{nm};\ m,n\in\mathbb{N}\}$, $\mathcal{F}_1=\mathcal{F}_o\cup\mathcal{F}_1'$, then $\mathcal{F}_1$ is a countable subset of $A(K)$, and for any $h\in\mathcal{F}_1$ we can find a sequence $(h_n')_{n\in\mathbb{N}}$, $h_n'\in A(K)$, $n\in\mathbb{N}$, such that

$$h^2\leq h_n',\ n\in\mathbb{N},\quad\text{and}\ \mu(h^2)=\inf_{n\in\mathbb{N}}\mu(h_o'\wedge\ldots\wedge h_n').$$

Let $\mathcal{F}_2'=\{h_n';\ h\in\mathcal{F}_1,\ n\in\mathbb{N}\}$ and $\mathcal{F}_2=\mathcal{F}_1\cup\mathcal{F}_2'$.

By induction, we can find an increasing sequence $(\mathcal{F}_n)_{n\in\mathbb{N}}$ of countable subsets of $A(K)$, such that for any $h\in\mathcal{F}_n$ there exists a sequence $(h'_n)_{n\in\mathbb{N}}$ in $\mathcal{F}_{n+1}$, such that $h^2 \leq h'_n$, $n \in \mathbb{N}$, and

$$\mu(h^2) = \inf_{n\in\mathbb{N}} \mu(h'_0 \wedge \ldots \wedge h'_n).$$

Let $\mathcal{F} = \bigcup_{n\geq 0} \mathcal{F}_n$. Then $\mathcal{F}$ is a countable subset of A(K) such that

1) D is $\mathcal{F}$-measurable;

2) for any $h \in \mathcal{F}$ there exists a sequence $(h'_n)_{n\in\mathbb{N}}$ in $\mathcal{F}$, such that $h^2 \leq h'_n$, $n \in \mathbb{N}$, and

(1) $$\mu(h^2) = \mu(\overline{h^2}) = \inf_n \mu(h'_0 \wedge h'_1 \wedge \ldots \wedge h'_n).$$

We shall now consider the affine continuous mapping $\theta : K \to \mathbb{R}^{\mathcal{F}}$, given by

$$\theta(x) = (h(x))_{h\in\mathcal{F}}, \qquad x \in K.$$

Since $R^{\mathcal{F}}$ is metrizable, we infer that $K_o = \theta(K)$ is a metrizable compact convex subset of $R^{\mathcal{F}}$. Of course, $\theta(D)$ is a Baire measurable subset of $\theta(K)$ and $D = \theta^{-1}(\theta(D))$.

We shall now consider the Radon probability measure $\mu_o = \theta_*(\mu)$ on K_o.

If we denote by p_h the projection in $R^{\mathcal{F}}$ (or its restriction to K_o) which corresponds to $h \in \mathcal{F}$, then we have

$$h = p_h \circ \theta, \qquad h \in \mathcal{F}.$$

Since $(p_h)_{h\in\mathcal{F}}$ is a total set of affine continuous real function on K_o, from Corollary 1 to Proposition 1.3 from [6] we infer that

(2) $$\text{ex } K_o = \{ y \in K_o;\ p_h^2(y) = \overline{p_h^2}(y),\ h \in \mathcal{F} \}.$$

From (1) we now infer that

$$\mu(h^2)=\mu_o(p_h^2)\leq\mu_o(\overline{p_h^2})\leq\mu(\overline{h^2}), \quad h\in\mathcal{F},$$

and, therefore,

(3) $$\mu_o(p_h^2)=\mu_o(\overline{p_h^2}), \quad h\in\mathcal{F}.$$

From (2) and (3) we infer that

$$\mu_o(\text{ex } K)=1,$$

and, therefore, μ_o is a Choquet maximal Radon probability measure on K_o.

We have ,therefore, the following <u>Approximation Theorem</u> .

<u>THEOREM 1</u>. <u>Let</u> $D\subset K$ <u>be a Baire measurable subset and</u> $\mu\in\mathcal{M}_+^1(K)$ <u>a Choquet maximal Radon probability measure on</u> K. <u>Then there exists a (metrizable) compact convex set</u> $K_o\subset\mathbb{R}^N$ <u>and an affine continuous surjective mapping</u> $\theta: K\longrightarrow K_o$, <u>such that</u>

a) $\theta(D)$ <u>is a Baire measurable subset of</u> K_o;

b) $\theta^{-1}(\theta(D))=D$;

c) $\theta_*(\mu)$ <u>is a Choquet maximal Radon probability measure on</u> K_o.

We can now prove the following

<u>COROLLARY</u>. <u>Let</u> $D\subset K$ <u>be a Baire measurable subset</u>. <u>Then, for any</u> $\varepsilon>0$, <u>there exists a compact extremal Baire measurable subset</u> $D_o\subset D$, <u>such that</u> $\mu(D)-\varepsilon<\mu(D_o)$.

<u>PROOF.</u> With the notations of the preceding Theorem, by Ulam's Theorem (see [1] ,Ch.1,Theorem 1.4), there exists a compact subset $A_o\subset\theta(D)\cap\text{ex } K_o$, such that $\theta_*(\mu)(\theta(D))-\varepsilon<\theta_*(\mu)(A_o)$.

Let $D_o = \theta^{-1}(A_o)$. Then $D_o \subset D$ and D_o is a compact extremal Baire measurable subset of K, such that

$$\mu(D_o) = \theta_*(\mu)(A_o) > \theta_*(\mu)(\theta(D)) - \varepsilon = \mu(D) - \varepsilon .$$

The Corollary is proved.

REMARK. The preceding Theorem obviously holds for any sequence $\{D_n\}_{n\geq 0}$ of Baire measurable subsets of K and any sequence $(\mu_n)_{n\geq 0}$ of Choquet maximal Radon probability measures .

THEOREM 2. Let $H \subset K$ be a G_δ- subset and $\mu \in \mathcal{M}_+^1(K)$ a Choquet maximal Radon probability measure on K. Then, for any $\varepsilon > 0$, there exists a compact extremal Baire measurable subset $D_1 \subset H$, such that $\mu(D_1) > \mu(H) - \varepsilon$.

PROOF. By Lemma 1 there exists a compact Baire measurable subset $D \subset H$, such that

$$\mu(D) > \mu(H) - \frac{\varepsilon}{2}.$$

Let K_o and θ correspond to D and μ, as in the preceding Theorem. Then we have

$$\theta_*(\mu)(\theta(D) \cap \text{ex } K_o) = \theta_*(\mu)(\theta(D)) = \mu(\theta^{-1}(\theta(D))) = \mu(D).$$

Since ex K_o is a Polish space, by Ulam's Theorem (see [1], Ch.1, Theorem 1.4) there exists a compact subset $D_o \subset \theta(D) \cap \text{ex } K_o$ such that

$$\theta_*(\mu)(D_o) > \theta_*(\mu)(\theta(D) \cap \text{ex } K_o) - \frac{\varepsilon}{2}.$$

If we denote $D_1 = \theta^{-1}(D_o)$, then the set D_1 defined in this manner has all the required properties. The Theorem is proved.

We recall that the Choquet topology on ex K is that for which $\{F \cap ex\ K;\ F \subset K$ compact, extremal$\}$ is the set of all closed subsets of the topology (see [2],Ch.II.2 ; [6] , p.27).

We shall specify by C-closed, C-open, etc., the various topological epithets corresponding to the Choquet topology.

Let $\mathcal{B}_o(K)$ be the σ-algebra of all Baire measurable subsets of K and $\mathcal{A}_o(ex\ K) = \{D \cap ex\ K;\ D \in \mathcal{B}_o(K)\}$.

For any Choquet maximal Radon probability measure $\mu \in \mathcal{M}_+^1(K)$ one can induce the <u>boundary measure</u> $\check{\mu} : \mathcal{A}_o(ex\ K) \to [0,1]$, given by

$$\check{\mu}(D \cap ex\ K) = \mu(D), \qquad D \in \mathcal{B}_o(K).$$

We have proved in [6] that for any C-closed subset $A \subset ex\ K$ we have

$$\check{\mu}^*(A) = \mu(F),$$

for any compact extremal subset $F \subset K$, such that $F \cap ex\ K = A$ (see [6] ,Proposition 1.11),

We shall now prove the following <u>Inner Regularity Theorem</u>.

<u>THEOREM 3</u>. <u>Let</u> $G \subset ex\ K$ <u>be any C-open subset of</u> ex K.<u>Then</u>

$$\check{\mu}_*(G) = \sup\{\check{\mu}(A);\ A \subset G, \text{ C-closed and } A \in \mathcal{A}_o(ex\ K)\}.$$

<u>PROOF</u>. Let $F \subset K$ be any compact extremal subset of K, such that $G = (ex\ K) \setminus F$. We then have

$$\check{\mu}_*(G) = 1 - \check{\mu}^*(F \cap ex\ K) = 1 - \mu(F),$$

the second equality being a consequence of Proposition 1.11 from [6].

If we denote $F' = Z(\overline{\chi}_F)$, then, by Proposition 3, we have $\mu(F') = \check{\mu}_*(G)$. Let $\varepsilon > 0$ be given. By Theorem 2, there exists a compact extremal Baire measurable subset $D_1 \subset F'$, such that

$$\mu(F') < \mu(D_1) + \varepsilon ;$$

we infer that $A_1 = D_1 \cap \text{ex } K$ is a C-closed $\mathcal{A}_0(\text{ex } K)$-measurable subset of G, such that

$$\check{\mu}_*(G) < \check{\mu}(A_1) + \varepsilon .$$

The Theorem is proved.

In [6] we proved that any C-Baire measurable subset of ex K (with respect to the Choquet topology; i.e., any set belonging to the smallest σ-algebra of subsets of ex K, containing all C-closed (C-G_δ)-subsets of ex K) is $\check{\mu}$-measurable (see [6], Theorem 1.5). We shall now prove that the boundary measure $\check{\mu}$ is inner regular on the σ-algebra $\mathcal{B}_0(\text{ex } K;C)$ of all C-Baire measurable subsets of ex K.

<u>THEOREM 4</u>. <u>For any</u> $B \in \mathcal{B}_0(\text{ex } K;C)$ <u>and any</u> $\varepsilon > 0$ <u>there exists a</u> C-<u>closed subset</u> $A \subset \text{ex } K$, <u>such that</u>

$$A \subset B,\ A \in \mathcal{A}_0(\text{ex } K) \text{ \underline{and} } \check{\mu}(B) - \varepsilon < \check{\mu}(A).$$

<u>PROOF</u>. a) Let $A \subset \text{ex } K$ be any C-closed (C-G_δ)-subset of ex K. Then A and $G = (\text{ex } K) \setminus A$ are $\check{\mu}$-measurable, and there exists an increasing sequence $(F_n)_{n \geq 0}$ of compact extremal subsets of K, such that $\bigcup_{n=0}^{\infty} (F_n \cap \text{ex } K) = G$. Let $F \subset K$ be any compact extremal subset of K, such that $F \cap \text{ex } K = A$.

Let $H = K \setminus (\bigcup_{n=0}^{\infty} F_n)$. Then H is a G_δ-subset of K; by Theorem 2, given $\varepsilon > 0$, there exists a compact extremal Baire measurable subset $D_1 \subset H$, such that $\mu(H) - \varepsilon < \mu(D_1)$.

On the other hand, from $F \cap F_n = \emptyset$, $n \in \mathbb{N}$, we infer that $F \subset H$ and

$$\mu(F) + \mu(F_n) = \check{\mu}^*(F \cap \text{ex } K) + \check{\mu}^*(F_n \cap \text{ex } K), \quad n \geq 0;$$

we obtain that

$$\mu(F)+\mu(\bigcup_{n\geq 0} F_n)=\check{\mu}^*(F \cap \text{ex } K)+\check{\mu}^*(G)=1,$$

and, therefore, we have

$$\mu(F) = \mu(H).$$

If we denote $A_1=D_1 \cap \text{ex } K$, then $A_1 \subset A$, A_1 is C-closed and $A_1 \in \mathcal{A}_0(\text{ex } K)$.

b) If $G \subset \text{ex } K$ is any open (C-F_σ)-subset of ex K, then it is $\check{\mu}$-measurable, by virtue of Theorem 1.5 from [6] and the set $A \subset G$ required by the Theorem exists by virtue of Theorem 3.

c) Let $\mathcal{B}_1$ be the set of all subsets B of ex K, such that B and $(\text{ex } K)\setminus B$ have the property required by Theorem 4. Then, by a) and b), any C-closed (C-G_δ)-subset of ex K belongs to $\mathcal{B}_1$ and, since $\mathcal{B}_1$ is easily shown to be a σ-algebra of subsets of ex K, we obviously have that $\mathcal{B}_0(\text{ex } K;\, C)\subset\mathcal{B}_1$. The Theorem is proved.

According to Theorem 1.5 from [6] we have

$$\mathcal{B}_0(\text{ex } K;\, C)\subset \mathcal{A}_0(\text{ex } K)^{\sim}_{\check{\mu}} ,$$

where the right-hand member is the completion of $\mathcal{A}_0(\text{ex } K)$ with respect to $\check{\mu}$.

4. The preceding results can be strengthened as follows.

Let us consider the σ-algebra $\mathcal{B}_1(\text{ex } K)$ of subsets of ex K, generated by all the sets of the form $D \cap \text{ex } K$, where $D \subset K$ is a compact extremal Baire measurable subset. Of course, we have

(*) $$\mathcal{B}_1(\text{ex } K)\subset \mathcal{A}_0(\text{ex } K),$$

and also

(**) $$\mathcal{B}_1(\text{ex } K)\subset \mathcal{B}(\text{ex } K;C),$$

where $\mathcal{B}(\text{ex } K;C)$ denotes the σ-algebra of all the Borel measurable subsets of ex K, with respect to the Choquet topology.

We shall denote by $\mathcal{B}_1(\text{ex } K)^{\sim}_{\check{\mu}}$ the Lebesgue completion of the σ-algebra $\mathcal{B}_1(\text{ex } K)$, with respect to the restriction of $\check{\mu}$ to $\mathcal{B}_1(\text{ex } K)$.

We have the following Regularity Theorem.

THEOREM 5. a) $\mathcal{B}_1(\text{ex } K)^{\sim}_{\check{\mu}} = \mathcal{A}_0(\text{ex } K)^{\sim}_{\check{\mu}}$.

b) For any $A \in \mathcal{A}_0(\text{ex } K)^{\sim}_{\check{\mu}}$ and any $\varepsilon > 0$ there exists a C-closed set $A_0 \in \mathcal{A}_0(\text{ex } K)$, such that $A_0 \subset A$ and $\check{\mu}(A) - \varepsilon < \mu(A_0)$.

PROOF. a) From (*) we immediately obtain that

(1) $$\mathcal{B}_1(\text{ex } K)^{\sim}_{\check{\mu}} \subset \mathcal{A}_0(\text{ex } K)^{\sim}_{\check{\mu}}.$$

Let now $A_1 \in \mathcal{A}_0(\text{ex } K)^{\sim}_{\check{\mu}}$ and $\varepsilon > 0$ be given. Then there exists an $A_0 \in \mathcal{A}_0(\text{ex } K)$, such that

(2) $$A_0 \subset A_1 \quad \text{and} \quad \check{\mu}(A_0) = \check{\mu}(A_1).$$

Let $D_0 \in \mathcal{B}_0(K)$ be a Baire measurable subset of K, such that $D_0 \cap \text{ex } K = A_0$. By the Corollary to Theorem 1, there exists a compact extremal Baire measurable subset $D \subset D_0$, such that $\mu(D_0) - \varepsilon < \mu(D)$. We then have :

$$\check{\mu}(A_0) - \varepsilon < \check{\mu}(D \cap \text{ex } K),$$

and $A = D \cap \text{ex } K \in \mathcal{B}_1(\text{ex } K)$.

By a standard argument we infer that there exists a set $A \in \mathcal{B}_1(\text{ex } K)$, such that

$$A \subset A_1 \quad \text{and} \quad \check{\mu}(A) = \check{\mu}(A_1),$$

where we have also taken into account (2).

A similar argument, applied to $\complement A_1$, yields a set $A' \in \mathcal{B}_1(\mathrm{ex}\ K)$, such that

$$A_1 \subset A' \quad \text{and} \quad \check{\mu}(A') = \check{\mu}(A_1).$$

We infer that $A_1 \in \mathcal{B}_1(\mathrm{ex}\ K)^{\sim}_{\check{\mu}}$ and, therefore, we have

(3) $$\mathcal{A}_0(\mathrm{ex}\ K)^{\sim}_{\check{\mu}} \subset \mathcal{B}_1(\mathrm{ex}\ K)^{\sim}_{\check{\mu}}.$$

From (1) and (3) we infer statement a) of the Theorem.

b) For any compact extremal Baire measurable subset $D \subset K$, property b) obviously holds for the set $A = D \cap \mathrm{ex}\ K$, with $A_0 = A$.

Let us now consider the set $(\mathrm{ex}\ K) \setminus A$. We have $(\mathrm{ex}\ K) \setminus A = (K \setminus D) \cap \mathrm{ex}\ K$, and $K \setminus D$ is a Baire measurable subset of K. By the Corollary to Theorem 1, there exists a compact extremal Baire measurable subset $D_0 \subset K \setminus D$, such that $\mu(K \setminus D) - \varepsilon < \mu(D_0)$. Then we have $A_0 = D_0 \cap (\mathrm{ex}\ K) \subset (\mathrm{ex}\ K) \setminus A$, and $\check{\mu}((\mathrm{ex}\ K) \setminus A) - \varepsilon < \check{\mu}(A_0)$. The set A_0 meets the requirements from statement b) of the Theorem.

Let now $\mathcal{B}'$ be the set of all subsets, $S \in \mathcal{B}_1(\mathrm{ex}\ K)$, of ex K, such that for any $\varepsilon > 0$ there exists compact extremal Baire measurable subsets $D_0, D_1 \subset K$, such that $D_0 \cap \mathrm{ex}\ K \subset S$, $D_1 \cap \mathrm{ex}\ K \subset (\mathrm{ex}\ K) \setminus S$ and $\check{\mu}(S) - \varepsilon < \check{\mu}(D_0 \cap \mathrm{ex}\ K)$, $\check{\mu}((\mathrm{ex}\ K) \setminus S) - \varepsilon < \check{\mu}(D_1 \cap \mathrm{ex}\ K)$. Then $\mathcal{B}'$ obviously is a σ-algebra of subsets of ex K, containing all the generators of $\mathcal{B}_1(\mathrm{ex}\ K)$, by virtue of the preceding argument. We first infer that $\mathcal{B}' = \mathcal{B}_1(\mathrm{ex}\ K)$ and then , by an easy argument, that property b) holds for any $A \in \mathcal{B}_1(\mathrm{ex}\ K)^{\sim}_{\check{\mu}}$. Part a) of the Theorem now concludes the proof.

5. In this Section we shall consider the case of the quasi-states space $E_o(\mathcal{C})$ of an arbitrary C^* - algebra $\mathcal{C}$; i.e.,

$$E_o(\mathcal{C})=\{f\in\mathcal{C}^*; \quad f\geq 0, \quad \|f\|\leq 1\},$$

endowed with the $\sigma(\mathcal{C}^*;\mathcal{C})$ - topology. Then $E_o(\mathcal{C})$ is a compact convex set, in $(\mathcal{C}^*;\sigma(\mathcal{C}^*;\mathcal{C}))$, and ex $E_o(\mathcal{C})= P(\mathcal{C})\cup\{0\}$,where $P(\mathcal{C})$ denotes the set of all pure states of $\mathcal{C}$.

Let $\mu\in\mathcal{M}_+^1(E_o(\mathcal{C}))$ be a maximal orthogonal Radon probability measure, such that $\|b(\mu)\|$ =1. By Henrichs' Theorem (see [3] , p.106; and also [6] , Theorem 3.10) μ is maximal for the Choquet order relation, and, therefore, the foregoing Theory can be applied to μ. We shall make extensive use of the results of [6] . According to Proposition 3.2 from [6] we have $\check{\mu}^*(\{0\})=0$ and , therefore , $\check{\mu}^*(P(\mathcal{C}))=1$.

Moreover, $\{0\}\subset P(\mathcal{C})\cup\{0\}$ is a C-closed subset of $P(\mathcal{C})\cup\{0\}$; hence, $P(\mathcal{C})$ is a C-open subset of $P(\mathcal{C})\cup\{0\}$.

According to Theorem 5.2 from [6] , the probability measure $\check{\mu}:\mathcal{A}_o(P(\mathcal{C})\cup\{0\})\to[0,1]$ can be extended to a probability measure

$$\tilde{\mu}:\mathcal{A}_1(P(\mathcal{C})\cup\{0\})\to[0,1],$$

defined on the σ-algebra $\mathcal{A}_1(P(\mathcal{C})\cup\{0\})$, generated by $\mathcal{A}_o(P(\mathcal{C})\cup\{0\})$ and $\mathcal{B}(P(\mathcal{C})\cup\{0\};C)$.

Since we have

$$\mathcal{B}_1(P(\mathcal{C})\cup\{0\})\subset\mathcal{B}(P(\mathcal{C})\cup\{0\};C),$$

we infer that we also have

$$\mathcal{A}_o(P(\mathcal{C}) \cup \{o\})_{\tilde{\mu}}^{\sim} = \mathcal{B}_1(P(\mathcal{C}) \cup \{o\})_{\tilde{\mu}}^{\sim} \subset$$
$$\subset \mathcal{B}(P(\mathcal{C}) \cup \{o\} ; C)_{\tilde{\mu}}^{\sim} ,$$

where we have also applied Theorem 5.

<u>THEOREM 6</u> . a) $\mathcal{A}_o(P(\mathcal{C}) \cup \{o\})_{\tilde{\mu}}^{\sim} \subset \mathcal{B}(P(\mathcal{C}) \cup \{o\} ; C)_{\tilde{\mu}}^{\sim}$.

b) <u>For any</u> $A \in \mathcal{B}(P(\mathcal{C}) \cup \{o\} ; C)_{\tilde{\mu}}^{\sim}$ <u>and any</u> $\varepsilon > 0$ <u>there exists a</u> C-<u>closed subset</u> $A_o \subset P(\mathcal{C}) \cup \{o\}$, <u>such that</u>

$$A_o \subset A \ \underline{\text{and}} \ \tilde{\mu}(A) - \varepsilon < \tilde{\mu}(A_o).$$

<u>PROOF.</u> In order to develop the proof, we have to recall some notations and results that we have used and obtained in [6] .

Namely, let $f_o = b(\mu)$, and let $\pi_{f_o} : \mathcal{C} \to \mathcal{L}(H_{f_o})$ be the corresponding cyclic representation, according to the GNS-construction; let $\xi^o_{f_o} \in H_{f_o}$ be the corresponding cyclic vector, and $\mathcal{C}_\mu \subset (\pi_{f_o}(\mathcal{C}))'$ the maximal abelian von Neumann algebra, corresponding to μ (see [6] , Theorem 3.4). Let $\mathcal{B} \subset \mathcal{L}(H_{f_o})$ be the C^* - algebra generated by $\pi_{f_o}(\mathcal{C})$ and $\mathcal{C}_\mu$, and let $\alpha \in \mathcal{M}_+^1(E(\mathcal{B}))$ be the central measure on $E(\mathcal{B})$, corresponding to the vector state $g_o = \omega_{\xi^o_{f_o}} | \mathcal{B} \in E(\mathcal{B})$. Since

$$\mathcal{C}_\mu \subset \mathcal{B} \quad \text{and} \quad \mathcal{B}' = \mathcal{C}_\mu ,$$

α is a maximal orthogonal measure on $E(\mathcal{B})$, representing g_o and, by C.F. Skau's Theorem, it is the greatest Radon probability measure on $E(\mathcal{B})$, representing g_o, for the Choquet order relation (see [6], Proposition 3.5 ; and also Chapter 5).

The mapping $\tau : E(\mathcal{B}) \longrightarrow E_0(\mathcal{C})$ is defined by

$\tau(g) = g \circ \pi_{f_0}$, $g \in E(\mathcal{B})$, whereas $\sigma = \tau \setminus P(B)$.

We have $\sigma(P(\mathcal{B})) \subset P(\mathcal{C}) \cup \{0\}$ (see [6], Chapter 3), and $\tau_*(\alpha) = \mu$, $\sigma_*(\check{\alpha}) = \check{\mu}$ (see [6], Lemma 3.8 and Proposition 3.1o).

i) For any compact extremal subset $F \subset E_0(\mathcal{C})$ the set $\tau^{-1}(F) \subset E(\mathcal{B})$ is a compact extremal subset of $E(\mathcal{B})$, and we have

$$\tilde{\mu}(F \cap (P(\mathcal{C}) \cup \{0\})) = \tilde{\alpha}(\sigma^{-1}(F \cap (P(\mathcal{C}) \cup \{0\}))) =$$
$$= \tilde{\alpha}(\tau^{-1}(F) \cap P(\mathcal{B})) = \check{\alpha}^*(\tau^{-1}(F) \cap \operatorname{supp} \check{\alpha}) =$$
$$= \alpha(\tau^{-1}(F) \cap F_0) = \alpha(\tau^{-1}(F)) = \tau_*(\alpha)(F) = \mu(F),$$

where we denoted by F_0 the smallest compact extremal subset of $E(\mathcal{B})$, containing $\operatorname{supp} \alpha$, whereas $\operatorname{supp} \check{\alpha} = F_0 \cap P(\mathcal{B})$.

Let us now remark that property b) obviously holds for any C-closed subset $A \subset P(\mathcal{C}) \cup \{0\}$. Let now $G = (P(\mathcal{C}) \cup \{0\}) \setminus F$ be any C-open subset of $P(\mathcal{C}) \cup \{0\}$, where $F \subset E_0(\mathcal{C})$ is any compact extremal subset of $E_0(\mathcal{C})$. By virtue of the Inner Regularity Theorem (see Theorem 3, above), there exists a C-closed subset $A_0 \subset G$, such that $\check{\mu}_*(G) - \varepsilon < \tilde{\mu}(A_0)$.

On the other hand, we have

$$\check{\mu}_*(G) = 1 - \check{\mu}_*(P(\mathcal{C}) \cup \{0\} \setminus G) = 1 - \mu(F) =$$
$$= 1 - \tilde{\mu}(F \cap (P(\mathcal{C}) \cup \{0\})) = \tilde{\mu}(G),$$

where we have taken into account equality (1).

ii) If we now denote by $\mathcal{B}'$ the set of all $A \in \mathcal{B}(P(\mathcal{C}) \cup \{0\}; C)^{\sim}_{\mu}$, such that property b)holds for A and for $(P(\mathcal{C}) \cup \{0\}) \setminus A$, it is easy to prove that $\mathcal{B}'$ is a σ-algebra. Since, by virtue of i), $\mathcal{B}'$ contains any C-closed subset of $P(\mathcal{C}) \cup \{0\}$, we infer that $\mathcal{B}(P(\mathcal{C}) \cup \{0\}; C) \subset \mathcal{B}'$ and, therefore, $\mathcal{B}' = \mathcal{B}(P(\mathcal{C}) \cup \{0\}; C)^{\sim}_{\mu}$, since $\mathcal{B}'$ is easily shown to be complete.

In this manner, property b) in the statement of the Theorem is proved. Part a) of the Theorem was already proved just before.

REMARK. The preceding theory can be viewed as a non-commutative extension of the theory of Radon measures.

BIBLIOGRAPHY

1. P.Billingsley. Convergence of Probability Measures. John Wiley and Sons, New York-London-Sydney-Toronto, 1969.
2. N.Boboc, Gh.Bucur. Conuri convexe de funcţii continue pe spaţii compacte. Ed.Acad.R.S.R., Bucureşti, 1976.
3. R.W.Henrichs. On decomposition theory for unitary representations of locally compact groups. Journal of Functional Analysis, vol.31, no.1, January 1979, p.101-114.
4. P.A.Meyer. Probability and Potentials, Blaisdell Publishing Company, Waltham, Toronto-London, 1966.
5. R.R.Phelps. Lectures on Choquet's Theorem. D.van Nostrand Co., Princeton-Toronto-New York-London, 1966.
6. S.Teleman. An introduction to Choquet theory with applications to reduction theory. INCREST Preprint Series in Mathematics, No.71/1980. Bucharest.

ON STATIONARY PROCESSES IN COMPLETE CORRELATED ACTIONS

by

Ilie VALUSESCU

1. Introduction. This paper contains some results concerning stationary processes in complete correlated actions, large part of this results beeing published in [3], [4], [6], Operator methods are the main tools in investigation which is made here.

In this first part, some notations and terminology which are used in what follows are established. The second section is devoted to the main results in discrete case, and in the last section the continuous parameter case is analysed.

As usualy, let us denote by $\mathbb{D}$ the open unit disc and by $\mathbb{T}$ the unit circle in the complex plane $\mathbb{C}$. Let $\mathcal{E}$ be a Hilbert space and $\mathcal{L}(\mathcal{E})$ the set of all linear bounded operators on $\mathcal{E}$. By an $\mathcal{L}(\mathcal{E})$-valued semispectral measure F on $\mathbb{T}$ one means a map $\sigma \longmapsto F(\sigma)$ from $\mathcal{B}(\mathbb{T})$ - the family of all Borel subsets of $\mathbb{T}$ into $\mathcal{L}(\mathcal{E})$, such that for any $a \in \mathcal{E}$, the Radon measure μ_a on $\mathbb{T}$ given by $\mu_a(\sigma)=(F(\sigma)a,a)$ is a positive one. A semispectral measure E is spectral if $E(\mathbb{T})=I$ and $E(\sigma_1 \cap \sigma_2)= =E(\sigma_1)E(\sigma_2)$. Via Naimark dilation theorem, for any $\mathcal{L}(\mathcal{E})$-valued semispectral measure F there exist a Hilbert space $\mathcal{K}$, an $\mathcal{L}(\mathcal{K})$--valued spectral measure E and a bounded operator $V: \mathcal{E} \mapsto \mathcal{K}$ such that

$$F(\sigma)=V^{*}E(\sigma)V, \qquad (\sigma\in\mathcal{B}(\mathbf{T})). \tag{1.1}$$

The triplet $[\mathcal{K},V,E]$ is called the <u>spectral dilation</u> of the semispectral measure F. If

$$\mathcal{K} = \bigvee_{\sigma\in\mathcal{B}(\mathbf{T})} E(\sigma)V\mathcal{E} , \tag{1.2}$$

then $[\mathcal{K},V,E]$ is the <u>minimal</u> spectral dilation of F.

Let $\mathcal{E}$ and $\mathcal{F}$ be Hilbert spaces and $\Theta(\lambda)$ an $\mathcal{L}(\mathcal{E},\mathcal{F})$--valued function on $\mathbf{D}$ given by

$$\Theta(\lambda)=\sum_{n=0}^{\infty} \lambda^{n}\Theta_{n} , \tag{1.3}$$

where Θ_n are linear bounded operators from $\mathcal{E}$ into $\mathcal{F}$. The triplet $\{\mathcal{E},\mathcal{F},\Theta(\lambda)\}$ is called an <u>operator valued analytic function</u>. An L^2-<u>bounded analytic function</u> is an operator valued analytic function with the properties that there exists a positive constant M such that for any $a\in\mathcal{E}$

$$\sup_{0\le r<1} \frac{1}{2\pi}\int_0^{2\pi} \|\Theta(re^{it})a\|^{2}dt \le M^{2}\|a\|^{2} , \tag{1.4}$$

or equivalently

$$\sum_{n=0}^{\infty} \|\Theta_n a\|^{2} \le M^{2}\|a\|^{2} . \tag{1.5}$$

If instead (1.4) we have

(1.6) $$\|\Theta(\lambda)\| \leq M ,$$

then $\{\mathcal{E}, \mathcal{F}, \Theta(\lambda)\}$ is a <u>bounded</u> analytic function. It is easy to see that (1.6) implies (1.4) but the converse is not true.

To any L^2-bounded analytic function $\{\mathcal{E}, \mathcal{F}, \Theta(\lambda)\}$ we can attach an $\mathcal{L}(\mathcal{E})$-valued semispectral measure F_Θ on $\mathbb{T}$ as follows. Let $V_\Theta : \mathcal{E} \longmapsto L^2(\mathcal{F})$ be the bounded operator defined by $(V_\Theta a)(\lambda) = \Theta(\lambda)a$. If $E^x_{\mathcal{F}}$ is the spectral measure corresponding to the shift operator on $L^2(\mathcal{F})$, then putting

(1.7) $$F_\Theta(\sigma) = V_\Theta^* E^x_{\mathcal{F}}(\sigma) V_\Theta \quad , \qquad (\sigma \in \mathcal{B}(\mathbb{T})),$$

one obtaines what we need.

Conversely, to any $\mathcal{L}(\mathcal{E})$-valued semispectral measure F whose the minimal spectral dilation is $[\mathcal{K}, V, E]$ we can attach an L^2-bounded analytic function $\{\mathcal{E}, \mathcal{F}, \Theta(\lambda)\}$ with the properties that $F_\Theta \leq F$ and for any L^2-bounded analytic function $\{\mathcal{E}, \mathcal{F}_1, \Omega(\lambda)\}$ with $F_\Omega \leq F$ we have $F_\Omega \leq F_\Theta$, via the factorization theorem [3] of semispectral measures. This L^2-bounded analytic function is called the <u>maximal</u> (L^2-bounded) function of the semispectral measure F.

An L^2-bounded analytic function $\{\mathcal{E}, \mathcal{F}, \Theta(\lambda)\}$ has a <u>scalar multiple</u>, if there exist a contractive analytic function $\{\mathcal{F}, \mathcal{E}, \Omega(\lambda)\}$ and a scalar function $\delta(\lambda) \neq 0$ in H^2 such that

(1.8) $$\Omega(\lambda)\Theta(\lambda) = \delta(\lambda) I_{\mathcal{E}} \quad , \quad \Theta(\lambda)\Omega(\lambda) = \delta(\lambda) I_{\mathcal{F}} \quad .$$

Let us consider the Lebesgue decomposition of the Radon measure $\mu_a(\sigma)=(F(\sigma)a,a)$ with respect to the normalized Lebesgue measure $\frac{1}{2\pi}dt$, $d\mu_a=\frac{1}{2\pi}h_a dt+d\mu_s$. We have the following

THEOREM 1. <u>The following assertions are equivalent</u>

(i) <u>There exists a positive function</u> $h\in L^1(dt)$ <u>such that for any</u> $a\in\mathcal{E}$, $\|a\|=1$ <u>one has</u> $h\leq h_a$ a.e. <u>and</u>

$$\int_0^{2\pi}\log h(t)dt > -\infty \quad . \tag{1.9}$$

(ii) <u>The maximal</u> L^2-<u>bounded function</u> $\{\mathcal{E},\mathcal{F}_1,\Theta_1(\lambda)\}$ <u>attached to</u> F <u>has a scalar multiple</u>.

(iii) <u>There exists an</u> L^2-<u>bounded analytic function</u> $\{\mathcal{E},\mathcal{F},\Theta(\lambda)\not\equiv 0\}$ <u>which has a scalar multiple and</u> $F_\Theta\leq F$.

For a complete proof see [6].

2. <u>Discrete time stationary processes</u>. Let us introduce the context of a complete correlated action, in which all properties of stationary processes here considered will be studied.

Let $\mathcal{E}$ be a Hilbert space and $\mathcal{H}$ a right $\mathcal{L}(\mathcal{E})$-module. The map $\mathcal{L}(\mathcal{E})\times\mathcal{H}\longmapsto\mathcal{H}$ given by $(Ah,h)=Ah$, where $Ah=h\cdot A$, is called the <u>action</u> of $\mathcal{L}(\mathcal{E})$ on $\mathcal{H}$. The map $\mathcal{H}\times\mathcal{H}\longmapsto\mathcal{L}(\mathcal{E})$ given by

$$(h,g)\longmapsto\Gamma[h,g] \tag{2.1}$$

is called a <u>correlation</u> of the action of $\mathcal{L}(\mathcal{E})$ on $\mathcal{H}$ if it is provided with the following properties

(i) $\Gamma[h,h] \geq 0, \quad \Gamma[h,h]=0 \implies h=0,$

(ii) $\Gamma[h,g]^{*}=\Gamma[g,h], \qquad\qquad (h,g\in\mathcal{H})$

(iii) $\Gamma[\sum_i A_i h_i, \sum_j B_j g_j]=\sum_{i,j} A_i^{*}\Gamma[h_i,g_j]B_j \quad ,$

where i and j takes a finit number of values.

The triplet $\{\mathcal{E}, \mathcal{H}, \Gamma\}$ as above is the <u>correlated action</u> of $\mathcal{L}(\mathcal{E})$ on $\mathcal{H}$. The space $\mathcal{E}$ is the <u>parameter space</u> and $\mathcal{H}$ is the <u>state space</u>.

Let us consider an exemple. Take $\mathcal{H}=\mathcal{L}(\mathcal{E},\mathcal{K})$ where $\mathcal{E}$ and $\mathcal{K}$ are Hilbert spaces. $\mathcal{H}$ is organized as an right $\mathcal{L}(\mathcal{E})$-module if for any $A\in\mathcal{L}(\mathcal{E})$ and $V\in\mathcal{L}(\mathcal{E},\mathcal{K})$ one takes $AV=VA$, where VA is understood in the usual sense of multiplication of operators. The correlation of the action of $\mathcal{L}(\mathcal{E})$ on $\mathcal{H}$ is given by

$$\Gamma[V_1,V_2]=V_1^{*}V_2 \ . \tag{2.2}$$

This correlated action $\{\mathcal{E}, \mathcal{H}, \Gamma\}$ is called the <u>operatorial model</u>. This will play a principal rol in the following and by the following theorem it is seen that an arbitrary correlated action can be imbedded in such a one.

THEOREM 2. <u>Let $\{\mathcal{E}, \mathcal{H}, \Gamma\}$ be a correlated action. There exist a Hilbert space $\mathcal{K}$ and an algebraic imbedding $h\longmapsto V_h$ of $\mathcal{H}$ in $\mathcal{L}(\mathcal{E},\mathcal{K})$ such that</u>

$$\Gamma[h_1,h_2]=V_{h_1}^{*}V_{h_2} \qquad (h_1,h_2\in\mathcal{H}). \tag{2.3}$$

<u>The subset of the elements</u>

(2.4) $\gamma_{(a,h)} = V_h a$, $(h \in \mathcal{H},\ a \in \mathcal{E})$

<u>is a dense subset in</u> $\mathcal{K}$.

<u>The imbedding is unique up to a unitary equivalence.</u>

Proof. Taking $\lambda = (a,h)$ and $\mu = (b,g)$, to the positive definite kernel $\gamma : \mathcal{E} \times \mathcal{H} \mapsto \mathbb{C}$ given by

(2.5) $$\gamma_{(a,h)}(b,g) = (\Gamma[g,h]a,b) .$$

one can attach a reproducing kernel Hilbert space $\mathcal{K}$. The imbedding given by $V_h a = \gamma_{(a,h)}$ verifies the request properties.

If $h \longmapsto V'_h$ is another imbedding of $\mathcal{H}$ in $\mathcal{L}(\mathcal{E},\mathcal{K}')$, then $XV'_h a = V_h a$ gives rise to a unitary operator X from $\mathcal{K}'$ onto $\mathcal{K}$ such that $XV'_h = V_h$.

The unique Hilbert space attached to $\{\mathcal{E},\mathcal{H},\Gamma\}$ as in above theorem is the <u>measuring space</u> of the correlated action.

One say that a correlated action $\{\mathcal{E},\mathcal{H},\Gamma\}$ is <u>complete</u> if the imbedding $h \longmapsto V_h$ of $\mathcal{H}$ into $\mathcal{L}(\mathcal{E},\mathcal{K})$ is onto.

If $\{\mathcal{E},\mathcal{H},\Gamma\}$ is a complete correlated action, it is easy to prove the following

THEOREM 3. <u>Let</u> $\mathcal{H}_1$ <u>be a right</u> $\mathcal{L}(\mathcal{E})$-<u>submodule in</u> $\mathcal{H}$ <u>and</u> $\mathcal{K}_1 = \bigvee_{x \in \mathcal{H}_1} V_x \mathcal{E}$. <u>For any</u> h <u>in</u> $\mathcal{H}$ <u>there exists a unique</u> h_1 <u>in</u> $\mathcal{H}$ <u>such that for any</u> $a \in \mathcal{E}$ <u>we have</u>

(2.6) $$V_{h_1} a \in \mathcal{K}_1 \quad \text{and} \quad V_{h-h_1} a \in \mathcal{K}_1^{\perp} .$$

<u>Moreover, we have</u>

(2.7) $$\Gamma[h-h_1,h-h_1]=\inf_{x\in\mathcal{H}_1}\Gamma[h-x,h-x],$$

<u>where the infimum is taken in the subset of positive operators in</u> $\mathcal{L}(\mathcal{E})$.

If one denotes $\mathcal{P}_{\mathcal{H}_1}$ the endomorphism of $\mathcal{H}$ given by $\mathcal{P}_{\mathcal{H}_1}h=h_1$, then we have $\mathcal{P}^2_{\mathcal{H}_1}h=\mathcal{P}_{\mathcal{H}_1}h$ and $\Gamma[\mathcal{P}_{\mathcal{H}_1}h,g]=\Gamma[h,\mathcal{P}_{\mathcal{H}_1}g]$. This is the reason why we can consider $\mathcal{P}_{\mathcal{H}_1}$ as a "projection" or, more precisely, a Γ-<u>orthogonal projection</u> of $\mathcal{H}$ onto $\mathcal{H}_1$.

A sequence $\{f_n\}_{-\infty}^{+\infty}$ of elements in $\mathcal{H}$ is a stationary process in the complete correlated action $\{\mathcal{E},\mathcal{H},\Gamma\}$, or a Γ-<u>stationary process</u>, if $\Gamma[f_n,f_m]$ depends only on the difference m-n. In the measuring space $\mathcal{K}$ of the complete correlated action $\{\mathcal{E},\mathcal{H},\Gamma\}$ one puts in evidence the following subspaces related to a process $\{f_n\}$: the <u>past of the process</u> $\mathcal{K}_n^f$ given by $\mathcal{K}_n^f=\bigvee_{k\le n}V_{f_n}\mathcal{E}$, the <u>remote past</u> $\mathcal{K}_{-\infty}^f=\bigcap_n\mathcal{K}_n^f$, and the space spaned by the process $\mathcal{K}_\infty^f=\bigvee_{-\infty}^{+\infty}V_{f_n}\mathcal{E}$. Also, if in $\mathcal{H}$ we take $\mathcal{H}_n^f=\{h\in\mathcal{H};\ h=\sum_{k\le n}A_kf_k,\ A_k\in\mathcal{L}(\mathcal{E})\}$ then $\mathcal{K}_n^f=\bigvee_{h\in\mathcal{H}_n^f}V_h\mathcal{E}$.

Two Γ-stationary processes $\{f_n\}$ and $\{g_n\}$ are <u>cross--correlated</u> if $\Gamma[f_n,g_n]$ depends only on the difference m-n.

Defining on the generators of $\mathcal{K}_\infty^f$ the operator $U_f(\sum_n V_{f_n}a_n)=\sum_n V_{f_{n+1}}a_n$, we obtain for the process $\{f_n\}$ a unitary operator U_f on $\mathcal{K}_\infty^f$ such that $V_{f_n}=U_f^nV_{f_o}$. This is so called <u>shift operator</u> of the process $\{f_n\}$. In a similar way an extended shift operator U_{fg} can be obtained for the cross-correlated processes f_n and g_n such that $U_{fg}|\mathcal{K}_\infty^f=U_f$ and $U_{fg}|\mathcal{K}_\infty^g=U_g$.

Let us give some definition of the processes which are used in the reminder of this section. A Γ-stationary process $\{g_n\}$ is a <u>white noise</u> if $\Gamma[g_n, g_m]=0$ for $n\neq m$.

A Γ-stationary process $\{f_n\}$ <u>containes</u> the white noise $\{g_n\}$ if $\{f_n\}$ and $\{g_n\}$ are cross-correlated, $\Gamma[f_n, g_m]=0$ for $m>n$, $V_g\mathcal{E}\subset\mathcal{K}_0^f$ and $\mathrm{Re}\,\Gamma[f_n-g_n, g_n]\geqslant 0$.

A proces $\{f_n\}$ is <u>deterministic</u> iff it containes no white noise.

The Γ-stationary process $\{f_n\}$ is a <u>moving average</u> of a white noise $\{g_n\}$ if $\{f_n\}$ containes $\{g_n\}$ and $\mathcal{K}_\infty^g=\mathcal{K}_\infty^f$.

If for a Γ-stationary process $\{f_n\}$ we take

$$(2.8)\qquad g_n=f_n-\mathcal{P}_{\mathcal{H}_{n-1}^f} f_n \quad ,$$

then one obtaines a Γ-stationary process $\{g_n\}$ which is called the <u>innovation</u> part of $\{f_n\}$. In fact, the innovation process $\{g_n\}$ is the maximal white noise contained in $\{f_n\}$.

THEOREM 4. (Wold decomposition). <u>The Γ-stationary process $\{f_n\}$ admits a unique decomposition of the form</u>

$$(2.9)\qquad f_n=u_n+v_n$$

<u>where $\{u_n\}$ is the moving average of the maximal white noise $\{g_n\}$ contained in $\{f_n\}$, $\{v_n\}$ is a deterministic process, and $\Gamma[u_n, v_m]=0$, for $n,m\in\mathbb{Z}$.</u>

A complete proof can be found in [4] or [6].

Acting on the process with some specific experiences we can obtain information about his future part, "knowing" the past of the process. In the case here analysed, the action is in the context of a complete correlated action, the results of experiences are measured by the metric of the attached measuring space. The present and the past $\mathcal{H}_o$ of the process can be interpret as the total information obtained acting on the process up to the moment n=0. To predict the next moment of the process, means to obtain the best information about f_1 in terms of the elements in $\mathcal{H}_o$.

If we take for a Γ-stationary process $\{f_n\}$ the elements of the form

$$(2.10) \qquad \hat{f}_1 = \mathcal{P}_{\mathcal{H}_o^f} f_1 = f_1 - g_1 \quad ,$$

then can be proved that

$$(2.11) \qquad \Gamma[f_1-\hat{f}_1, f_1-\hat{f}_1] = \inf_{h \in \mathcal{H}_o^f} \Gamma[f_1-h, f_1-h] \; ,$$

where the infimum is taken in the set of the positive operators in $\mathcal{L}(\mathcal{E})$.

Such a way, $\hat{f}_1$ contains the best information about f_1, acting on the process up to the present moment. This is the reason why $\hat{f}_1$ is called the <u>predictible part</u> of f_1 and $\Delta[f] = \Gamma[f_1-\hat{f}_1, f_1-\hat{f}_1]$, the <u>prediction error operator</u>.

Due to the fact that the correlation function of the process $\{f_n\}$, $n \longmapsto \Gamma(n) = \Gamma[f_k, f_{k+n}]$ is a complete positive operator valued function on $\mathbb{Z}$, there exists a unique $\mathcal{L}(\mathcal{E})$--valued semispectral measure F on $\mathbb{T}$ such that

$$(2.12) \qquad \Gamma(n) = \int_0^{2\pi} e^{-int} dF(t) \quad .$$

This F is called the spectral distribution of the process $\{f_n\}$. Let $\{\mathcal{E}, \mathcal{F}, \Theta(\lambda)\}$ be the maximal L^2-bounded analytic function of the semispectral measure F. This $\Theta(\lambda)$ is also called the maximal function of the process.

Under a boundedness condition on the spectral distribution F of the process $\{f_n\}$, similar to that imposed by Wiener and Masani in the matrix case [7], the predictible part can be obtained using a linear filter consisting in succesive actions up to the present moment. The coeficients of the filter will be determined in terms of the coeficients of maximal function attached to $\{f_n\}$.

Let us sketch the way to obtain the filter of prediction. The boundedness condition on F is

$$(2.13) \qquad \frac{1}{2\pi} c\,dt \leq F \leq \frac{1}{2\pi} c^{-1} dt \quad ,$$

where c is a positive constant.

Firstly can be proved [4] that F verifies the condition (2.13) if and only if $\{\mathcal{E}, \mathcal{F}, \Theta(\lambda)\}$ is a bounded analytic function which has a bounded inverse, the semispectral measure attached to $\Theta(\lambda)$ is $F_\Theta = F$, $\dim \mathcal{E} = \dim \mathcal{F}$, $\Delta[f] = \Theta(0)^* \Theta(0)$, and there exists an identification of $\{\mathcal{E}, \mathcal{F}, \Theta(\lambda)\}$ with an invertible bounded analytic function $\{\mathcal{E}, \mathcal{E}, \Phi(\lambda)\}$ such that

$$(2.14) \qquad \Phi(0) = \Delta[f]^{1/2} \quad .$$

Let g_n be the maximal white noise contained in $\{f_n\}$. If we put

$$(2.15) \qquad h_n = \Delta[f]^{-1/2} g_n \quad ,$$

then $\{h_n\}$ is white noise process with the properties that $\Gamma[h_n,h_n]=I_{\mathcal{E}}$. The process $\{h_n\}$ is so called the normalized innovation process of $\{f_n\}$.

Taking account of the obove identification of $\Theta(\lambda)$, we can consider the maximal function to be $\{\mathcal{E},\mathcal{E},\Theta(\lambda)\}$ and $\mathcal{K}=L^2(\mathcal{E})$, $\mathcal{K}_+=L^2_+(\mathcal{E})$, $V_{f_o}=\Theta|\mathcal{E}$, $(V_{f_o}a)(t)=\Theta(e^{it})a$, and the shift operator of the process U to be the multiplication by e^{-it} on $L^2(\mathcal{E})$. So we can see our processes $\{f_n\}$, $\{g_n\}$ and $\{h_n\}$ as operators from $\mathcal{E}$ into $L^2(\mathcal{E})$ as follows: $f_n=e^{-int}\Theta(e^{it})$, $g_n=e^{-int}\Theta(0)$, $h_n=e^{-int}$.

Taking the Taylor expansions of the maximal function $\{\mathcal{E},\mathcal{E},\Theta(\lambda)\}$ and its inverse $\{\mathcal{E},\mathcal{E},\Omega(\lambda)\}$ as follows

$$(2.16)\qquad \Theta(\lambda)=\Delta[f]^{1/2}+\sum_{k=1}^{\infty}\Theta_k\lambda^k\ ,\ \Omega(\lambda)=\Delta[f]^{-1/2}+\sum_{k=1}^{\infty}\Omega_k\lambda^k\ ,$$

then is simply to verify that

$$(2.17)\qquad f_n=\sum_{k=0}^{\infty}\Theta_k h_{n-k}\quad\text{and}\quad h_n=\sum_{k=0}^{\infty}\Omega_k f_{n-k}\ .$$

Such a way, the predictible part $\hat{f}_n$ of f_n is given by

$$\hat{f}_n=\mathcal{P}_{\mathcal{H}^f_{n-1}}f_n=f_n-g_n=\sum_{k=0}^{\infty}\Theta_k h_{n-k}-\Delta[f]^{1/2}h_n=\sum_{k=1}^{\infty}\Theta_k h_{n-k}=$$

$$=\sum_{k=1}^{\infty}V_{h_{n-k}}\Theta_k=\sum_{k=1}^{\infty}V_{\sum_s\Omega_s f_{n-k-s}}\Theta_k=\sum_{k=1}^{\infty}\sum_{s=0}^{\infty}V_{f_{n-k-s}}(\Omega_s\Theta_k)=$$

$$=\sum_{k=1}^{\infty}\sum_{s=0}^{\infty}\Omega_s\Theta_k f_{n-k-s}=\sum_{p=0}^{\infty}\sum_{s=0}^{\infty}\Omega_s\Theta_{p+1}f_{n-k-s}=$$

$$=\sum_{j=0}^{\infty}\Big(\sum_{p+s=j}\Omega_s\Theta_{p+1}\Big)f_{(n-1)-j}=\sum_{j=0}^{\infty}\Big(\sum_{p=0}^{j}\Omega_{j-p}\Theta_{p+1}\Big)f_{(n-1)-j}=$$

$$=\sum_{j=0}^{\infty} E_j f_{(n-1)-j} .$$

Therefore the predictible part $\hat{f}_n$ of f_n is obtained using the linear filter $\{E_j\}_0^{\infty}$, so called the linear predictor, or the Wiener filter for prediction,

(2.18) $$\hat{f}_n = \sum_{j=0}^{\infty} E_j f_{(n-1)-j} .$$

3. Continuous parameter stationary processes. Let us consider the stationary process $(f_t)_{-\infty<t<+\infty}$ in the complete correlated action $\{\mathcal{E}, \mathcal{H}, \Gamma\}$, i.e. $\Gamma[f_t, f_s]$ depends only on the difference s-t. In the measuring space of $\{\mathcal{E}, \mathcal{H}, \Gamma\}$, the past $\mathcal{K}_t^f$, the remote past $\mathcal{K}_{-\infty}^f$ and the space spaned by the process $\mathcal{K}_{\infty}^f$ can be considered. Defining on the generators of $\mathcal{K}_{\infty}^f$ the operators $U_t(\sum_s V_{f_s} a_s) = \sum_s V_{f_{s+t}} a_s$, one obtaines a unique group of unitary operators $(U_t)_{-\infty<t<+\infty}$ on $\mathcal{K}_{\infty}^f$ such that

(3.1) $$V_{f_t} = U_t V_{f_o} .$$

This group is the so called shift group of the process (f_t).

The Γ-stationary process (f_t) is called continuous if the corresponding shift group (U_t) is a continuous one parameter group of unitary operators on $\mathcal{K}_{\infty}^f$, i.e. $U_o = I_{\mathcal{K}_{\infty}^f}$, $U_{t_1+t_2} = U_{t_1} U_{t_2}$ and U_t converges weakly to the identity operator on $\mathcal{K}_{\infty}^f$ for t converging to zero.

Let U be the cogenerator of the shift group (U_t) of (f_t). Puting

(3.2) $$V_{f'_n}=U^n V_{f_o}$$

one obtaines a discrete parameter Γ-stationary process $\{f'_n\}$ which has U as the shift operator. The process $\{f'_n\}$ obtained as above is called the discrete parameter process associated with (f_t). As in [2] can be proved that $\mathcal{K}_o^{f'}=\mathcal{K}_o^f$, $\mathcal{K}_{\infty}^{f'}=\mathcal{K}_{\infty}^f$ and $\mathcal{K}_{-\infty}^{f'}=\mathcal{K}_{-\infty}^f$. Therefore a continuous Γ-stationary process (f_t) is deterministic if and only if the associated discrete parameter process is a deterministic one.

If $\{g'_n\}$ is the maximal white noise contained in $\{f'_n\}$, then

(3.3) $$V_{g_t}=U_t V_{g'_o}$$

give rise to a continuous Γ-stationary process (g_t) cross--correlated with (f_t), the past and present of (g_t) is given by

(3.4) $$\mathcal{K}_t^g=\mathcal{K}_t^f \ominus \mathcal{K}_{-\infty}^f$$

and (see [6]) can be proved that

(3.5) $$P_{\mathcal{K}_s^g} V_{g_t}=e^{s-t} V_{g_s}, \qquad (-\infty< s< t<+\infty),$$

and the corresponding correlation function is given by

(3.6) $$\Gamma_g(t)=e^{-|t|}\,\Gamma_g(0), \qquad (-\infty<t<+\infty).$$

Let us consider the process (ξ_t) given by

$$(3.7)\qquad V_{\xi_t} = \frac{1}{\sqrt{2}}\left[V_{g_t} - V_{g_o} + \int_o^t V_{g_s}\, ds\right], \qquad (-\infty < t < +\infty).$$

This process will play the role of the *differential innovations* of the continuous Γ-stationary process (f_t) and it is proved [6] that (ξ_t) has increments which are Γ-stationary under the group (U_t), i.e.,

$$(3.8)\qquad U_t V_{\xi_b - \xi_a} = V_{\xi_{b+t}} - V_{\xi_{a+t}}, \qquad (a,b,t \in (-\infty,+\infty)),$$

also (ξ_t) has Γ-orthogonal increments, i.e. if $-\infty < a < b \leq c < d < < +\infty$, then $\Gamma[\xi_b - \xi_a, \xi_d - \xi_c] = 0$, and for any real a,b one has $\Gamma[\xi_b - \xi_a, \xi_b - \xi_a] = |b-a|\,\Gamma_g(0)$, for $-\infty < a < c < b < d < +\infty$ we have $\Gamma[\xi_b - \xi_a, \xi_d - \xi_c] = \Gamma[\xi_b - \xi_c, \xi_b - \xi_c] = (b-c)\,\Gamma_g(0)$.

THEOREM 5. (Continuous Wold decomposition). *Let (f_t) be a continuous Γ-stationary process and $\{g'_n\}$ the maximal white noise contained in the discrete process $\{f'_n\}$ associated with (f_t). If (g_t) is the continuous Γ-stationary process corresponding to $\{g'_n\}$ by* (3.3), *and (ξ_t) the process with Γ-orthogonal increments defined by* (3.7) *then (f_t) admits a unique decomposition of the form*

$$(3.9)\qquad f_t = u_t + v_t,$$

where (u_t) is a moving overage, i.e. $V_{u_t} = \int_o^\infty c(s)\, dV_{\xi_{t-s}}$, *with* $c \in L^2[0,\infty)$ *and* $\mathcal{K}_t^u = \bigvee_{s \leq t} V_{u_s} \xi = \mathcal{K}_t^g$, $t \in (-\infty,+\infty)$, *and (v_t) is a deterministic process with* $\mathcal{K}_t^v = \mathcal{K}_{-\infty}^f$.

Proof. Let us remark that for any complex valued function $c \in L^2(-\infty,+\infty)$ the integral $\int_{-\infty}^{+\infty} c(s)V_{\xi_s} a$ exists. This yelds an expression for the past and present of the continuous process (g_t) in terms of the orthogonal increments. Namely, for any real t, $\mathcal{K}_t^g$ is the set of all integrals of the form $\int_{-\infty}^{+\infty} c(s)dV_{\xi_s} a$, where $a \in \mathcal{E}$ and $c \in L^2(-\infty,t)$.

We have only to put

$$u_t = \mathcal{P}_{\mathcal{H}_t^g} f_t \tag{3.10}$$

and

$$v_t = (I - \mathcal{P}_{\mathcal{H}_t^g}) f_t \quad . \tag{3.11}$$

The uniqueness is proved similar to the proof of theorem 6.8 of [2].

Let us remark that (u_t) and (v_t) have the same shift group, namely the shift group of (f_t).

Between the Wold decomposition for the discrete case and the continuous case there exist the following correspondence.

THEOREM 6. *The moving average part* $\{u'_n\}$ *and the deterministic part* $\{v'_n\}$ *of the Wold decomposition of the discrete process* $\{f'_n\}$ *associated with the* Γ*-stationary continuous process* (f_t) *are the discrete processes associated with the moving average part* (u_t) *and the deterministic part* (v_t)*, respectively, of the Wold decomposition of the continuous process* (f_t).

<u>Proof</u>. By the fact that $\mathcal{K}_{-\infty}^{f}=\mathcal{K}_{-\infty}^{f'}$ and $V_{f_o'}=V_{f_o}$ it follows that

$$V_{v_o}=P_{\mathcal{K}_{-\infty}^{f}}V_{f_o}=P_{\mathcal{K}_{-\infty}^{f'}}V_{f_o'}=V_{v_o'} \quad ,$$

hence $v_o=v_o'$. Consequently we have also

$$V_{u_o}=V_{f_o}-V_{v_o}=V_{f_o'}-V_{v_o'}=V_{u_o'}$$

and thus we have the desired result.

Let (U_t) be the shift group of the continuous process (f_t). By Stone theorem there exists a unique $\mathcal{L}(\mathcal{K}_{\infty}^{f})$-valued spectral measure $\mathbf{E}$ on the real line such that

$$(3.12) \qquad U_t=\int_{-\infty}^{+\infty} e^{-itx}dE(x) \ .$$

If we take

$$(3.13) \qquad \Gamma=V_{f_o}^{*}\,EV_{f_o} \quad ,$$

then one obtaines an $\mathcal{L}(\mathcal{E})$-valued semispectral measure on the real line, which is the <u>spectral distribution</u> of the continuous process (f_t).

In that follows some spectral properties of a continuous parameter Γ-stationary process are established corresponding to properties in the discrete parameter case.

Taking the map of the real line onto the unit circle given by

$$(3.14) \qquad e^{i\theta}=(x-i)(x+i)^{-1}$$

(or equivalently, $\theta=-2\operatorname{arc\,ctg} x$), to the spectral distribution F of (f_t) corresponds a semispectral measure F_1 on $\mathbb{T}$ which is just the spectral distribution of the associated discrete parameter process $\{f'_n\}$.

Let $L^2_R(\mathcal{E})$ be the space of the square integrable functions on the real line into $\mathcal{E}$, and $H^2_\Delta(\mathcal{E})$ be the space of all analytic function on the upper half plane Δ with values in $\mathcal{E}$ such that

$$\sup_{y>0} \int_{-\infty}^{+\infty} \| f(x+iy) \|^2 dx < \infty . \tag{3.15}$$

It is known [5] that the space $L^2(\mathcal{E})$ is transformed onto $L^2_R(\mathcal{E})$, if for any g in $L^2(\mathcal{E})$ we put associate f in $L^2_R(\mathcal{E})$ by

$$f(x)=(x+i)^{-1}g\left(\frac{x-i}{x+i}\right) . \tag{3.16}$$

In a similar way a correspondence between $H^2(\mathcal{E})$ and $H^2_\Delta(\mathcal{E})$ is realized.

As in the disc case one defines an L^2-bounded analytic function $\{\mathcal{E}, \mathcal{F}, S(z)\}$ on Δ by

$$\sup_{y>0} \frac{1}{\pi} \int_{-\infty}^{+\infty} \| S(x+iy)a \|^2 dx \leq M^2 \| a \|^2 . \tag{3.17}$$

Via the above identification, there exists a one-to-one correspondence between the L^2-bounded function $\Theta(\lambda)$ on the unit disc $\mathbb{D}$ and the L^2-bounded function $S(z)$ on the half plane Δ. If $\Theta(\lambda)$ is the maximal function of the discrete process $\{f'_n\}$ associated with (f_t), then the corresponding $S(z)$ is called the <u>maximal function of the continuous process</u> (f_t).

THEOREM 7. *Let* (f_t) *be a continuous* Γ*-stationary process,* F *its spectral distribution and* $\{\mathcal{E}, \mathcal{F}, S(z)\}$ *the attached maximal function. If* $f_t = u_t + v_t$ *is the corresponding Wold decomposition, then*

(i) F_S *is the spectral distribution of the nondeterministic part* (u_t).

(ii) *The process* (f_t) *is nondeterministic if there exists a positive function* h *in* $L^1(\frac{dx}{1+x^2})$ *such that* $h \leq h_a$, *where* h_a *is the derivative of* $d(F(x)a,a)$ *with respect to* $\frac{dx}{1+x^2}$ *and*

$$\int_{-\infty}^{+\infty} \frac{\log h(x)}{1+x^2} dx > -\infty . \tag{3.18}$$

The condition become necessary if the maximal function of the associated discrete process has a scalar multiple.

Proof. By the above identification between $L^2(\mathcal{E})$ and $L^2_R(\mathcal{E})$, taking account of Theorem 1, the proof of theorem is obvious.

REFERENCES

[1]. HELSON, H. and LOWDENSLAGER, D., *Prediction theory and Fourier series in several variables*, Part I Acta Math. 99(1958), 165-202; Part II Acta Math.106(1961), 175-213.

[2]. MASANI, P. and ROBERTSON, J., *The time domain analysis of a continuous parameter weakly stationary stochastic processes*, Pacific J.of Math. vol.12(1962) no.4, 1361-1378.

[3]. SUCIU, I. and VALUŞESCU, I., Factorization of semispectral measures, Rev.Roum.Math.Pures et Appl. 21(1976), no.6, 773-793.

[4]. SUCIU, I. and VALUŞESCU, I., Factorization theorems and prediction theory, Rev.Roum.Math.Pures et Appl. 23(1978), no.9, 1393-1423.

[5]. SZ.-NAGY, B. and FOIAŞ, C., Harmonic Analysis of Operators on Hilbert Space, Acad.Kiadó Budapest, North Holland Comp. Amsterdam, London, 1970.

[6]. VALUŞESCU, I., Operator methods in prediction theory (Roumanian) Part I, St.Cerc.Mat. 33(1981), no.3, 343-401; Part II, St.Cerc.Math. 33(1981), no.4, 467-492.

[7]. WIENER, R. and MASANI, P., The prediction theory of multivariate stochastic processes, Part I, Acta Math. 98(1957) 111-150; Part II, Acta Math. 99(1958), 93-139.